Comparative Biology and Evolutionary Relationships of Tree Shrews

ADVANCES IN PRIMATOLOGY

THE PRIMATE BRAIN
Edited by Charles R. Noback and William Montagna

MOLECULAR ANTHROPOLOGY: Genes and Proteins in the Evolutionary Ascent of the Primates
Edited by Morris Goodman and Richard E. Tashian

SENSORY SYSTEMS OF PRIMATES
Edited by Charles R. Noback

NURSERY CARE OF NONHUMAN PRIMATES
Edited by Gerald C. Ruppenthal

COMPARATIVE BIOLOGY AND EVOLUTIONARY RELATIONSHIPS OF TREE SHREWS
Edited by W. Patrick Luckett

Comparative Biology and Evolutionary Relationships of Tree Shrews

Edited by

W. Patrick Luckett

Creighton University School of Medicine
Omaha, Nebraska

Plenum Press · New York and London

Library of Congress Cataloging in Publication Data

Main entry under title:

Comparative biology and evolutionary relationships of tree shrews.
(Advances in primatology)
Includes bibliographical references and index.
1. Tupaiidae–Evolution. 2. Mammals–Evolution. I. Luckett, Winter Patrick.
II. Series.
QL737.P968C65 599.3'3 80-19824
ISBN 978-1-4684-1053-2

ISBN 978-1-4684-1053-2 ISBN 978-1-4684-1051-8 (eBook)
DOI 10.1007/978-1-4684-1051-8

Softcover reprint of the hardcover 1st edition 1980
A Division of Plenum Publishing Corporation
227 West 17th Street, New York, N.Y. 10011

Contributors

Percy M. Butler
Department of Zoology
Royal Holloway College
Englefield Green, Surrey
England

C. B. G. Campbell
Department of Medical Neurosciences
Division of Neuropsychiatry
Walter Reed Army Institute of Research
Washington, D. C.

Matt Cartmill
Department of Anatomy
Duke University Medical Center
Durham, North Carolina

John E. Cronin
Department of Anthropology
Peabody Museum
Harvard University
Cambridge, Massachusetts

Howard Dene
Department of Anatomy
Wayne State University School of Medicine
Detroit, Michigan

Gerrell Drawhorn
Department of Anthropology
University of California, Davis
Davis, California

Morris Goodman
Department of Anatomy
Wayne State University School of Medicine
Detroit, Michigan

Louis L. Jacobs
Department of Geology
Museum of Northern Arizona
Flagstaff, Arizona

W. Patrick Luckett
Department of Anatomy
Creighton University
Omaha, Nebraska

R. D. E. MacPhee
Department of Anatomy
Duke University Medical Center
Durham, North Carolina

Genji Matsuda
Department of Biochemistry
Nagasaki University School of Medicine
Nagasaki, Japan

Michael J. Novacek
Department of Zoology
San Diego State University
San Diego, California

William Prychodko
Department of Biology
Wayne State University
Detroit, Michigan

Vincent M. Sarich
Departments of Anthropology and Biochemistry
University of California
Berkeley, California

Frederick S. Szalay
Department of Anthropology
Hunter College, CUNY
New York, New York

Preface

Tree shrews are small-bodied, scansorial, squirrel-like mammals that occupy a wide range of arboreal, semi-arboreal, and forest floor niches in Southeast Asia and adjacent islands. Comparative aspects of tree shrew biology have been the subject of extensive investigations during the past two decades. These studies were initiated in part because of the widely accepted belief that tupaiids are primitive primates, and, as such, might provide valuable insight into the evolutionary origin of complex patterns of primate behavior, locomotion, neurobiology, and reproduction. During the same period, there has been a renewed interest in the methodology of phylogenetic reconstruction and in the use of data from a variety of biological disciplines to test or formulate hypotheses of evolutionary relationships. In particular, interest in the comparative and systematic biology of mammals has focused on analysis of phylogenetic relationships among Primates and a search for their closest relatives. Assessment of the possible primate affinities of tree shrews has comprised an important part of these studies, and a considerable amount of dental, cranioskeletal, neuroanatomical, reproductive, developmental, and molecular evidence has been marshalled to either corroborate or refute hypotheses of a special tupaiid-primate relationship. These contrasting viewpoints have resulted from differing interpretations of the basic data, as well as alternative approaches to the evolutionary analysis of data.

The present volume was organized in order to evaluate the possible evolutionary relationships of tree shrews to primates and other eutherian mammals. Such analysis is deemed an essential prerequisite for the use of tree shrews as possible models for the study of the evolutionary origin of various primate organ systems. Assessment of the mammalian affinities of tree shrews is limited because of the sparseness of the tupaiid fossil record (restricted at present to the recently discovered Miocene specimens described by Jacobs in this volume). In order to evaluate the possible mammalian affinities of tupaiids, it is essential to attempt reconstruction of the morphotypic condition for each organ system analyzed in Tupaiidae, Primates, Dermoptera, Chiroptera, Lipotyphla, Macroscelididae, and other eutherian taxa, because such analyses facilitate the recognition of convergent evolution of individual traits in these systems. Reconstruction of the eutherian morphotype for selected features was also undertaken by many contributors, and these hypothetical reconstructions serve as focal points for identifying areas of disagreement in phylogenetic analyses.

Contributors to the volume were requested to consider two interrelated questions during their assessments of the possible evolutionary relationships of tree shrews. (1) Are there uniquely derived biological attributes shared solely by tupaiids and primates? (2) If tupaiids are not cladistically primates, are they more closely related phyletically to primates, dermopterans, and chiropterans (in a superordinal taxon Archonta) than they are to any other eutherians? Following evaluations of the available dental, cranial, postcranial, neuroanatomical, reproductive, developmental, and molecular evidence, the contributors were in general agreement that tupaiids do not appear to share any uniquely derived features with living or fossil primates that would warrant inclusion of tree shrews in a monophyletic order Primates. Available biological data suggest, instead, that tree shrews have evolved independently since at least the early Tertiary, despite the fact that corroborating fossil evidence is lacking. As proposed previously by Butler, recognition of the separate ordinal status of tree shrews (as the order Scandentia) appears to be the best manner of expressing such a long and independent evolutionary history.

There was less agreement among contributors concerning the possible superordinal relationships of Scandentia. The available molecular and postcranial evidence provides some support for a modified archontan hypothesis of phylogenetic affinities among Scandentia, Primates, and Dermoptera, but data evaluated from other organ systems provide little, if any, corroboration for this postulate. As emphasized by Szalay and Drawhorn in this volume, however, the available biological evidence does not corroborate any alternative hypotheses of superordinal affinities of Scandentia. Future investigations of tupaiid biology should provide additional data for testing the archontan hypothesis.

Even if tree shrews are not considered to be members of the order Primates, these fascinating mammals will remain the subject of extensive study by primate biologists. Analyses of the evolutionary relationships of tree shrews during the past two decades have been a major stimulus to renewed investigations into the origin and phylogeny of Primates. Moreover, assessments of possible tupaiid-primate affinities have focused attention on the wide range of molecular and soft anatomical evidence that can be used to supplement, rather than replace, the more traditional dental and skeletal data for phylogenetic reconstruction.

On behalf of the contributors, I wish to thank Mr. Kirk Jensen for his encouragement and continued support of this project. Special thanks are also due to Miss Mary Markytan and Mrs. Edith Witt for their invaluable assistance in preparing this volume for publication. This book was composed by Western

Typesetting Company, Kansas City, Missouri, and the editor acknowledges the assistance of Mr. Jerry Germany and Mr. Alex Kreicbergs in completing this phase of the project.

W. P. L.

Contents

I. Systematics

II. Cranioskeletal System and Dentition

IV. Reproductive System

V. Molecular Evolution

Systematics I

The Suggested Evolutionary Relationships and Classification of Tree Shrews

1

W. PATRICK LUCKETT

1. Introduction

The tree shrews (family Tupaiidae) of Southeast Asia are small, scansorial, squirrel-like mammals which occupy a range of arboreal, semi-arboreal, and forest floor niches. They are insectivorous/omnivorous and predominantly diurnal, although one species (*Ptilocercus lowii*) is crepuscular or nocturnal. The only monographic description of all genera and species of the family was that by Lyon (1913). Most taxonomists have accepted Lyon's subdivision of Tupaiidae into two subfamilies: (1) the diurnal subfamily Tupaiinae, containing five genera (*Tupaia, Anathana, Dendrogale, Lyonogale,* and *Urogale*); and (2) the crepuscular (or nocturnal) subfamily Ptilocercinae, containing the single genus *Ptilocercus*. The geographic distribution of the family extends from India to the Philippines, and from southern China to Java, Borneo, Sumatra, and Bali, including the many islands within this region. This distribution corresponds closely with the limits of the zoogeographic Oriental Region.

The earliest written account of tree shrews to reach the Western World was that of William Ellis, a surgeon's mate who maintained a naturalist's journal during the Pacific voyage of the British ship *Discovery*. On January 20, 1780,

W. PATRICK LUCKETT • Department of Anatomy, Creighton University, Omaha, Nebraska 68178 U.S.A.

the *Discovery* dropped anchor at Pulo Condore, a small island off the southern coast of what is now Vietnam, and Ellis reported the sighting of a few monkeys and squirrels on the island. He designated one of these "squirrels" as *Sciurus dissimilis,* but it is evident from his description and accompanying sketch[1] that the specimen was actually a tree shrew. Ellis's account was generally unknown to biologists until it was republished by Gray (1860). Diard (1820) described a new species of "shrew" from Penang, Malaya, as *Sorex glis*, although he admitted that it differed from other shrews in its dentition, squirrel-like habits, and in the possession of a cecum. Raffles (1822) subsequently recognized the distinctness of the diurnal squirrel-like shrews of Malaya and Sumatra, and he proposed the generic name *Tupaia* for them. Ironically, "tupai" is the Malay word for both tree shrews and squirrels.

Uncertainties concerning the taxonomic affinities of tree shrews originated with these original descriptions and have continued to the present day. One of the earliest mammalian classifications which incorporated tree shrews was that of Gray (1825); they were included as the "subfamily Tupaina" within the family Talpidae. It should be noted, however, that Gray's family Talpidae corresponded essentially to recent concepts of the order Insectivora (excluding Macroscelididae). During the first half of the 19th century, reports on tree shrews were restricted principally to the description of new species and genera (*Ptilocercus, Dendrogale*), and to the extension of their known geographic distribution. Peters (1864) was one of the first investigators to consider the question of "natural" relationships within Insectivora. He divided extant Insectivora into two major (unnamed) groups, based on the presence or absence of an intestinal cecum, and these groups were further subdivided on the basis of a few cranioskeletal features. Subsequently, Haeckel (1866) adopted Peters' criteria for subdividing Insectivora, and he proposed the suborder Menotyphla for insectivores with a cecum (modern families Tupaiidae and Macroscelididae), and the suborder Lipotyphla for taxa which lack a cecum (modern families Soricidae, Talpidae, Erinaceidae, Tenrecidae, Solenodontidae, and Chrysochloridae).

Although most systematists during the latter half of the 19th century adopted Darwin's concept of descent with modification as the theoretical basis for explaining taxonomic relationships, evolutionary theory had little immediate effect on the construction of mammalian classifications. Haeckel's (1866) typological classification of Insectivora and other mammals, adopted subsequently by Huxley (1872), was based on the presence or absence of a few key characters, despite the fact that both authors were staunch advocates of Darwin's theory. Haeckel (1866) clustered the Insectivora with the Prosimiae, Simiae, Chiroptera, and Rodentia in a superordinal taxon Discoplacentalia, based on their common possession of a single feature, a discoidal chorioallantoic placenta. Further hypotheses concerning the possible affinities of tree shrews with Insectivora or other Eutheria have reflected the changing trends in mammalian systematic philosophy during the post-Darwinian period.

[1]Ellis's original sketch was reproduced by Lyon (1913) as Plate 1 in his monograph.

Mivart (1868) examined a variety of cranioskeletal features in Insectivora, and he concluded that extant forms should be separated into nine "natural" families, including his newly proposed family Tupaiidae for tree shrews. However, Mivart was uncertain concerning the superfamilial relationships among insectivoran families. Weber (1904, 1928) and Gregory (1910) summarized much of the available morphological evidence relevant to the evolutionary relationships among extant and fossil mammals, and both assessed the phylogenetic relationships of Tupaiidae to other Eutheria, in particular, to Macroscelididae, Primates, and Lipotyphla. Both authors argued for the close relationship of Tupaiidae and Macroscelididae in a higher taxon Menotyphla, although they disagreed on the taxonomic rank of Menotyphla.

Gregory (1910) emphasized that Menotyphla appears to be widely divergent from other Insectivora, and he suggested that it should be elevated to ordinal rank. Furthermore, Gregory indicated (p. 272) that "in many characters the *skull of the Tupaiidae differs widely from that of any Lipotyphlous Insectivore* and approaches the lemuroid type." Although Gregory did not include the Tupaiidae or Menotyphla within his order Primates, he concluded that the many similarities in cranioskeletal features among Tupaiidae, Macroscelididae, Lipotyphla, Dermoptera, Chiroptera, and Primates precluded a sharp taxonomic separation between Insectivora and Primates.

Formal allocation of the Tupaiidae to the order Primates was suggested initially by Carlsson (1922), although several earlier investigators (Gray, 1848; Haeckel, 1866; Huxley, 1872; Leche, 1885; Van Kampen, 1905; Gregory, 1910; Kaudern, 1911) reported similarities in certain cranioskeletal and reproductive features between Tupaiidae and lemuriform Primates. Carlsson (1922) examined numerous features of the cranioskeletal system, dentition, musculature, skin, gastrointestinal tract, nervous system, and reproductive system in several species of *Tupaia*, and these were compared to similar features in representative Prosimii (principally *Lemur*), Macroscelididae, and Lipotyphla (principally Erinaceidae). Her analysis revealed the presence of many shared character states in *Tupaia* and Malagasy lemuriforms which did not occur in Macroscelididae or Erinaceidae. Carlsson concluded that the family Tupaiidae is more closely related genealogically to the Prosimii than to Macroscelididae or Lipotyphla, and she suggested that this relationship should be expressed by classifying Tupaiidae within the Prosimii. Strong support for Carlsson's hypothesis of a special tupaiid-prosimian relationship was provided by Le Gros Clark's (1924a, 1924b, 1925, 1926, 1932) detailed examination of the myology, brain, skull, and postcranial skeleton of *Tupaia* and *Ptilocercus*. These and additional data which provide evidence for the primate affinities of the Tupaiidae have been summarized by Le Gros Clark (1959, 1971).

Early hypotheses of tupaiid-macroscelidid or tupaiid-primate affinities (Fig. 1) were based on the identification of shared similarities, but in many cases the theoretical basis for the origin of these similarities was not evaluated. Even those investigators (Gregory, 1910; Le Gros Clark, 1925, 1926) who speculated upon the relative primitive or advanced nature of shared similarities among taxa resorted ultimately to a concept of overall (phenetic) resemblance

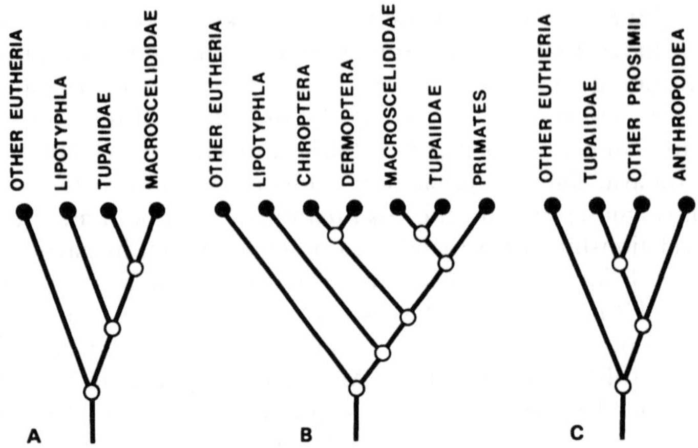

Fig. 1. Several hypotheses of Tupaiidae affinities with other Eutheria. A. Adapted from Peters (1864) and Weber (1928). B. Adapted from Gregory (1910). C. Adapted from Le Gros Clark (1959, 1971).

in assessing phylogenetic relationships. Similarities among organisms may be the result of *homology* (inheritance from a common ancestor) or *homoplasy* (convergent or parallel similarities not inherited from a common ancestor), and it is essential to distinguish between these types of similarities when assessing phylogenetic relationships.

2. *Principles of Phylogenetic Reconstruction*

During the past two decades, there has been renewed interest in the theoretical basis for phylogenetic reconstruction, stimulated in part by the publication of Hennig's (1966) *Phylogenetic Systematics*. There is general agreement among evolutionary biologists that phylogeny can not be observed directly in the fossil record; therefore, the construction and testing of phylogenetic hypotheses should incorporate observations derived from a wide range of characters in both extant and fossil organisms. Such phylogenetic hypotheses can be corroborated or falsified by the analysis of additional characters, but they can not be *proved*. Phylogenetic analysis of characters is recognized as a necessary prerequisite for the construction of biological classifications by students of both "evolutionary" and "phylogenetic" schools of taxonomy, although they disagree concerning the extent to which phylogenetic hypotheses should be reflected in classifications (see Simpson, 1961, 1975; Hennig, 1966; Mayr, 1974; Bock, 1977; Bonde, 1977; Szalay, 1977b).

Essential tasks in the phylogenetic analysis of characters include: (1) determination of the homologous or homoplastic nature of shared similarities among organisms, (2) identification of all possible "character states" of homologous features, and their arrangement in a transformation series (*morphocline*), and (3) assessment of the relative primitiveness or derivedness of each character state in a hypothesis concerning the pattern and direction of evolutionary change of characters (determination of morphocline polarity). Homologous character states are recognized on the basis of detailed phenetic similarities (both structural and functional), and, whenever feasible, ontogenetic studies should be used to test hypotheses of homology. As an example, the suggested homology (Gregory, 1910; Saban, 1956/1957) between the entotympanic bulla of tupaiids and the petrosal bulla of lemuriform primates, proposed on the basis of phenetic resemblances in adults, is not corroborated by ontogenetic studies (see Cartmill and MacPhee, this volume).

Homologous character states inherited from a common ancestor provide evidence of *patristic* relationships (Cain and Harrison, 1960) among taxa, but determination of the branching sequence (*cladistic* relationships) of phylogeny requires further analysis of shared homologies. Cladistic relationships are defined in terms of relative recency of common ancestry. Two taxa are more closely related cladistically to each other than to any other taxa when it can be demonstrated that they share derived homologous character states (synapomorphies) which were inherited from a relatively recent common ancestor not shared with other taxa (Hennig, 1966). The development of hypotheses concerning the direction of evolutionary change in character states of a transformation series (morphocline polarity) is the most difficult and most important aspect of character analysis, and these hypotheses of character phylogeny form the only valid test of phylogenetic hypotheses about taxa (Bock, 1977).

Methodologies for the assessment of relative primitiveness or derivedness of character states vary, and there is considerable disagreement concerning the validity of some of these methods (see the contrasting views of Bock, 1977 and Bonde, 1977). A majority of investigators have utilized the frequency of distribution of character states among taxa (the "commonality" principle) as a measure of their polarity. Character states which are most widely distributed within the taxa being compared, as well as in more distantly related taxa ("outgroup" comparisons), are assumed to represent the relatively primitive (plesiomorphous) condition, whereas the relative rarity of a character state is suggestive of its derived (apomorphous) condition. Analysis of polarities using the commonality principle may be relatively simple when there are only two character states for a feature and when the rare (presumably derived) condition is a unique feature. However, for many features there may be three or more character states, without any one state being rare or unique. In such instances, it is essential to consider all available data (including ontogenetic character precedence, analysis of form-function relationships, and stratigraphic position) when assessing morphocline polarity. The arrangement of character states in a transformation series (a character phylogeny) serves as a hypothesis for the sequence of evolutionary change in a feature. Such a char-

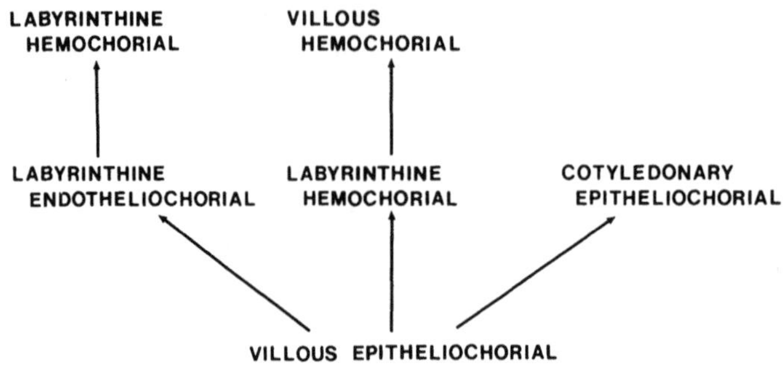

Fig. 2. Character phylogeny for the eutherian chorio-allantoic placenta, based primarily on ontogenetic data. Note the differences in the suggested evolutionary pathways for the attainment of several types of hemochorial placenta (data derived from Luckett, 1974, 1975).

acter phylogeny for the eutherian chorioallantoic placenta is illustrated in figure 2 (data derived from Luckett, 1974, 1975).

Character analyses are utilized to test hypotheses of phylogenetic relationships among taxa. Two taxa are considered to be related cladistically and to form a higher monophyletic (= holophyletic) taxon when it can be demonstrated that they share derived, homologous character states (synapomorphies). Shared primitive character states (symplesiomorphies) provide no evidence of special cladistic relationships among taxa, because these features may have been inherited unchanged from a more distant ancestor. Character analysis of features which have been used to formulate hypotheses of tupaiid phylogenetic relationships reveals that many of these shared similarities between tupaiids and other taxa are symplesiomorphous eutherian features. No synapomorphies are recognized among the shared similarities discussed by Peters (1864), Haeckel (1866), or Weber (1928) for the menotyphlan affinities of Tupaiidae and Macroscelididae (Figs. 3, 4; Tables 1, 2). Plesiomorphy and apomorphy are relative concepts, and plesiomorphous character states of one taxon may be relatively derived within a higher taxonomic category. For instance, a molariform trigonid on P_4 is plesiomorphous for Tupaiidae, but within Eutheria it is a relatively apomorphous character state (character 6 in Fig. 4 and Table 2).

Both paleontological and neontological data can be utilized in character analysis for determining the branching sequences of a phylogeny, but only paleontological data can provide the essential time dimension for assessment of possible ancestor-descendant relationships among taxa. Unquestioned early Tertiary fossil tupaiids are unknown at the present time (McKenna, 1963; Van Valen, 1965; Butler, this volume; Jacobs, this volume; Novacek, this volume), and it is unlikely that they will be recognized solely on the basis of dental morphology. Ideally, phylogenetic hypotheses should include information on both the cladistic and anagenetic (ancestor-descendant) relationships among

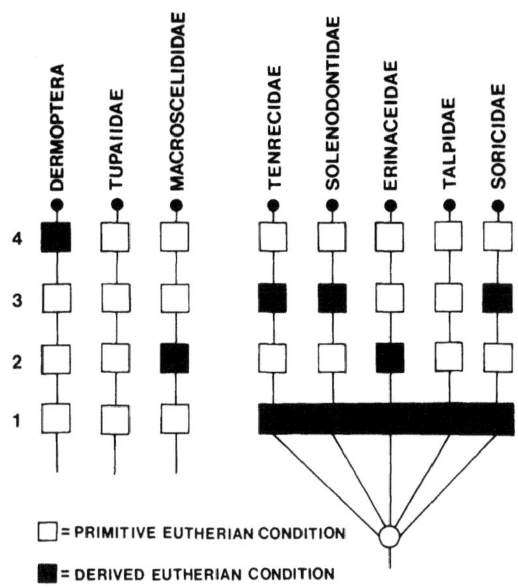

Fig. 3. Character analysis of features utilized by Peters (1864) for establishing the menotyphlan-lipotyphlan dichotomy of Insectivora. In this and subsequent figures, relatively primitive (plesiomorphous) character states are indicated by an open square, and derived (apomorphous) character states are indicated by black squares. Shared, derived, homologous character states (synapomorphies) between taxa are represented by black rectangles. Note that the monophyly of the Lipotyphla is suggested by only a single synapomorphy in this scheme (character 1), and that no synapomorphies are present in the Menotyphla (here also including Dermoptera). See Table 1 for the list of characters evaluated.

organisms, particularly when there is sufficient fossil evidence to warrant the latter. Hypotheses of ancestor-descendant relationships can be falsified readily by identifying apomorphous character states which occur in the suggested fossil ancestor but not in the descendant (Szalay, 1977a; Bock, 1977). Although it is questionable whether ancestor-descendant hypotheses can be corroborated, Szalay (1977a) has emphasized the value of morphotype reconstruction in narrowing the search for the approximate fossil ancestor of a taxon.

Table 1. Character analysis of features utilized by Peters (1864) in distinguishing between menotyphlous and lipotyphlous Insectivora (see Fig. 3)

Primitive	Derived
1. Cecum present	1. Cecum absent
2. Tibia and fibula unfused	2. Tibia and fibula fused distally
3. Zygomatic arch complete	3. Zygomatic arch incomplete
4. Ulna complete distally	4. Ulna slender, reduced distally

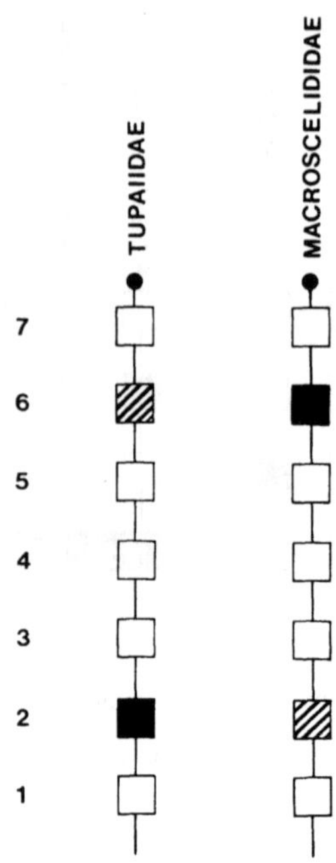

Fig. 4. Character analysis of features used by Weber (1928) to support the monophyly of Menotyphla (intermediately derived character states are indicatd by obliquely-lined squares). Note that a majority of the shared resemblances are plesiomorphous eutherian features. See Table 2 for the list of characters evaluated.

Table 2 Character analysis of features utilized by Weber (1928) to support monophyly of Menotyphla (see Fig. 4)

Primitive	Derived
1. Puboischiac symphysis complete	1. Puboischiac symphysis incomplete
2. Entotympanic does not form major portion of bulla	2. Entotympanic forms major portion of bulla
3. Zygomatic arch complete	3. Zygomatic arch incomplete
4. Upper molars tribosphenic	4. Upper molars dilambdodont
5. Upper canines caniniform[1]	5. Upper canines premolariform[1]
6. P_4 premolariform	6. P_4 molariform[2]
7. Cecum present	7. Cecum absent

[1]Butler (personal communication) suggests that the polarity of this character may be reversed. However, this reversal would have no effect on assessment of tupaiid-macroscelidid affinities, since both character states occur in both families.

[2]P_4 trigonid molariform-intermediately derived character state.

Some of the supposed similarities between tupaiids and macroscelidids (Table 2) listed by Weber (1928) are not shared by all members of each family (premolariform upper canines), or else they have been attributed incorrectly to one taxon (dilambdodont molars in macroscelidids). Similar problems are encountered in assessing tupaiid-primate affinities. It is essential to reconstruct the ancestral state (*morphotype*) of each character within a taxon, prior to evaluating its possible phylogenetic relationships with other taxa. The plesiomorphous character state of each feature within a taxon is considerd to represent the morphotype of that feature. Morphotype reconstruction suggests that a premolariform cusp pattern on the canines was an unlikely attribute of the ancestral tupaiids and macroscelidids, and that the occurrence of this character state in some genera of both families is probably the result of convergent evolution (see Butler, this volume, for further assessment of the dental evidence for tupaiid affinities).

Tests for the recognition of homology and synapomorphy have a low level of resolution (Bock, 1977), and it is essential that hypotheses of phylogenetic relationships be tested by character analysis of a wide variety of features in both extant and fossil taxa. Additional character analyses may provide further corroboration of phylogenetic hypotheses; this is Hennig's (1966) "principle of reciprocal illumination." In some cases, however, analysis of additional traits may lead to contradictory hypotheses of phylogenetic relationships. Such incongruences are commonly the result of differing viewpoints concerning the homologous or convergent nature of shared derived attributes. Some method of *a posteriori* character weighting is therefore essential whenever conflicting phylogenetic hypotheses result from multiple character analyses. Hecht and Edwards (1976) have formalized a method for character state weighting in which greatest value is placed on the identification of shared and derived homologous character states that are unique and innovative. In contrast, minimal weight is given to shared, derived states which are the result of character loss, especially when there are no data to indicate the ontogenetic pathway for such loss. The five weighting categories proposed by Hecht and Edwards are not absolute; instead, they reflect to varying degrees our knowledge of ontogeny and form-function relationships of characters, and the possible interrelationship of these features in a functionally integrated character complex.

3. Sources of Potential Error in Assessing Tupaiid Phylogenetic Relationships

Character analysis of features utilized by Carlsson (1922) and Le Gros Clark (1959, 1971) to support a hypothesis of tupaiid-primate affinities (Fig. 5 and Table 3) serves to identify several sources of error which are common in phylogenetic reconstruction. Most of these have been discussed previously (Van Valen, 1965; Campbell, 1966, 1974; Martin, 1968b).

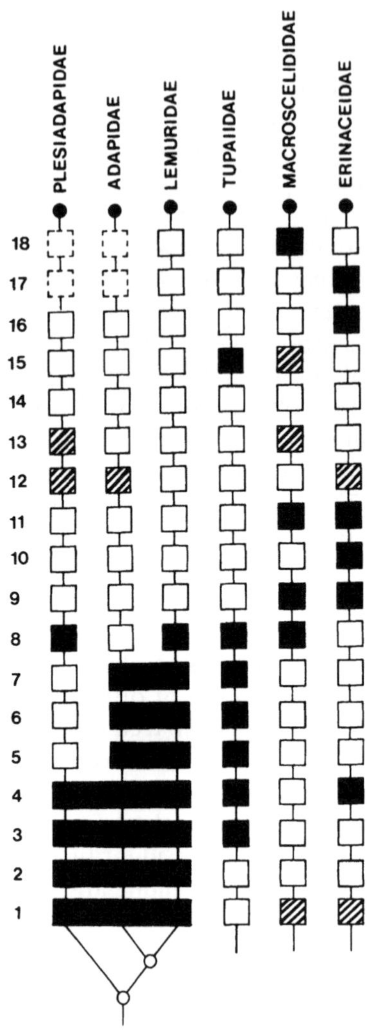

Fig. 5. Cladogram or character phylogeny of features utilized by Carlsson (1922) to support hypothesis of tupaiid-lemurid affinities. Intermediately derived traits are indicatd by obliquely-lined squares. The primate families Plesiadapidae and Adapidae are included in the analysis, although their characters were not discussed by Carlsson (character states of traits 17 and 18 are unknown in these fossil families). The character analysis supports the monophyly of the Primates, but not the inclusion of Tupaiidae within the same clade. See Table 3 for the characters evaluated.

(1) *Failure to reconstruct the morphotype of taxa prior to assessing their possible phylogenetic relationships.* In evaluating the affinities of Tupaiidae, Carlsson (1922) restricted her comparison principally to *Tupaia* as the representative tupaiid, *Lemur* as the representative lemuriform, and *Erinaceus* as a represen-

tative lipotyphlan insectivore. Such limited comparisons also characterized the studies of tupaiid relationships by Evans (1942), Saban (1956/1957), and Le Gros Clark (1959,1971). There are cranioskeletal and dental differences among the genera of Tupaiidae (as emphasized previously by Gregory, 1910 and Lyon, 1913), and it is essential to reconstruct the morphotype of the tupaiid ancestor by character analysis, prior to assessing their phylogenetic relationships to other Eutheria. Likewise, morphotype reconstruction must be carried out on Primates, Macroscelidea, Insectivora, and other Eutheria for valid analysis of tupaiid affinities. From the foregoing discussion it should be evident that morphotype reconstruction of character states in ancestral eutherians is a necessary prerequisite for assessing phylogenetic relationships among higher taxa such as Tupaiidae, Primates, Macroscelidea, Dermoptera, and Insectivora. Unfortunately, such analyses are rarely presented in studies of mammalian phylogeny, with the notable exceptions of Gregory (1910) and Winge (1941).

Analysis of the available cranioskeletal, soft anatomical, and molecular data corroborates the hypothesis that the family Tupaiidae is a monophyletic

Table 3. Character analysis of features utilized by Carlsson (1922) to support hypothesis of tupaiid-lemurid affinities (see Fig. 5)

Primitive	Derived
1. Petrosal wing of bulla minute or absent	1. Petrosal forms virtually entire bulla
2. Foramen rotundum confluent with sphenoidal fissure	2. Foramen rotundum distinct from sphenoidal fissure
3. Prominent fibular facet on calcaneus	3. Fibular facet reduced or absent
4. Sustentacular facet of astragalus separate from navicular facet	4. Sustentacular facet continuous with navicular facet
5. Ectotympanic exposed at least partially at lateral margin of bulla	5. Ectotympanic ring "intrabullar"
6. Postorbital processes of frontal and jugal absent	6. Postorbital bar complete
7. Jugal (zygomatic) foramen absent	7. Jugal (zygomatic) foramen present
8. Fibular and tibial cristae of astragalar trochlea subequal	8. Fibular crista of trochlea higher than tibial crista
9. Postglenoid process of skull well developed	9. Postglenoid process reduced or absent
10. Entepicondylar foramen of humerus present	10. Entepicondylar foramen absent
11. Tibia and fibula unfused	11. Tibia and fibula fused distally
12. Greater trochanter projects beyond head of femur	12. Greater trochanter subequal in height with head of femur
13. Ectotympanic annular or nearly so	13. Ectotympanic expanded laterally
14. Foramen ovale not covered by bulla	14. Foramen ovale covered by bulla
15. Entotympanic component of bulla minute or lacking	15. Entotympanic forms virtually entire bulla
16. Free os centrale in wrist	16. Os centrale fused with other wrist bones
17. Cecum present	17. Cecum absent
18. Seminal vesicles present	18. Seminal vesicles absent

taxon (see below), and it is the morphotypic character states of the family which should be utilized in assessing their phylogenetic relationships with Primates and other Eutheria. Among the shared similarities of *Tupaia* and *Lemur* listed by Carlsson (1922), three features differ in *Tupaia* and *Ptilocercus* (Characters 2, 14, 16 in Table 3 and Fig. 5). In each case, the character state in *Ptilocercus* is judged to be relatively plesiomorphous and thus to represent the tupaiid morphotype (see Table 4). In the present analysis, characters 14 and 16 are considered to be shared plesiomorphous features of tupaiids and primates, whereas character 2 is plesiomorphous for tupaiids but apomorphous for primates. None of these provide corroborating evidence for special tupaiid-primate affinities.

(2) *Failure to utilize available fossil evidence in assessing character state polarity and morphotype reconstruction.* Character states in fossil primates were not utilized by Carlsson (1922), Saban (1956/1957), or Le Gros Clark (1959, 1971) in assessing tupaiid-primate affinities, despite the fact that considerable evidence on character polarity in adapids was provided by Gregory (1920), and subsequently for plesiadapids by Van Valen (1965). Characters 6 and 7 (Table 3) exhibit apomorphous states among Tupaiidae, Lemuridae, and Adapidae, while a more plesiomorphous character state is retained by Plesiadapidae. Unless the Tupaiidae are considered to be more closely related cladistically to the common adapid-lemurid stock that they are to the ancestral primates, the shared and derived features of tupaiids and lemurids are probably the result of convergent evolution. Furthermore, character weighting according to the scheme of Hecht and Edwards (1976) suggests that the acquisition of a petrosal bulla (character 1 in Table 3) in all fossil and extant primates is a uniquely derived (autapomorphous) feature within Eutheria, and thus it should be weighted heavily in assessing the branching sequence among tupaiids, plesiadapids, and primates.

(3) *Failure to explore the possibility that shared similarities may be the result of homoplastic (convergent) evolution.* Several shared similarities which have played a major role in the hypothesis of a special tupaiid-primate relationship appear to be the result of convergent evolution between tupaiids and lemuriforms. These include the erroneous assertion that the auditory bulla is constructed in a similar fashion in tupaiids and lemuriforms (Carlsson, 1922; Saban, 1956/57). As emphasized by Van Valen (1965) and Cartmill and MacPhee (this volume), the primate bulla develops from the petrosal, whereas that of tupaiids is formed by the entotympanic. Consequently, the various components of the bulla in tupaiids and primates are analyzed separately in table 3 (characters 1, 5, 13, 15). The "intrabullar" position of the ectotympanic (character 5) in tupaiids and lemuriforms is clearly convergent, because the bullar composition in the two taxa is nonhomologous. Form-function analyses also suggest that convergent evolution led to the occurrence of a laminated lateral geniculate nucleus in the thalamus of a variety of arboreal, gliding, and flying eutherians, including tupaiids, primates, dermopterans, and megachiropterans (Campbell, 1974, this volume; Kaas et al., 1978).

(4) *Failure to distinguish between shared plesiomorphous and apomorphous character states in assessing phylogenetic relationships.* Many of the shared similarities listed for tupaiids and lemuriforms (Carlsson, 1922; Saban, 1956/57; Le Gros Clark, 1959, 1971) are considered to be symplesiomorphous eutherian features (characters 9–14, 16–18 in Table 3 and Fig. 5). As such, these characters provide no evidence to corroborate a hypothesis of special tupaiid-primate relationship.

4. Assessment of Cladistic Relationships among Genera of Tupaiidae

Lyon (1913) presented a character key for features of the dentition, skull, and pelage which can be used to distinguish between the subfamilies Tupaiinae and Ptilocercinae, and to identify the five genera of Tupaiinae. Few authors since Lyon have examined character states in all genera of tupaiids, and this has hindered further analysis of intrafamilial relationships. Most recent comparisons among tupaiid genera have been restricted primarily to *Tupaia* and *Ptilocercus* (see especially Le Gros Clark, 1959, 1971), with the tacit assumption that character states in *Tupaia* are representative of the Tupaiinae.

The validity of Lyon's subfamilial division of the Tupaiidae was questioned by Davis (1938) in his study on the anatomy of *Dendrogale*. He noted that there were numerous cranioskeletal and soft anatomical similarities between *Dendrogale* and *Ptilocercus*, and Davis concluded that subfamilial separation of *Ptilocercus* from other tupaiids was unwarranted on the basis of these resemblances. More recently, Steele (1973) has provided a phenetic analysis of 43 dental traits in five of the six genera of tree shrews (no specimens of *Anathana* were studied), utilizing an unweighted pair group method of cluster analysis. Although Steele was cautious in drawing taxonomic conclusions from his analysis, several of his observations are of considerable interest in assessing the intrafamilial relationships of tupaiids. Twenty of the 43 traits examined served to distinguish between the subfamilies Tupaiinae and Ptilocercinae, thus supporting Lyon's (1913) subdivision of the family. Relationships among the genera of Tupaiinae are unclear from Steele's analysis, although he emphasized that *Lyonogale* and *Urogale* are clustered together on the basis of some dental similarities. This would seem to support the contention (Lyon, 1913) that *Lyonogale* and *Urogale* were derived from a recent common ancestor distinct from *Tupaia*, and it would be consistent with Lyon's separation of *Lyonogale* and *Tupaia* at the generic level.

Unfortunately, none of the above authors have speculated on the relatively plesiomorphous or apomorphous nature of shared similarities among tupaiid genera. Character analysis of many of the dental, cranial, and other anatomical features studied by Lyon (1913), Le Gros Clark (1925, 1926, 1971), Davis (1938),

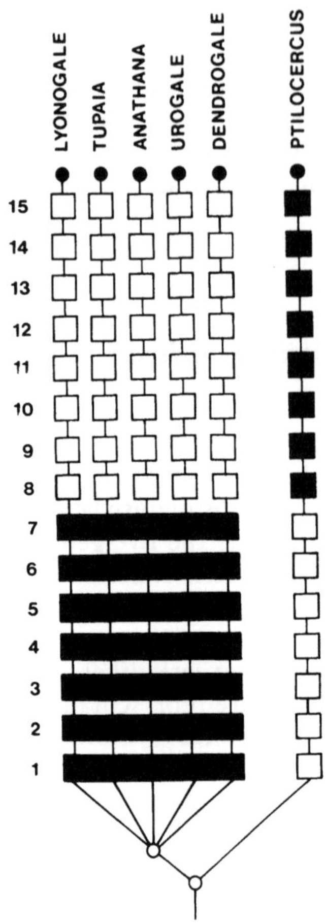

Fig. 6. Character phylogeny of selected features within the family Tupaiidae. The monophyly of the subfamilies Tupaiinae and Ptilocercinae is corroborated by this analysis. See Table 4 for the characters evaluated.

Verma (1965), Steele (1973), and Butler (this volume) provides strong corroboration for Lyon's subdivision of Tupaiidae into two monophyletic subfamilies Tupaiinae and Ptilocercinae (Fig. 6 and Table 4). Most of the shared similarities cited by Davis (1938) for *Dendrogale* and *Ptilocercus* (including a minute malar foramen, double rooted upper canines, absence of posterior palatal vacuities, and scaphoid and lunate unfused) are symplesiomorphous eutherian or tupaiid features. As such, these provide no evidence for the pattern of branching among *Ptilocercus*, *Dendrogale*, and *Tupaia*, whereas apomorphous character states shared by *Dendrogale* and other Tupaiinae (characters 1–7 in Fig. 6 and Table 4) corroborate the hypothesis of tupaiine monophyly. Subfamilial dis-

Table 4. Character analysis of selected features within the family Tupaiidae (see Fig. 6)

Primitive	Derived
1. Foramen rotundum confluent with sphenoidal fissure	1. Foramen rotundum separate from sphenoidal fissure
2. Foramen ovale well in front of bulla	2. Foramen ovale partially covered by bulla
3. Supraorbital foramen absent	3. Supraorbital foramen present
4. Mesostyle absent on upper molars	4. Mesostyle present on upper molars
5. Metacone lower than paracone on M^1	5. Metacone higher than paracone on M^1
6. Hypoconulid not displaced	6. Hypoconulid displaced lingually
7. Preprotocone cingulum present	7. Preprotocone cingulum lost
8. Tail bushy or close-haired	8. Tail naked and scaly basally; tufted distally
9. Facial wing of lacrimal moderately developed	9. Facial wing of lacrimal reduced
10. Buccal cingulum absent on lower molars	10. Buccal cingulum present on lower molars
11. Paracone with transverse shearing crest (paracrista)	11. Paracone displaced bucally; paracrista nearly longitudinal
12. P^3_3 not reduced	12. P^3_3 reduced
13. I_2 subequal to I_1	13. I_2 twice as large as I_1
14. I^2 without posterior cusp	14. I^2 with distinct posterior cusp and crest
15. Upper canine caniniform	15. Upper canine premolariform

tinction of Ptilocercinae and Tupaiinae is also supported by the available (although incomplete) immunological data (Sarich and Cronin, 1976; Cronin and Sarich, this volume; Dene et al., this volume).

Steele's (1973) phenetic clustering of *Urogale* and *Lyonogale* on the basis of dental similarities is not supported by the character analysis of the present study (Fig. 7). With the exception of an elongated snout and large body size (apparently associated with their semiterrestrial habits), these two genera exhibit no shared features which are not also found in *Tupaia* or other tupaiines. Moreover, closer inspection of the dental features examined by Steele reveals no shared character states which are unique to *Urogale* and *Lyonogale*; instead, all their shared similarities are widely distributed along the Tupaiinae.

The character analysis of the present study, as well as that of Butler (this volume), fails to clarify the nature of sister-group relationships among the genera of Tupaiinae (Fig. 7 and Table 5). The occurrence of an enlarged zygomatic (malar) foramen and posterior palatal vacuities in *Tupaia* and *Lyonogale* are shared derived character states which suggest a recent common ancestor. The immunological studies of Dene et al. (1976, this volume) also support the hypothesis that *Lyonogale* and *Tupaia* are more closely related to each other than either is to *Urogale*. As emphasized by Martin (1968a), suggestions for the return of *Lyonogale* and *Anathana* to the genus *Tupaia* (Fiedler, 1956) seem premature without a detailed character analysis of dental, cranioskeletal, and pelage traits in all tupaiid genera.

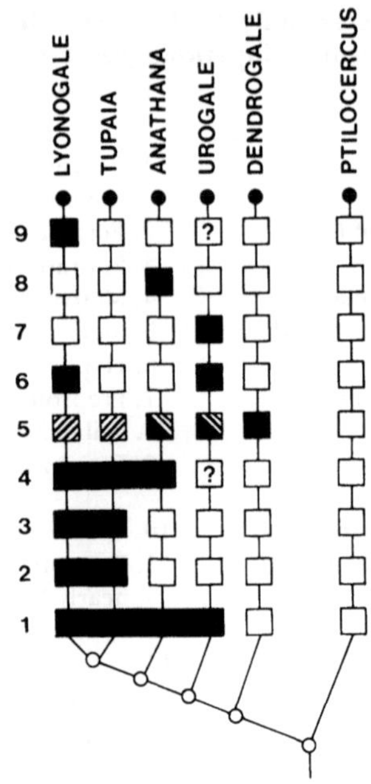

Fig 7. Character phylogeny of selected features within the subfamily Tupaiinae. Relationships within Tupaiinae are unclear from the analysis, although characters 2 and 3 support the sister group relationship of *Tupaia* and *Lyonogale*. See Table 5 for the characters evaluated. Note: For character 5, ■ = hypocone lost; ▧ = hypocone enlarged; ▨ = hypocone reduced.

Table 5. Character analysis of selected features which exhibit variation within the subfamily Tupaiinae (see Fig. 7)

Primitive	Derived
1. Upper canine double rooted	1. Upper canine single rooted
2. Zygomatic (malar) foramen small	2. Zygomatic (malar) foramen large
3. Posterior palatal vacuities absent	3. Posterior palatal vacuities present
4. Scaphoid and lunate unfused	4. Scaphoid and lunate fused
5. Hypocone moderately developed	5. Hypocone lost or enlarged
6. Snout moderately developed	6. Snout greatly elongated
7. I^2 subequal to I^1; I_3 moderately developed	7. I^2 larger than I^1; I_3 reduced and functionless
8. Entoconid somewhat lower than hypoconid	8. Entoconid elevated relative to hypoconid
9. Cecum present	9. Cecum absent

5. *Phylogenetic Relationships of Tupaiidae to Other Eutheria*

Since the time of Haeckel's (1866) association of Tupaiidae and Macroscelididae in a separate suborder Menotyphla of the Insectivora, numerous hypotheses have been proposed concerning the possible phylogenetic relationships of Tupaiidae with other Eutheria. Testing of these hypotheses has been handicapped by the absence of early Tertiary fossil tupaiids. Analyses of the skull, dentition, postcranial skeleton, nervous system, reproductive system, and serum proteins have played a major role in assessing the possible affinities of tree shrews, and the evidence from each of these systems is presented in detail elsewhere in this volume. Comparative aspects of social, sexual, and maternal behavior in several species of *Tupaia* and *Lyonogale* in captivity have been discussed by Sorenson (1970) and Martin (1968a, 1975), and D'Souza (1974) has presented a preliminary report on the behavior and ecology of *T. minor* in the field. Although extremely interesting from a biological viewpoint, these behavioral studies have contributed little to our understanding of tupaiid affinities.

A. Possible Affinities of Tupaiidae with Macroscelididae or Extant Insectivora

The initial association of Tupaiidae with the order Insectivora, and with the Macroscelididae in particular, was based upon their common retention of symplesiomorphous eutherian features. Character analyses of numerous cranioskeletal features by Novacek (1977, this volume), as well as those of the present study (Figs. 5, 8), provide no evidence to support a special phylogenetic relationship between Tupaiidae and Macroscelididae. Immunodiffusion studies also indicate a lack of special affinity between Tupaiidae and Macroscelididae (Dene et al., 1976, this volume). Neither anatomical nor molecular data provide evidence for a special relationship between Tupaiidae and any family of extant lipotyphlous insectivorans.

B. Tupaiid-Leptictid Affinities

Several authors (McDowell, 1958; Van Valen, 1965; McKenna, 1966) have discussed the possibility of a special relationship between tupaiids and the Paleocene-Oligocene family Leptictidae, based in part on their common possession of an entotympanic bulla. However, Novacek's (1977, this volume) recent character analysis of numerous dental, cranial, and postcranial features in leptictids, tupaiids, macroscelidids, lipotyphlans, and primates does not corroborate a hypothesis of tupaiid-leptictid affinities. Many of the shared

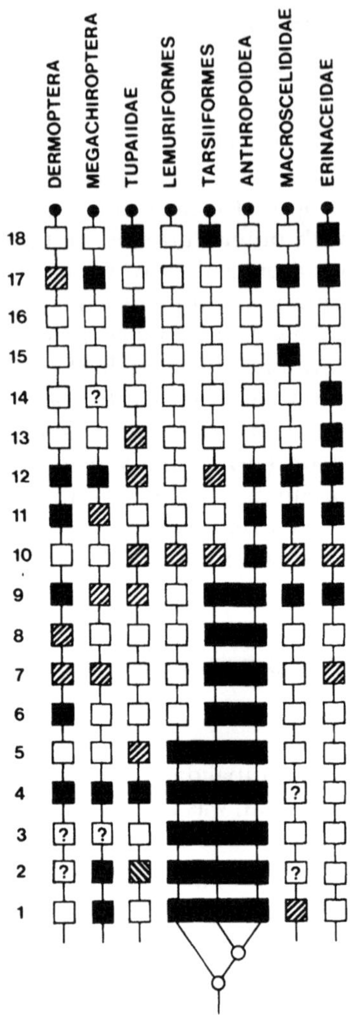

Fig. 8. Character phylogeny of fetal membrane and other anatomical features of tupaiids, primates, and selected eutherians, including those taxa which have been assigned to the Archonta by various authors. Intermediately derived character states are indicated by obliquely-lined squares. Note that the monophyly of extant Primates is corroborated by this analysis, whereas possible relationships of tupaiids and other archontans to primates are unclear. See Table 6 for the characters evaluated.

similarities between the two families appear to be plesiomorphous eutherian features and therefore provide no evidence of their possible affinities. Novacek (1977) has emphasized the relatively remote phyletic position of Leptictidae within the Eutheria, although the possibility remains for a close leptictid-macroscelidid relationship.

C. *Tupaiid-Mixodectid Affinities*

Szalay (1969) suggested that the Paleocene family Mixodectidae probably exhibits affinities with Tupaiidae, based on several shared similarities in their dentition. More recently, Szalay (1977b) formally allocated the Tupaiidae and Mixodectidae to his order Scandentia (see Fig. 9b), apparently on the basis of shared dental similarities. Butler's (this volume) analysis of these dental features does not corroborate Szalay's hypothesis of special affinities between tupaiids and mixodectids. In the absence of postcranial or adequate cranial remains of mixodectids, it seems premature to suggest a possible phylogenetic relationship between tupaiids and mixodectids.

D. *Tupaiid-Primate Affinities*

Tree shrews were characterized by Simpson (1945, p. 183) as "either the most primate-like insectivores or the most insectivore-like primates." Simpson's decision to include Tupaiidae within the infraorder Lemuriformes was influ-

Table 6. Character analysis of fetal membranes and other anatomical features in Tupaiidae and selected Eutheria (see Fig. 8)

Primitive	Derived
1. Antimesometrial orientation of embryonic disc at implantation	1. Orthomesometrial orientation of embryonic disc at implantation
2. Uncrossed corticospinal tract in ventral funiculus[1]	2. Crossed corticospinal tract in lateral funiculus[1]
3. Triradiate calcarine sulcus complex absent	3. Triradiate calcarine sulcus complex present
4. Lateral geniculate nucleus not clearly laminated	4. Lateral geniculate nucleus of thalamus laminated
5. Large olfactory bulbs	5. Olfactory bulbs reduced
6. Choriovitelline placenta present	6. Choriovitelline placenta absent
7. Allantoic vesicle large	7. Allantoic diverticulum reduced
8. Mesodermal body stalk absent	8. Mesodermal body stalk permanent
9. Chorioallantoic placenta epitheliochorial	9. Chorioallantoic placenta hemochorial
10. Uterus duplex	10. Uterus simplex
11. Blastocyst attachment at paraembryonic pole	11. Blastocyst attachment at embryonic pole
12. Primordial amniotic cavity absent; amniogenesis by folding	12. Primordial amniotic cavity present; amniogenesis by cavitation
13. Definitive yolk sac free, reduced	13. Definitive yolk sac incompletely inverted
14. Cecum present	14. Cecum absent
15. Seminal vesicles present	15. Seminal vesicles absent
16. Gland-free, endometrial pads absent	16. Bilateral, gland-free, endometrial pads present
17. Sublingua present	17. Sublingua absent
18. Os penis present	18. Os penis absent

[1]It is unclear whether any of the various character states of this feature can be considered more primitive.

enced considerably by the earlier studies of Carlsson (1922) and Le Gros Clark (1924a, 1924b, 1925, 1932); however, he presented no discussion of the morphological data which support such a relationship. Le Gros Clark (1959, 1971) summarized most of the evidence for the primate affinities of Tupaiidae. In particular, he emphasized the similar evolutionary "trends" in tupaiids and primates for construction of the orbito-temporal region and auditory bulla, elaboration of the visual pathways of the eye and associated brain centers, reduction of the olfactory apparatus, development of a "dental comb" and similar cusp pattern of the teeth, ossicles of the middle ear, the serrated sublingua of the tongue, and certain features of placentation. On the basis of these and additional features, Le Gros Clark (1971, p. 71) concluded that tree shrews "show in their total morphological pattern such a remarkable number of resemblances to the lemurs, and particularly the Lemuriformes, that in spite of their primitive characters their affinities with the latter seem reasonably certain." All of these shared similarities have been re-examined during the past two decades, and phylogenetic analysis of these features suggests that a special tupaiid-primate relationship is unlikely.

Simpson's (1945) characterization of tupaiids as insectivore-like primates has had a pronounced effect on subsequent studies of tupaiid affinities, and many investigators have limited their comparisons of traits in tree shrews with those of lemuriform primates and "representative" insectivorans (e.g., Saban, 1956/1957). Van Valen (1965) has criticized such a restricted analysis of tupaiid affinities, and he emphasized the necessity of determining character states in a wide range of Eutheria when assessing the phylogenetic relationships of tree shrews. As a result of his studies, Van Valen concluded that most, if not all, similarities of the skull and dentition between tupaiids and extant primates are the result of shared primitive retentions or convergent evolution. This hypothesis has been corroborated by character analyses of numerous soft anatomical features (Figs. 5, 8), in particular, the nervous system (Campbell, 1966, 1974, this volume), reproductive systems (Martin, 1968a, 1968b, 1975), placentation (Hill, 1965; Luckett, 1969, 1974), and myology (Campbell, 1974). Similar conclusions have been reached as the result of further studies of dental, cranial, and postcranial features in tupaiids, primates, and other eutherians (Butler, this volume; Cartmill and MacPhee, this volume; Novacek, this volume).

E. Ordinal Status of Tupaiidae

The family Tupaiidae does not appear to share any uniquely derived character states with Primates, Macroscelidea, Dermoptera, Chiroptera, or Insectivora (= Lipotyphla), and Butler (1972) has suggested that tupaiids be placed in a separate order Scandentia[2]. Butler's ordinal concept was adopted subsequently by McKenna (1975) and Szalay (1977b) in their classifications of mammals. The earliest recommendation for ordinal distinctness of tree shrews

[2]Scandentia was originally utilized as a family name for tree shrews by J. A. Wagner in 1855 (see Gregory, 1910, for a reproduction of Wagner's classification of mammals).

was apparently that by Straus (1949), although he presented no evidence to support the separation of his order Tupaioidea from Primates. Tupaioidea was also employed as an ordinal designation by Haines and Swindler (1972) and Goodman (1975), following a consideration of the neuroanatomical and molecular evidence, respectively. Although there are no precise rules for the derivation of ordinal names, the International Code of Zoological Nomenclature has recommended that the ending-*oidea* be adopted for the names of superfamilies, and this recommendation has been followed in virtually all recent mammalian classifications. In order to avoid ambiguity between the superfamilial and ordinal ranking of tree shrews in different classifications, it is suggested here that the name Scandentia be adopted as the ordinal designation for the group.

Some investigators (e.g. Van Valen, 1965) who have recognized the phyletic distinctness of tree shrews from other eutherians have preferred to retain the family Tupaiidae within the order Insectivora. This difference of opinion over taxonomic rank is the result of alternative philosophies concerning the degree to which the cladistic or patristic aspects of a phylogenetic hypothesis should be expressed in a formal classification. Thus, some taxonomists (Simpson, 1945; Romer, 1966) have included within the Insectivora those Cretaceous-early Tertiary and Recent eutherians which are not related clearly to other orders and which have retained a considerable number of primitive eutherian features. Bulter (1972) has emphasized that "waste-basket" taxa such as Insectivora are only temporary solutions for tidying up the classificiation of mammals, and that sooner or later they must be sorted out. Analyses of cranioskeletal features in extant lipotyphlans (Butler, 1956, 1972; McDowell, 1958; Novacek, unpublished manuscript) lend support to Gregory's (1910) assertion that Lipotyphla is a "natural" (= monophyletic) group. As a result of these and other studies, Butler (1972) has recommended separate ordinal status for Lipotyphla, Scandentia, and Macroscelidea. The findings of the present study, as well as the available skeletal, molecular, and soft anatomical data, provide further support for Butler's proposal. The remaining Cretaceous-early Tertiary eutherians whose affinities are uncertain have been allocated temporarily by Butler (1972) to the paraphyletic or "waste-basket" order Proteutheria.

F. The Archonta Hypothesis

Gregory (1910, p. 322) hypothesized that "the orders Menotyphla, Dermoptera, Chiroptera, and Primates have had a common origin, possibly from some Upper Cretaceous family resembling in many characters the Tupaiidae," and he proposed a superorder Archonta to embrace these four taxa. Although a number of traits shared by Tupaiidae and Lemuriformes were cited by Gregory, he presented no discussion of characters states possessed by all members of his superorder Archonta. Support for Gregory's hypothesis was provided by Butler's (1956) review of cranial features in Lipotyphla, Leptictidae,

Menotyphla, Primates, and Dermoptera. Butler suggested that there was a major phylogenetic dichotomy between Lipotyphla and the Archonta, and he postulated that archontans were derived from early erinaceomorphs. Although Butler discussed the relative primitive or derived nature of most of the character states presented, many of his common archontan features are eutherian symplesiomorphs, and others are of uncertain phylogenetic significance. More recently, McKenna (1975) has endorsed a modified version of the Archonta (excluding Macroscelidea), but he presented no discussion of shared derived characters which would unite his grandorder Archonta.

Szalay (1977b) has suggested that the Archonta (minus Macroscelidea) is a monophyletic taxon (Fig. 9b), based on the identification of several synapomorphies in the foot, with particular emphasis on the astragalocalcaneal complex. He postulated that the earliest archontans were arboreal forms which were similar to *Ptilocercus* postcranially, and that the origin of Archonta was from an unknown terrestrial stock of adapisoricid erinaceoids. No discussion of tarsal features in Chiroptera was presented by Szalay, although he emphasized that similarities in patagial morphology between Chiroptera and Dermoptera provide support for the hypothesis of a sister-group relationship between the two orders. An extensive analysis of postcranial features in archontans, lipotyphlans, macroscelideans, and leptictids (Novacek, this volume) indicates that few, if any, of the tarsal synapomorphies of archontans are uniquely derived. (For a contrasting view, however, see Szalay and Drawhorn, this volume.)

An alternative hypothesis of archontan relationships (Fig. 9a) has been postulated on the basis of molecular evidence (Goodman, 1975; Dene et al., this volume). The available immunodiffusion data provide no evidence for a special relationship between Chiroptera and other archontans, and thus the Chiroptera as well as Macroscelidea are excluded from the Archonta in Good-

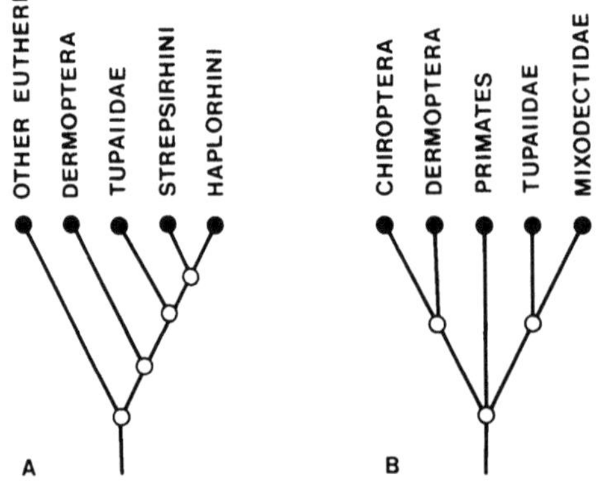

Fig. 9. Two alternative hypotheses of cladistic relationships among taxa of the Archonta. A. Adapted from Goodman (1975). B. Adapted from Szalay (1977b).

man's (1975) scheme. The albumin plus transferrin immunological distance data analysed by Sarich and Cronin (1976) are also consistent with Goodman's hypothesis of the Archonta, although differences are evident in the branching sequences and resulting classifications derived by the two teams of investigators.

The several hypotheses of archontan phylogenetic relationships have not been adopted by all mammalian systematists. Simpson (1945) believed that Gregory's (1910) Archonta is an "unnatural" group, although he did favor a close tupaiid-primate affinity. Simpson considered it to be unlikely that primates are more closely related to bats than to insectivorans; instead, he postulated an early dichotomy of most eutherians into the cohorts Unguiculata and Ferungulata. The Unguiculata included the orders Insectivora, Chiroptera, Primates, Dermoptera, Edentata, and Pholidota, as well as the extinct orders Tillodontia and Taeniodonta. Although Simpson believed that these diverse orders have much in common, he presented no list of diagnostic features of the cohort Unguiculata; instead, he acknowledged that their living representatives have retained numerous primitive eutherian features. There appears to be little, if any, evidence to corroborate the concept of a monophyletic clade which corresponds to Simpson's Unguiculata.

Martin (1968b) has suggested that the taxa included in Gregory's (1910) Archonta are characterized by the retention of numerous symplesiomorph eutherian features and by the development of specializations for arboreal life. Martin also postulated that the ancestors of viviparous (= therian) mammals were probably arboreal; therefore, differences between Archonta and Lipotyphla may be the result of the secondary adaptation of lipotyphlans to terrestrial habits. Clearly, analysis of possible relationships among archontans requires the concomitant consideration of the nature of the postcranial and locomotor adaptations of the earliest therians and eutherians (for further discussion, see Jenkins, 1974; Szalay and Decker, 1974; Cartmill, 1974; Novacek, this volume; Szalay and Drawhorn, this volume).

6. *Discussion*

Resurgence of interest in the comparative biology of tree shrews during the past two decades was related in part to the belief that these animals might serve as a model for the "insectivorous" mammals which gave rise to the order Primates. This implication is particularly evident in a number of studies on the nervous, reproductive, and locomotor systems of tupaiids. However, caution should be exercised in using living species as models for the study of evolutionary change in lineages. Concepts such as "primitive species" or "living fossils" are invalid for extant organisms; all living species exhibit a mosaic distribution of primitive and derived character states. Therefore, the use of tupaiids as models for the study of evolutionary change in primates or other eutherians should be preceded by character analysis and morphotype reconstruction for each organ system and taxon investigated. Such analysis supports the use of scansorial tupaiids as a model for the study of form-function relationships which may have been associated with the arboreal adaptations of early primates (Jenkins, 1974; Szalay, 1977b). However, it remains uncertain whether tupaiids and primates share any uniquely derived postcranial traits which might corroborate their monophyletic relationship (Novacek, this volume).

The hypothesis of a monophyletic relationship between Tupaiidae and Primates has not been corroborated by phylogenetic analysis of the available anatomical and molecular data. It is possible, however, that future discovery of early Tertiary tupaioids might provide new evidence for reconsidering the validity of tupaiid-primate affinities. Previous assessments of possible phylogenetic relationships between tupaiids and primates have been handicapped by the absence of a clear diagnosis of the order Primates. Thus, Le Gros Clark (1959, 1971) characterized the order by a series of "evolutionary trends" which included: (1) plesiomorphous eutherian features, such as pentadactyl limbs and retention of the clavicle; (2) apomorphous features which characterize some, but not all, primates, such as the replacement of claws by nails; and (3) a series of "progressive trends" which serve to characterize the Anthropoidea rather than the Primates as a whole. In contrast, other investigators (Martin, 1968b; Szalay, 1973, 1975) have urged that Primates be diagnosed solely by their shared, derived character states. As a result of their retention of numerous primitive eutherian traits, it has been difficult to detect dental and cranioskeletal features which are autapomorphous for Cretaceous-Recent Primates. One such attribute which characterizes the known skulls of Paleocene-Recent Primates is an auditory bulla composed almost entirely by the petrosal (McKenna, 1966; Szalay, 1973, 1975). Other derived character states shared by all primates include features which occur sporadically in other eutherian taxa (Table 7); however, joint possession of this suite of synapomorphies is unique to Primates. Soft anatomical character states that are shared by all extant primates presumably occurred in their last common ancestor, and it is parsi-

Table 7. Eutherian apomorphous character states diagnostic of the order Primates

1. Auditory bulla formed solely by petrosal, with ectotympanic ring "intrabullar" or marginal
2. Median branch of primitive internal carotid artery lost in adult[1]
3. Astragalocalcaneal complex of foot adapted for habitual inversion[2]
4. Major corticospinal tract in lateral funiculus of spinal cord[3]
5. Triradiate calcarine sulcus complex
6. Lateral orientation of embryonic disc during implantation[4]

[1]Also occurs in Tupaiidae, Lipotyphla, and Macroscelidea.
[2]Also occurs in Tupaiidae and Dermoptera.
[3]Also occurs in Chiroptera and Carnivora.
[4]Also occurs in Tenrecinae and some families of Chiroptera.

monious to assume that these conservative traits also characterized plesiadapiform primates. Phylogenetic analysis of additional characters in fossil and extant primates and other eutherians should result in the identification of new synapomorphies for diagnosis of the order Primates.

The hypothesis of a possible superordinal relationship which includes Scandentia, Primates, and Dermoptera requires further testing. Those investigators who have speculated on superordinal relationships within Eutheria (Gregory, 1910; Simpson, 1945; McKenna, 1975; Szalay, 1977b) have acknowledged the tenuous nature of such hypotheses, and most of these hypotheses have not been rigorously tested by character analyses. To date, the best evidence for supporting the archontan hypothesis is that provided by postcranial and immunological data. Caution should be exercised, however, in using quantitative measurements, such as immunological distance or brain-body weight ratios, for assessing phylogenetic relationships or delimiting taxonomic boundaries. In most of these cases, it is not possible to clearly distinguish between the relative amount of shared similarity which may be due to symplesiomorphy, synapomorphy, or homoplasy. On the other hand, amino acid sequence data are more amenable to reconstruction of presumed ancestral and derived character states, and the continued accumulation of protein sequences should provide valuable evidence for testing hypotheses of superordinal relationships. At present, there are extensive sequence data for anthropoideans, whereas only a few sequences are available for tarsier, strepsirhines, tupaiids, bats, and hedgehogs, and none have been reported for other lipotyphlans, macroscelidids, or dermopterans (see Dene et al., this volume).

Most studies of tupaiid biology have focused upon their possible higher taxonomic relationships to primates or other eutherians, while intrafamilial variation and affinities have received relatively little attention since Lyon's (1913) major review. The available anatomical and molecular data corroborate the tupaiine-ptilocercine dichotomy, whereas cladistic relationships within the Tupaiinae are unclear from analysis of the cranial and dental features discussed by Lyon (1913), Steele (1973), and Butler (this volume). In addition to further analyses of dental and cranioskeletal traits for evaluating affinities among the genera and species of tupaiines, a concerted effort should be made to extend the immunological studies discussed by Dene et al. and Cronin and Sarich (this volume). Another source of potentially valuable data for clarification of intrafamilial relationships is the analysis of chromosomal banding patterns. Most karyological studies of tupaiids have been limited to reports of chromosomal numbers and fundamental numbers for *Urogale* and several species of *Tupaia* (Arrighi et al., 1969). As yet, these data have provided little evidence for evaluating tupaiid affinities, even within the genus *Tupaia*. On the other hand, modern banding techniques allow a greater degree of certainty concerning the homology of various chromosomal arms between species, and thus may be of considerable value in assessing relationships at the species and generic level. Chromosomal banding patterns have been determined for *Tupaia chinensis* (Mandahl, 1976) but are unknown for other species of tupaiids.

Insight into the cladistic relationships among genera and species of tupaiids might also result from an analysis of their zoogeographic distribution. Paleogeographic, floral, and faunal evidence indicates that Borneo, Sumatra, and Java were continuous intermittently with the Asian mainland during the Pleistocene (Darlington, 1957), and it is probable that the islands of Indonesia have been isolated for only about 10,000–50,000 years. It is equally possible that speciation occurred among the island populations of tupaiines as a result of their relatively recent geographic isolation. The islands of Palawan and Mindanao in the Philippines were also continuous with, or in closer proximity to, Borneo during the Pleistocene, and this would have facilitated the eastward spread of tupaiines into these islands. *Urogale* is restricted zoogeographically to Mindanao, and analysis of chromosomal banding patterns and serological data among *Urogale* and the tupaiine genera of Borneo (*Tupaia, Lyonogale, Dendrogale*) might clarify their interrelationships.

7. Summary

During the past 200 years, studies of tupaiid systematic relationships have resulted in several alternative hypotheses for their possible affinities with insectivorans, macroscelidids, primates, leptictids, and mixodectids. In most cases, these suggested relationships have been based on concepts of overall (phenetic) similarity. Character analysis indicates that a majority of the shared similarities used to support these phylogenetic hypotheses may be the result of primitive retentions or convergence. In contrast, tree shrews share few, if any, uniquely derived character states with primates or other eutherians. This apparently remote phyletic position provides support for Butler's (1972) suggestion that Tupaiidae should be allocated to a separate order Scandentia. Alternative taxonomic schemes which would be consistent with current hypotheses of eutherian phylogeny are the inclusion of Tupaiidae within an admittedly paraphyletic order Insectivora, or else to list them simply as Eutheria, *incertae sedis*.

Superordinal affinities of Scandentia remain unclear, although the available immunological data do support a modified hypothesis of archontan affinities for Scandentia, Primates, and Dermoptera. In contrast, phylogenetic analysis of cranioskeletal and soft anatomical features provides little evidence for uniquely derived (autapomorphous) traits shared by archontans. The possibility persists that many shared similarities of the postcranial skeleton may be the result of convergent evolution (or primitive retention) in scansorial or arboreal tupaiids, primates, and dermopterans. Further assessment of the superordinal relationships of Scandentia awaits the discovery of cranial and postcranial remains of early Tertiary tupaioids, dermopterans, mixodectids, and microsyopids.

ACKNOWLEDGMENT

The writer wishes to thank his colleagues, in particular, Drs. P. Butler, M. Cartmill, R. MacPhee, and M. Novacek, for their careful reading and constructive criticism of this manuscript and its character analyses. This of course does not imply their complete agreement with the hypotheses presented. The illustrations were prepared by Dr. Nancy Hong.

8. References

Arrighi, F. E., Sorenson, M. W., and Shirley, L. R. 1969. Chromosomes of the tree shrews (Tupaiidae). *Cytogenetics* 8:199–208.

Bock, W. J. 1977. Foundations and methods of evolutionary classification, pp. 851–895. *In* M. K. Hecht, P. C. Goody, and B. M. Hecht (eds.). *Major Patterns in Vertebrate Evolution.* Plenum Press, New York.

Bonde, N. 1977. Cladistic classification as applied to vertebrates, pp. 741–804. *In* M. K. Hecht, P. C. Goody, and B. M. Hecht (eds.) *Major Patterns in Vertebrate Evolution.* Plenum Press, New York.

Butler, P. M. 1956. The skull of *Ictops* and the classification of the Insectivora. *Proc. Zool. Soc. Lond.* 126: 453–481.

Butler, P. M. 1972. The problem of insectivore classification, pp. 253–265. *In* K. A. Joysey and T. S. Kemp (eds.). *Studies in Vertebrate Evolution.* Oliver and Boyd, Edinburgh.

Cain, A. J., and Harrison, G. A. 1960. Phyletic weighting. *Proc. Zool. Soc. Lond.* 135: 1–31.

Campbell, C. B. G. 1966. The relationships of the tree shrews: The evidence of the nervous system. *Evolution* 20: 276–281.

Campbell, C. B. G. 1974. On the phyletic relationships of the tree shrews. *Mammal Rev.* 4: 125–143.

Carlsson, A. 1922. Uber die Tupaiidae und ihre Beziehungen zu den Insectivora und den Prosimiae. *Acta Zool.* 3: 227–270.

Cartmill, M. 1974. Pads and claws in arboreal locomotion, pp. 45–83. *In* F. A. Jenkins, Jr. (ed). *Primate Locomotion.* Academic Press, New York.

Darlington, P. J., Jr. 1957. *Zoogeography: The Geographic Distribution of Animals.* John Wiley and Sons, Inc., New York.

Davis, D. D. 1938. Notes on the anatomy of the treeshrew *Dendrogale*. *Field Mus. Nat. Hist., Zool. Series* 20: 383–404.

Dene, H. T., Goodman, M., and Prychodko, W. 1976. Immunodiffusion evidence on the phylogeny of the primates, pp. 171–195. *In* M. Goodman and R. E. Tashian (eds.). *Molecular Anthropology.* Plenum Press, New York.

Diard, M. 1820. Literary and philosophical intelligence. *Asiatic J. Month. Reg.* 10: 477–478.

D'Souza, F. 1974. A preliminary field report of the lesser tree shrew *Tupaia minor*, pp. 167–182. *In* R. D. Martin, G. A. Doyle, and A. C. Walker (eds.). *Prosimian Biology.* Gerald Duckworth and Co., London.

Evans, F. G. 1942. The osteology and relationships of the elephant shrews (Macroscelididae). *Bull. Amer. Mus. Nat. Hist.* 80: 85–125.

Fiedler, W. 1956. Ubersicht über das System der Primates, pp. 1–266. *In* H. Hofer, A. H. Schultz, and D. Starck (eds.). *Primatologia,* Vol. 1. S. Karger, Basel.

Goodman, M. 1975. Protein sequence and immunological specificity: Their role in phylogenetic studies of primates, pp. 219–248. *In* W. P. Luckett and F. S. Szalay (eds.). *Phylogeny of the Primates.* Plenum Press, New York.

Gray, J. E. 1825. An outline of an attempt at the disposition of Mammalia into tribes and families, with a list of the genera apparently appertaining to each tribe. *Ann. Philos.* 10: 337–344.

Gray, J. E. 1848. Description of a new genus of insectivorous Mammalia, or Talpidae, from Borneo. *Proc. Zool. Soc. Lond.* 1848: 23–24.

Gray, J. E. 1860. Early notice of the *Tapaia* found in Pulo Condore. *Ann. Mag. Nat. Hist.* 5: 71.

Gregory, W. K. 1910. The orders of mammals. *Bull. Amer. Mus. Nat. Hist.* 27: 1–524.

Gregory, W. K. 1920. On the structure and relationships of *Notharctus*, an American Eocene primate. *Mem. Amer. Mus. Nat. Hist.* 3: 49–243.

Haeckel, E. 1866. *Generelle Morphologie der Organismen.* G. Reimer, Berlin.

Haines, D. E., and Swindler, D. R. 1972. Comparative neuroanatomical evidence and the taxomomy of the tree shrews (*Tupaia*). *J. Human Evol.* 1: 407–420.

Hecht, M. K., and Edwards, J. L. 1976. The determination of parallel or monophyletic relationships: The proteid salamanders—a test case. *Amer. Nat.* 110: 653–677.

Hennig, W. 1966. *Phylogenetic Systematics.* Univ. of Illinois Press, Urbana.

Hill, J. P. 1965. On the placentation of *Tupaia*. *J. Zool.* 146: 278–304.

Huxley, T. H. 1872. *A Manual of the Anatomy of Vertebrated Animals.* D. Appleton and Co., New York.

Jenkins, F. A., Jr. 1974. Tree shrew locomotion and the origins of primate arborealism, pp. 85–115. *In* F. A. Jenkins, Jr. (ed.). *Primate Locomotion.* Academic Press, New York.

Kaas, J. H., Huerta, M. F., Weber, J. T., and Harting, J. K. 1978. Patterns of retinal terminations and laminar organization of the lateral geniculate nucleus of primates. *J. Comp. Neur.* 182: 517–554.

Kaudern, W. 1911. Studien über die männlichen Geschlechtsorgane von Insectivoren und Lemuriden. *Zool. Jahrb.* 31: 1–106.

Leche, W. 1885. Uber die Säugethiergattung *Galeopithecus*. *Kongl. Svenska Vet. Akad. Handl.* 21: 1–92.

Le Gros Clark, W. E. 1924a. The myology of the tree-shrew (*Tupaia minor*). *Proc. Zool. Soc. Lond.* 1924: 461–497.

Le Gros Clark, W. E. 1924b. On the brain of the tree-shrew (*Tupaia minor*). *Proc. Zool. Soc. Lond.* 1924: 1053–1074.

Le Gros Clark, W. E. 1925. On the skull of *Tupaia*. *Proc. Zool. Soc. Lond.* 1925: 559–567.

Le Gros Clark, W. E. 1926. On the anatomy of the pen-tailed tree-shrew (*Ptilocercus lowii*). *Proc. Zool. Soc. Lond.* 1926: 1179–1309.

Le Gros Clark, W. E. 1932. The brain of the Insectivora. *Proc. Zool. Soc. Lond.* 1932: 975–1013.

Le Gros Clark, W. E. 1959. *The Antecedents of Man.* Edinburgh University Press, Edinburgh.

Le Gros Clark, W E. 1971. *The Antecedents of Man,* 3rd ed. Edinburgh University Press, Edinburgh.

Luckett, W. P. 1969. Evidence for the phylogenetic relationships of tree shrews (family Tupaiidae) based on the placenta and foetal membranes. *J. Reprod. Fert.*, Suppl. 6: 419–433.

Luckett, W. P. 1974. The comparative development and evolution of the placenta in Primates. *Contrib. Primat.* 3: 142–234.

Luckett, W. P. 1975. Ontogeny of the fetal membranes and placenta: Their bearing on primate phylogeny, pp. 157–182. *In* W. P. Luckett and F. S. Szalay (eds.). *Phylogeny of the Primates.* Plenum Press, New York.

Lyon, M. W., Jr. 1913. Treeshrews: An account of the mammalian family Tupaiidae. *Proc. U. S. Nat. Mus.* 45: 1–188.

Mandahl , N. 1976. G- and C-banded chromosomes of *Tupaia chinensis* (Mammalia, Primates). *Hereditas* 83: 131–134.

Martin, R.D. 1968a. Reproduction and ontogeny in tree-shrews (*Tupaia belangeri*), with reference to their general behaviour and taxonomic relationships. *Z. Tierpsychol.* 25: 409–532.

Martin, R. D. 1968b. Towards a new definition of Primates. *Man* 3: 377–401.

Martin, R. D. 1975. The bearing of reproductive behavior and ontogeny on strepsirhine phylogeny, pp. 265–297. *In* W. P. Luckett and F. S. Szalay (eds.). *Phylogeny of the Primates.* Plenum Press, New York.

Mayr, E. 1974 Cladistic analysis or cladistic classification? *Z. Zool. Syst. Evolut.-Forsch.* 12: 94–128.

McDowell, S. B., Jr. 1958. The Greater Antillean insectivores. *Bull. Amer. Mus. Nat. Hist.* 115: 113–214.

McKenna, M. C. 1963. New evidence against tupaioid affinities of the mammalian family Anagalidae. *Amer. Mus. Novitates* 2158: 1–16.

McKenna, M. C. 1966. Paleontology and the origin of Primates. *Folia Primat.* 4: 1–25.

McKenna, M. C. 1975. Toward a phylogenetic classification of Mammalia, pp. 21–46. *In* W. P. Luckett and F. S. Szalay (eds.). *Phylogeny of the Primates.* Plenum Press, New York.

Mivart, St. G. 1868. Notes on the osteology of the Insectivora. *J. Anat. Physiol.* 2: 117–154.

Novacek, M. J. 1977. Evolution and relationships of the Leptictidae (Eutheria: Mammalia). Ph. D. Thesis, University of California, Berkeley.

Peters, W. 1864. Uber die Säugethiergattung *Solenodon*. *Abhandl. König. Akad. Wissensch. Berlin*, pp. 1–22.

Raffles, T. S. 1822. Descriptive catalogue of a zoological collection, made on account of the honourable East India Company, in the island of Sumatra and its vicinity. *Trans. Linn. Soc. Lond.* 13: 239–274.

Romer, A. S. 1966. *Vertebrate Paleontology,* 3rd ed. University of Chicago Press, Chicago.

Saban, R. 1956/1957. Les affinitiés du genre *Tupaia* Raffles 1821, d'aprés les charactères morphologiques de la tête osseuse. *Ann. Paléont.* 42: 169–224; 43: 1–44.

Sarich, V. M., and Cronin, J. E. 1976. Molecular systematics of the primates, pp. 141–170. *In* M. Goodman and R. E. Tashian (eds.). *Molecular Anthropology.* Plenum Press, New York.

Simpson, G. G. 1945. The principles of classification and a classification of mammals. *Bull. Amer. Mus. Nat. Hist.* 85: 1–350.

Simpson, G. G. 1961. *Principles of Animal Taxonomy.* Columbia University Press, New York.

Simpson, G. G. 1975. Recent advances in methods of phylogenetic inference, pp. 3–19. *In* W. P. Luckett and F. S. Szalay (eds.). *Phylogeny of the Primates.* Plenum Press, New York.

Sorenson, M. W. 1970. Behavior of tree shrews, pp. 141–194. *In* L. A. Rosenblum (ed.). *Primate Behavior,* Vol. I. Academic Press, New York.

Steele, D. G. 1973. Dental variability in the tree shrews (Tupaiidae). *Symp. 4th Int. Congr. Primatol.* 3: 154–179.

Straus, W. L., Jr. 1949. The riddle of man's ancestry. *Quart. Rev. Biol.* 24: 200–223.

Szalay, F. S. 1969. Mixodectidae, Microsyopidae, and the insectivore-primate transition. *Bull. Amer. Mus. Nat. Hist.* 140: 193–330.

Szalay, F. S. 1973. New Paleocene primates and a diagnosis of the new suborder Paromomyiformes. *Folia Primat.* 19: 73–87.

Szalay, F. S. 1975. Where to draw the nonprimate-primate taxonomic boundary. *Folia Primat.* 23. 158–163.

Szalay, F. S. 1977a. Ancestors, descendents, sister groups and testing of phylogenetic hypotheses. *Syst. Zool.* 26: 12–18.

Szalay, F. S. 1977b. Phylogenetic relationships and a classification of the eutherian Mammalia, pp. 315–374. *In* M. K. Hecht, P. C. Goody, and B. M. Hecht (eds.). *Major Patterns in Vertebrate Evolution.* Plenum Press, New York.

Szalay, F. S., and Decker, R. L. 1974. Origins, evolution, and function of the tarsus in late Cretaceous Eutheria and Paleocene primates, pp. 223–259. *In* F. A. Jenkins, Jr. (ed.). *Primate Locomotion.* Academic Press, New York.

Van Kampen, P. N. 1905. Die Tympanalgegend des Säugetierschädels. *Morph. Jahrb.* 34: 322–722.

Van Valen, L. 1965. Treeshrews, primates, and fossils. *Evolution* 19: 137–151.

Verma, K. 1965. Notes on the biology and anatomy of the Indian tree-shrew, *Anathana wroughtoni*. *Mammalia* 29: 289–330.

Weber, M. 1904. *Die Säugetiere.* G. Fischer, Jena.

Weber, M. 1928. *Die Säugetiere,* 2nd ed., Vol. 2. G. Fischer, Jena.

Winge, H. 1941. *The Interrelationships of the Mammalian Genera.* Vol. 1. C. A. Reitzels Forlag, Copenhagen.

Cranioskeletal System and Dentition

II

Cranioskeletal Features in Tupaiids and Selected Eutheria as Phylogenetic Evidence

2

MICHAEL J. NOVACEK

1. Introduction

Anyone hoping to illustrate current difficulties with mammalian systematics might well recruit the problem of tree shrew relationships as an heuristic example. Despite the numerous, and often excellent, taxonomic and comparative studies on the subject, the phyletic position of tupaiids within the Eutheria is far from satisfactorily ascertained. The relationships of these animals have been investigated through studies of a broad spectrum of morphological and biochemical evidence, but osteological features are emphasized in many published interpretations (see Luckett, this volume).

Herein, I review several hypotheses on tree shrew affinities through comparisons of cranioskeletal features in selected eutherian groups. Such an effort is hardly novel, but may be optimistically viewed as a contribution in light of its scope and methodology. Many considerations of tupaiid phylogeny focus on their alleged relationships with primates or a few insectivore groups. As Campbell (1974) rightly points out, this limitation in scope of comparisons has a detrimental effect on the utility of the evidence for phylogenetic investigation. Coverage in this study is extended to the major families of lipotyphlan

MICHAEL J. NOVACEK • Department of Zoology, San Diego State University, San Diego, California 92182

insectivores, leptictids, dermopterans, and bats, as well as the elephant shrews and primates with which tupaiids are most frequently compared. Unfortunately, the effectiveness of such analysis varies markedly from group to group, and in some cases (e.g., Chiroptera) morphological characterizations are based on crude estimates of the true variation. However, this approximate and speculative approach seems warranted, simply because tupaiid affinities are best considered in the context of as many proposed relationships as are practical to investigate.

Because sister group relationships are only meaningful when based on synapomorphy (shared, derived characters resulting from common inheritance), the establishment of character polarity or character morphoclines is the crucial starting point for phylogenetic analysis. The difficulties encountered with this approach are often severe (see discussion below), but can be alleviated when considering isolated features as functionally and ontogenetically related components of a "form-function complex." Regions of the mammalian skeleton particularly accessible to this approach include the orbital wall, the auditory region, the appendicular skeleton, and the proximal tarsus. Evidence for the polarities of these skeletal features is reviewed extensively elsewhere (Novacek, 1977b), but discussions are modified here in reference to the problem of tupaiid affinities. The character analysis to follow is largely confined to features in the above listed skeletal regions, although some attention is given to miscellaneous skull and postcranial features cited in previous studies of tupaiids (Table 5, Fig. 23). Dental features are not considered herein. I am in agreement with Butler (this volume) that the dentition provides only limited and occasionally suspect information on tupaiid affinities. Readers are referred to Butler's excellent accounting of this system.

2. *Methodology*

The systematic procedure adopted in this study is basically in accord with Hennig's (1950) cladistic or phylogenetic systematics. I shall not belabor the reader with an explanation of the cladistic methodology, as such information is readily available in related (but not always corroborative!) treatments of Hennig (1966, 1975); Bonde (1974); Brundin (1966); Kavanaugh (1972), and many others. Well known criticisms of cladistics are found in papers by Darlington (1970), Bock (1973), Hull (1970), Mayr (1974), and Sokal (1975). Viewpoints, pro and con, are available in articles particularly common to recent volumes of Systematic Zoology.

Eschewing a general outline of principles, brief remarks will be provided on several topics of systematic analysis germane to the development of this study. These matters have certainly been considered by others, but a sufficient diversity of opinion exists, even among workers who share similar attitudes in taxonomy, to warrant their mention. The first of these topics concerns the distinction between cladograms and phylogenetic trees. Here I follow the general characterizations stated by Eldredge (1977). Cladograms are branching sequences arranging taxa according to the nested hierarchical pattern of synapomorphic characters. They do not distinguish between ancestor-descendant relationships and sister group relationships. Thus, a cladogram uniting A and B as most closely related sister groups may mean that they share a common ancestor (X) that underwent a dichotomous speciation event, or it may mean that A is the ancestor of B or that B is the ancestor of A. Accordingly, common ancestors in cladograms are always hypothetical and no taxa are placed at the nodes of various branches.

Phylogenetic trees are naturally more robust and more inferential than cladograms. They place taxa in a time framework where at least two types of phylogenetic relationships (ancestor-descendant and sister group) are recognized (Eldredge, 1977). Trees may reflect hypotheses concerning varying rates of evolution or times of speciation by placing taxa and inferred dichotomies on a geochronologic scale. Arguments currently rage over the recognition and testability of ancestor-descendant relationships (Szalay, 1977a; Englemann and Wiley, 1977), evolutionary rates, multichotomous speciation, and related matters. But it is important to emphasize that these topics pertain to the development of phylogenetic trees, not cladograms.

No cladogram can be constructed without the establishment of polarity of its supportive characters. The determination of morphocline polarity, or the distinction between primitive and derived states of a character, is the fundamental problem in systematics. Criteria often cited for the purpose of recognizing and testing morphoclines include 1) commonality or frequency in the distribution of character states, 2) out-group comparisons, 3) analysis of character covariance in relation to a "form-function complex," 4) study of ontogenetic transformations, and 5) reference to the relative geochronologic age of taxa with certain character states. There exists a variety of opinion concerning the relative utility of these criteria in morphocline analysis. Some authors favor commonality and out-group comparisons (Schaeffer *et al.*, 1972; Englemann and Wiley, 1977; Hecht, 1976) while others advocate the effectiveness of a fossil record for direct evidence of character transformation (Simpson, 1975; Szalay, 1977a), or the use of ontogenetic studies (Nelson, 1973). The importance of functional analysis and the recognition of form-function complexes have also received emphasis (Bock, 1977). Such diversity in attitudes merely underscores the difficulty and challenge of polarity study. But it is evident that any hypothesis for character transformation is clearly superior if it has resisted falsification from as many of the above listed criteria as can be possibly investigated. Accordingly, the distinction between primitive and derived character states constitutes a hypothesis, as Hecht (1976, p. 340) remarks, "...to be continually reevaluated and tested." The tenuous nature of a hypothesis for character polarity judged from a single criterion is self evident.

The assessment of character transformation and the recognition of derived characters are doubtless the "core" procedures in cladistic methodology. However, these do not constitute the final phase of analysis. The phylogenetic nature of the derived features (or apomorphs) must be ascertained. Bock

(1977) makes a distinction between apomorphs of two kinds: synapomorphs are homologous apomorphs which stem phylogenetically from the same feature in the immediate common ancestor of any two groups sharing the derived feature; convergent features are non-homologous apomorphs which do not derive from the same feature in the immediate common ancestor of the two groups. Most taxonomists are cognizant of the difficulties in distinguishing convergence from synapomorphy, and a number of procedures for demarcation have been suggested. One commonly advocated practice is to simply accept the most parsimonious solution by favoring the cladistic arrangement which allows the fewest instances of convergence to explain the distribution of apomorphs (see Platnick, 1977, for a recent discussion). A criticism leveled against this approach stems from the argument that evolution is not necessarily parsimonious, and therefore we cannot assume that the cladistic hypothesis invoking the lowest number of independent or convergent acquisitions will necessarily represent the true phylogeny (Hecht, 1976). One can hardly dispute this point, but its relevance to phylogenetic methodology seems open to question; it is difficult to imagine systematic analysis or any scientific procedure which does not rely on parsimony. However, as a matter of practicality, parsimony alone may not permit a clear choice between phyletic alternatives because the number of convergent events may be equal in different arrangements, or the characters themselves do not seem readily comparable. In such cases, the worker may attempt to develop a larger set of characters or seek a means of weighting the existent characters. The latter procedure has been discussed by Hecht and Edwards (1976) and Hecht (1976) who attempted to establish guidelines for weighting on the basis of information content of each character. These authors advocate a ranking of characters (from lowest to highest valency) in the following order: 1) characters and character states involving a simple loss of structure; 2) derived characters involving reduction or simplification or loss with developmental information; 3) derived characters due to growth and developmental processes, dependent on size, age, hormonal, physiological, and allometric factors; 4) characters which are components of a highly integrated functional complex; 5) innovative characters and character complexes. I agree that attempts to weight characters in such manner are constructive, but I find the above ranking system overly formalistic and I doubt that it can be applied rigorously (i.e., assigning values of 1 to 5 or 1 to 10^4) with much benefit in many phylogenetic problems. Hecht (1976) admits that this weighting system is arbitrarily defined and there is much potential for overlap. It is clear that certain characters assigned higher weights might be phyletically more ambiguous than those assigned to lower ranking categories. For example, the presence of an ossified entotympanic element in the auditory bulla of many eutherians is an innovative character logically relegated to the highest ranking category (5). However, the uncertainty over the evolution of this feature, and its distribution within eutherian mammals, suggests the possibility of independent acquisition in several lineages (Novacek, 1977c). Certainly the entotympanic bulla could not be weighted more strongly than features comprising a highly integrated functional complex (category 4). These considerations lead me to question whether the particular system suggested by Hecht and Edwards (1976) will enable one to choose between alternative phylogenetic hypotheses where character distributions preclude the effective use of parsimony. Examples of such problems are discussed below in reference to tupaiid affinities.

3. Character Analysis

Character state distributions for cranioskeletal features in tupaiids and selected eutherian groups are presented in Figs. 4, 12, 13, 18, and 23. Descriptions and comparative studies of the tupaiid skeleton are provided in many papers, including some in this volume (e.g., Cartmill and MacPhee; Butler; Szalay and Drawhorn). The major purpose of this section is to discuss the basis for the advocated polarity of the characters listed. Features bearing on tupaiid affinities will be considered in light of this more general problem.

A. Intra-Group Variation

An aspect of character analysis often ignored or concealed is the nature and definition of the taxa being compared. Here the problem is two-fold. The first difficulty is purely one of definition (e.g., what does the author mean by "Insectivora"?). The second, and more serious difficulty, concerns the characterization of a taxon which satisfactorily accounts for its within-group variation. Brief remarks on these matters in reference to the groups considered in this paper follow.

1. Tupaiidae

At least since Gregory's (1910) classic review, taxonomists have recognized the basic ptilocercine-tupaiine division within the Tupaiidae. Oddly, *Ptilocercus* was frequently ignored in subsequent considerations of tupaiid affinities (see Campbell, 1974, for general discussion). Ptilocercines lack many of the specialized "primate-like" features present in tupaiines, and information on the skeleton of this subfamily is necessary for accurate characterization of the tupaiid morphotype. Character states for both subfamilies are separately considered here.

Both tupaiid subfamilies share a significant number of derived skeletal features (e.g., Fig. 12), but it is noteworthy that indisputable fossil tupaiids are virtually unknown (however, see Jacobs, this volume). This is disconcerting when one considers plausible the very remote divergence time for the family. The construction of the cranioskeletal morphotype for tupaiids should be considered in light of this dearth of historical information.

2. *Primates*

Unlike tupaiids, primates are represented by a relatively good fossil record. Such information in combination with the continuing ferment of taxonomic work permits somewhat better qualified statements on the morphotypical features of the primate clades. But if knowledge begets uncertainty I predict that more objection will be raised concerning my characterizations of primates than other, less adequately understood, taxa!

The categorical breakdown of primates used here follows the major divisions recognized by Szalay (1975), although Plesiadapiformes Simons (1972) is substituted for Szalay's (1975) Paromomyiformes.

3. *Insectivora*

Perhaps no eutherian group has engendered as much confusion as insectivores (see Dawson, 1967; Butler, 1972; Novacek, 1976). Historically this category has served as a "waste basket" for many poorly understood taxa. Herein I follow McKenna (1975) in restricting the formal usage of the Insectivora to the Lipotyphla[1], a group recognized to include the Recent Erinaceidae (hedgehogs), Talpidae (moles), Soricidae (shrews), Solenodontidae (solenodons), Tenrecidae (tenrecs), and Chrysochloridae (golden moles). These taxa comprise an apparently monophyletic, though very broadly defined, grouping based on several shared derived characters (cited by Butler, 1972). Such characters include: the absence or marked reduction of a jugal bone, the expansion of the maxillary in the wall of the orbit, specialized snout muscles, the weak pubic symphysis, and the absence of a medial internal carotid artery.

A number of fossil groups, including the Adapisoricidae, Micropternodontidae, Nesophontidae, Nyctitheriidae, Apternodontidae, Geolabididae, and Plesiosoricidae can with varying degrees of confidence be allied with living lipotyphlans, but the evidence for such allocations is mainly dental (see Van Valen, 1967; Krishtalka, 1976a,b; Butler, 1972; Novacek, 1976, and others cited therein). A few of these families are, however, represented by excellent cranial material, and the argument for their affinity with Recent families seems quite tenable. Unfortunately, the virtual lack of skeletal remains exclusive of the dentition for the primitive adapisoricids precludes consideration of a fossil group which might provide critical information on the morphotypical condition for insectivores.

[1]I see no compelling reason against equating Butler's (1972) Lipotyphla with Insectivora and thus retaining the latter term in the systematic literature. The Insectivora would accordingly have a more restricted definition than under former usage, but such action would serve to demonstrate the doubtful ordinal affinities of the non-lipotyphlan groups (e.g., apatemyids, pantolestids, and palaeoryctids) customarily allied with the Lipotyphla for the lack of a better arrangement. These taxa are effectively Eutheria *incertae sedis* and they should be recognized as such pending further study.

4. Leptictidae

This archaic eutherian group has been traditionally regarded as an ancestral stem for the Insectivora and many other eutherian clades, but this patristic reputation has been questioned in more recent papers (McKenna, 1975; Szalay, 1968, 1977b; Butler, 1972). Elsewhere (Novacek, 1977a,b) I argue that leptictids comprise an isolated clade having very little affinity with *bona fide* insectivores. Advanced leptictids are represented by complete skeletons, but such material is limited for primitive taxa. However, the available material suggests rather conservative evolution of auditory, orbital, and foot structure. This pattern, coupled with relatively low diversity (six North American and four European genera) facilitates morphological characterization of the family.

5. Dermoptera

All dermopteran features compared here are based on information on the sole living genus *Cynocephalus*. An argument for the affinities of the Early Tertiary Plagiomenidae with dermopterans (see Szalay, 1969; Rose and Simons, 1977) seems plausible, but knowledge of the former is virtually restricted to dental features.

6. Macroscelididae

Much information on the osteology of elephant shrews can be found in Evans' (1942) comprehensive review of the family. That author noted the failings of earlier studies to account adequately for the fundamental differences between the subfamilies Macroscelidinae and Rhynchocyoninae. Nonetheless, Evans' (1942) considerations of macroscelidid affinities were largely based on features unique to *Rhynchocyon*, a taxon obviously specialized in many cranioskeletal features. This somewhat erroneous procedure formed the basis of several subsequent treatments (see Novacek, 1977b, for review). The characters considered here are based on observations of all macroscelidid genera. In agreement with Evans, I find *Petrodromus* to be relatively more primitive than other members of the family, in most features exclusive of the appendicular skeleton.

The traditional concept of a close tupaiid-macroscelidid affinity at the subordinal or ordinal rank of Menotyphla (Haeckel, 1866; Gregory, 1910; Butler, 1956; McDowell, 1958) has been rejected by many authors (Simpson, 1945; Patterson 1965; Butler, 1972; McKenna, 1975; Szalay, 1977b). Although there are undoubted fossil members of the Macroscelididae (Butler and Hopwood, 1957), the alleged affinities of certain fossil taxa are open to question (Patterson, 1965; Sigé, 1974). Information on the macroscelidid morphotype is almost exclusively based on study of the Recent forms.

7. *Chiroptera*

Understanding of the morphological variation and phylogeny of bats is dismally poor. The characters cited by Miller (1907) in the major comprehensive review of chiropteran taxonomy are largely adequate for generic and familial diagnoses, but are limited to systems which shed little light on the affinities of the order. The features cited herein for mega- and microchiropterans are crude estimates of morphotypical characters based on first-hand observations of the following taxa: *Megaloglossus, Scotonycteris, Pteropus, Hypsignathus, Eidolon, Rousettus, Epomops* (Pteropodidae); *Natalus* (Natalidae); *Nycteris* (Nycteridae); *Pteronotus, Artibeus, Macrotus, Glossophaga, Carollia, Leptonycteris, Sturnira* (Phyllostomatidae); *Balantiopteryx, Taphozous* (Emballonuridae); *Pizonyx, Eptesicus, Myotis, Lasiurus, Miniopterus, Plecotus, Antrozous* (Vespertilionidae); *Tadarida, Eumops* (Molossidae): *Rhinolophus* (Rhinolophidae); *Desmodus* (Desmondontidae).

Additionally, I have considered features described for *Icaronycteris* (Jepsen, 1970) and other fossil taxa. This limited range of comparisons does not inspire confidence, particularly in the case of Microchiroptera, but optimistically provides a basis for some preliminary assessments of bat character variation.

B. *The Orbital Region*

The complex of features represented in the orbital wall of the skull has traditionally received emphasis in considerations of insectivore and primate affinities (e.g. Butler, 1972). Unfortunately, the polarity of these features is not adequately understood, largely because few workers attempt to relate details of this region to each other. Thus, one is hard pressed to ascertain the significance of an extra foramen or an expanded bone. A satisfactory treatment of this problem is not intended here. However, some initial speculation facilitates comparisons of orbital features and assessments of their polarities. Figure 1 is a flow chart expressing the interrelationships of orbital characters as a function of (sometimes opposing) selection for increased olfactory, optic, and cerebral development and greater efficiency in the jaw apparatus. The diagram is not a cause for dogmatism, yet it serves some heuristic purpose. It seems reasonable, for example, to attribute the posterior migration of the sinus canal in some taxa to the expansion of the alisphenoid. The latter development might be correlated with expansion of the cerebral cavity and development of the eye. Some modifications are mutually exclusive; Muller (1934) attributed the enlarged maxilla in the orbital wall of certain insectivores to expansion of the nasal cavity. With increased development of the eye or the neopallium, both the maxilla and the lacrimal could be crowded out of the orbit, a modification cited by Gregory (1910, 1920) for certain mammalian groups. The above speculations constitute largely untested but testable hypotheses. Perhaps evidence will be most effectively found through ontogenetic studies (see Cartmill

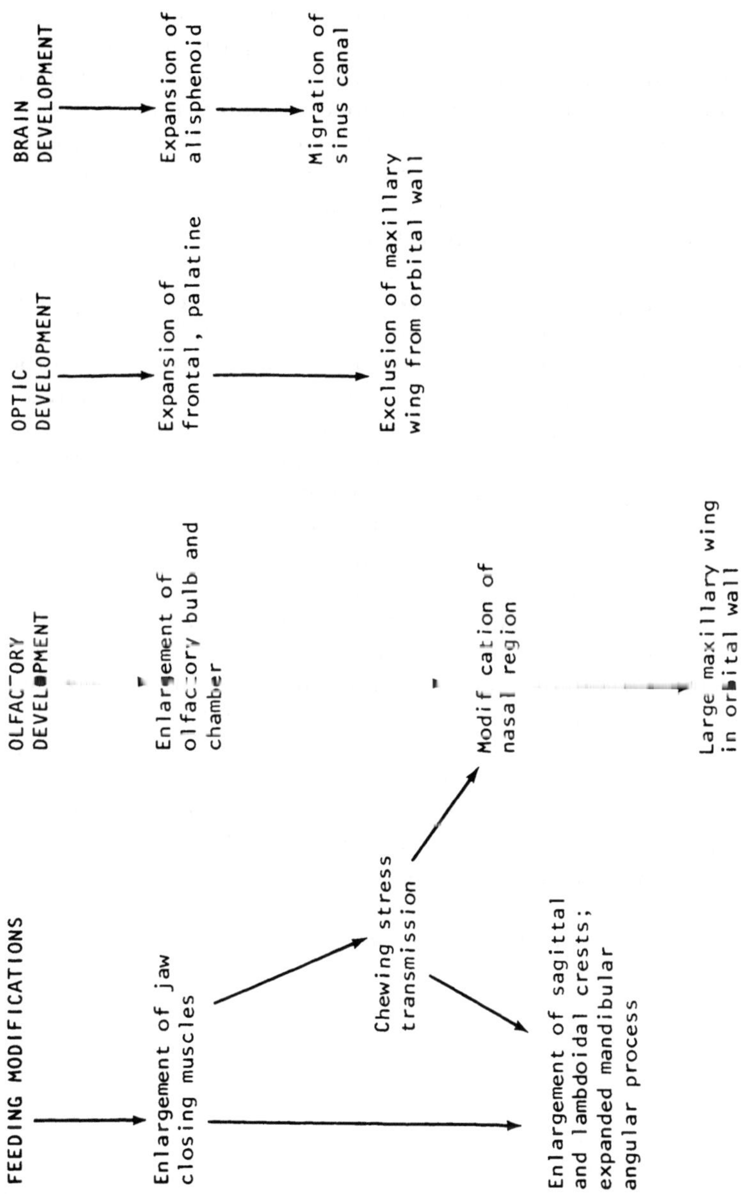

Fig. 1. Hypothetical flow diagram for interrelationships of cranial features. See text for explanation.

and MacPhee, this volume). Some of the basic relationships postulated in Figure 1 are discussed below.

The aspect of the orbital region eliciting the most attention in mammalian systematics is the condition of the bony mosaic (i.e., the relative sizes and positions of the major bony elements of the orbital wall). Particular interest has focused on the relationships of the maxilla, palatine, and lacrimals. The condition in *Tupaia* is shown in Figure 2, contrasted with the pattern observed in some insectivores, macroscelidids, leptictids, and primates. Both tupaiines and ptilocercines have a large orbital component of the palatine which forms a broad contact with the lacrimal and excludes the maxilla from contact with the frontal (Fig. 4, Table 1). This combination of features is common to chiropterans, some primates, macroscelidids, and dermopterans, but a markedly contrasting condition is observed in lipotyphlan insectivores. The latter show a pronounced expansion of the maxilla in the orbit and a confinement of the palatine to the ventral portion of the orbital wall (Fig. 2).

Several authors have considered the problem of polarity in the orbital mosaic, mostly in relation to studies of primate and tupaiid affinities. Le Gros Clark (1959, 1971) remarked that the situation in daubentoniids and indriids, where the palatine is excluded from contact with the lacrimal by the expansion of the maxilla (Fig. 2) is the "primitive" condition (1959) or "common" condition (1971) in mammals, citing as evidence the similar pattern in some insectivores. Le Gros Clark further believed that the affinities of primates with tupaiids were supported by the common existence of enlarged palatines, palatine-lacrimal, and palatine-frontal contacts in lemuriform lemuroids and tupaiids. But given Le Gros Clark's initial assumptions, this interpretation suggests that either 1) the earliest primates simply retained the broad maxillary-frontal contact present in the most primitive mammals, and the lemuriform-tupaiid similarity is a case of secondary parallelism, or 2) the lemuriform-tupaiid similarity is a synapomorphic character linking the most primitive primates and tree shrews, while the daubentoniid-indriid condition represents a secondary reversion to the primitive condition (see also discussion in Campbell, 1974, p. 128–131). Cartmill (1975) has contributed some recent remarks concerning this problem. Although he did not explicitly state his conclusions concerning the primitive condition in mammals, he noted the widespread occurrence of a large palatine bone in the orbit and its intrusion between the maxillary and the frontal in several primates and other eutherian lineages. However, Cartmill favored a frontal-maxillary contact as the primitive character for primates because 1) this condition is found in adapids (Le Gros Clark, 1934; Piveteau, 1957) and in *Plesiadapis* (Russell, 1964); 2) the palatine does not contact the lacrimal in (primitive) cheirogalines; 3) an enlarged maxillary contacting the frontal is widely distributed among Malagasy lemurs; and 4) the first-order affinities of primates are not with tree shrews or other Menotyphla, but with soricids and erinaceotans in which a broad maxilla-frontal contact is present. The last of these arguments is convincing only if a large number of independent synapomophies between these groups can be recruited. In this regard, a strong case falters, but it remains to be decided whether the erinaceotan-lemuriform condition is apomorphic for eutherians.

Van Valen (1965) and Butler (1956) conclude that such a condition is derived. In reference to the tupaiid condition, Van Valen (1965, p. 143) states: "The widely emphasized extension of the palatine to the lacrimal and the existence of a relatively long fronto-palatine suture are clearly both primitive retentions." Van Valen claimed that his interpretations were based on comparisons with leptictids, palaeoryctids, hyaenodontids, and diverse Recent placentals. Butler (1956, p. 470) expressed agreement with Van Valen, remarking that tupaiids, macroscelidids, as well as marsupials, retain the primitive mammalian condition described in the above quote. My survey of the variation of this feature in mammals leads me to roughly the same conclusion. It is of interest to note the presence of a broad lacrimal contact and the exclusion of the maxilla from the orbital wall in *Morganucodon* (Kermack and Kielan-Jaworowska, 1971), polyprotodont marsupials, tupaiids, macroscelidids, and some artiodactyls and carnivores (Haines, 1950). The condition in tupaiids is therefore basically primitive for eutherians.

Despite the considerable interest in maxillary and palatine bones, little in the recent literature is concerned with the transition from a "reptilian" grade,

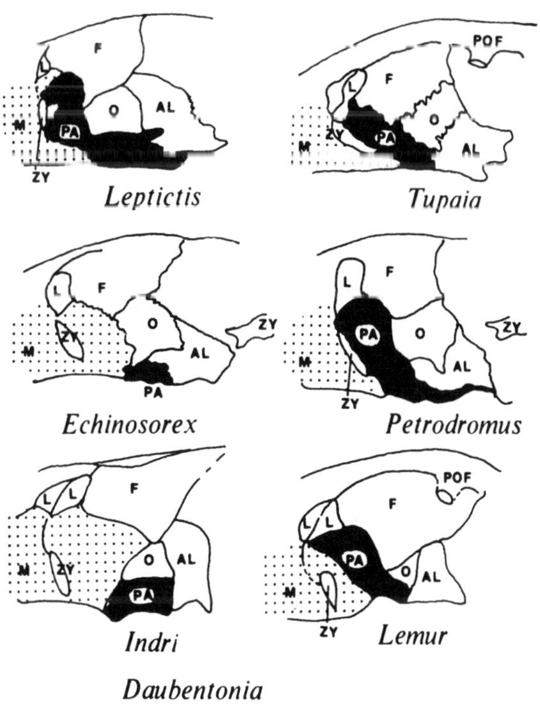

Fig. 2. Diagrams showing variation in the bony mosaic of the orbital wall for several eutherian taxa (after Evans, 1942; Butler, 1956; Campbell, 1974). For abbreviations, see pp. 88–89.

the basic modifications in mammals, and their possible adaptations. For such considerations it is necessary to refer to Muller's (1934) important study of evolution of the orbital bones in mammals. Muller suggested that the reptilian ancestors of mammals had only a rostral contribution of the palatine in the wall of the orbit, and its medial extension represented a later development. This interpretation was partially based on Fuch's (1910) comparisons of the braincase wall between reptiles and mammals. Muller reasoned that evolution of the palatine in part reflects the transition from trobibasic skull in reptiles to a platybasic mammalian cranium, and that the expansion of the rostral portion of the palatine in the orbit and its broad contact with the lacrimal and frontal were due to the marked widening and posterior extension of the nasopharyngeal duct. As a logical extension of this trend, further expansion of the nasal capsule caudally was accompanied by shift of the maxillary bone posteriorly and dorsally and its encroachment on the palatine.

Muller's (1934) hypothesis seems reasonable, but is perhaps oversimplified. I believe that an enlarged maxilla in the orbit could only result if expansion of the nasal capsule was not accompanied by marked increase in the size of the eye and possibly the neopallium. This would explain the mosaic present in various insectivores with enlarged olfactory bulbs but generally with small brains and small eyes. In leptictids the size of the optic foramen and the orbitosphenoid bone suggests that the eye in these animals was perhaps somewhat better developed than in primitive erinaceoids (see Butler, 1956, and remarks below), but the comparable development of the nasal capsule may be related to the slight intrusion of the orbital portion of the maxilla. The variable development of olfactory, optic, and cerebral specializations may have contributed to the complexity of patterns observed in primates and several other mammalian groups. Posterior extension of the maxilla may have other correlations, as noted by Muller. Primary among these are a shortening of the skull, the posterior shift of the molars, and the expansion of muscles inserting on or originating from the maxilla (e.g. *Elephas*).

Despite the attendant complexities, it seems reasonable to suggest a possible mode of change in the relationships of the palatine, maxilla, and neighboring elements in the orbital wall for at least the conditions present in various primates, insectivores, leptictids, macroscelidids, and tupaiids. Figure 3 is a series of block diagrams representing these conditions, interposed with the events which may have led to their derivation. The essentials of the model are explained in the caption and figure. I emphasize that this scheme is presented as a working hypothesis. The correlations and events are not particularly obvious in the case of primates. More careful analysis of covariation and ontogenetic descriptions are obviously needed.

Figures 2 and 3 also illustrate various positions and developments of the lacrimal bone among certain mammalian taxa. A prelacrimal crest sharply isolates the major portion of the lacrimal from the snout in leptictids, and the facial portion of this bone is either small or vestigial, being exposed only on the rim of the zygoma (Fig. 2). The pars facialis of the lacrimal is also small in *Ptilocercus*, Macroscelidinae, and erinaceids (Fig. 4, Table 1). A large portion of

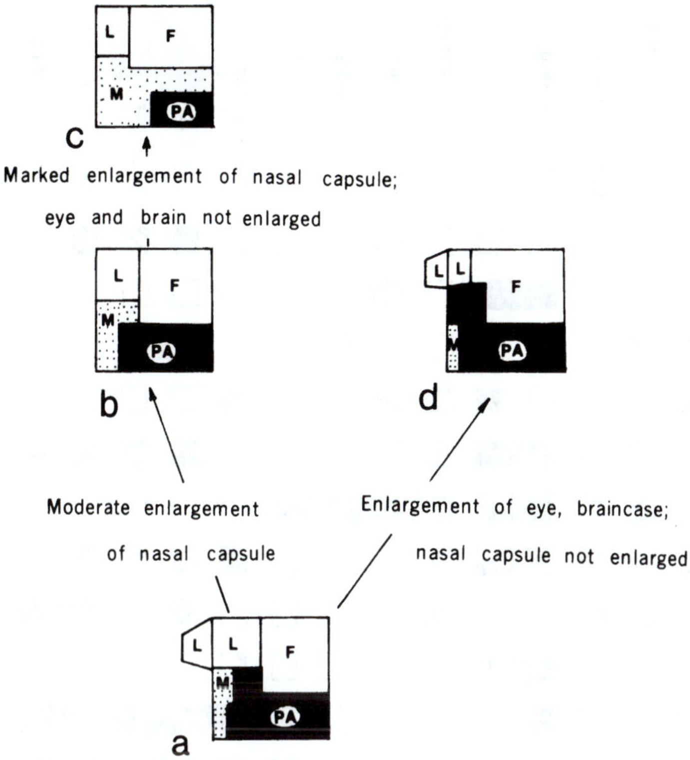

Fig. 3. Block diagrams representing the bony mosaic of the orbital wall showing several eutherian conditions with hypothesized events interposed: (a) primitive eutherian condition with large palatine and palatine-lacrimal contact; (b) leptictid condition where moderate enlargement of nasal capsule results in slight intrusion of the maxilla; (c) insectivore (lipotyphlan) condition where orbital wing of maxilla is very large and palatine is confined to floor of the orbit; (d) condition in some primates where palatine orbital wing is expansive and maxilla is virtually excluded from orbit. For abbreviations see pp. 88–89.

the lacrimal extends far out on the face in tupaiines *(Urogale)*, *Rhynchocyon*, and plesiadapid primates. A facial exposure of the lacrimal was regarded as a primitive mammalian character by some workers (Williston, 1925; Salomon, 1930), but Muller (1934) questioned the validity of this interpretation. Gregory (1920) was of the opinion that a well developed lacrimal was an unspecialized mammalian feature, and its continued reduction in size and confinement in the orbit were the result of encroachment of neighboring elements. Van Valen (1965) and Butler (1956) concur with this view.

It should be noted that Gregory (1920) recognized modifications of the lacrimal due to many factors in different mammalian lineages. He attributed reduction of this bone to reduction of the olfactory region (some lorisids and *Tarsius*), dorsad displacement of the zygomatic arch and anterodorsal shifting of the masseter (*Vombatus*, rodents, and typotheres), retraction of the nasals (tapirs, elephants, and *Brachyerus*), and expansion of the maxilla *(Erinaceus)*.

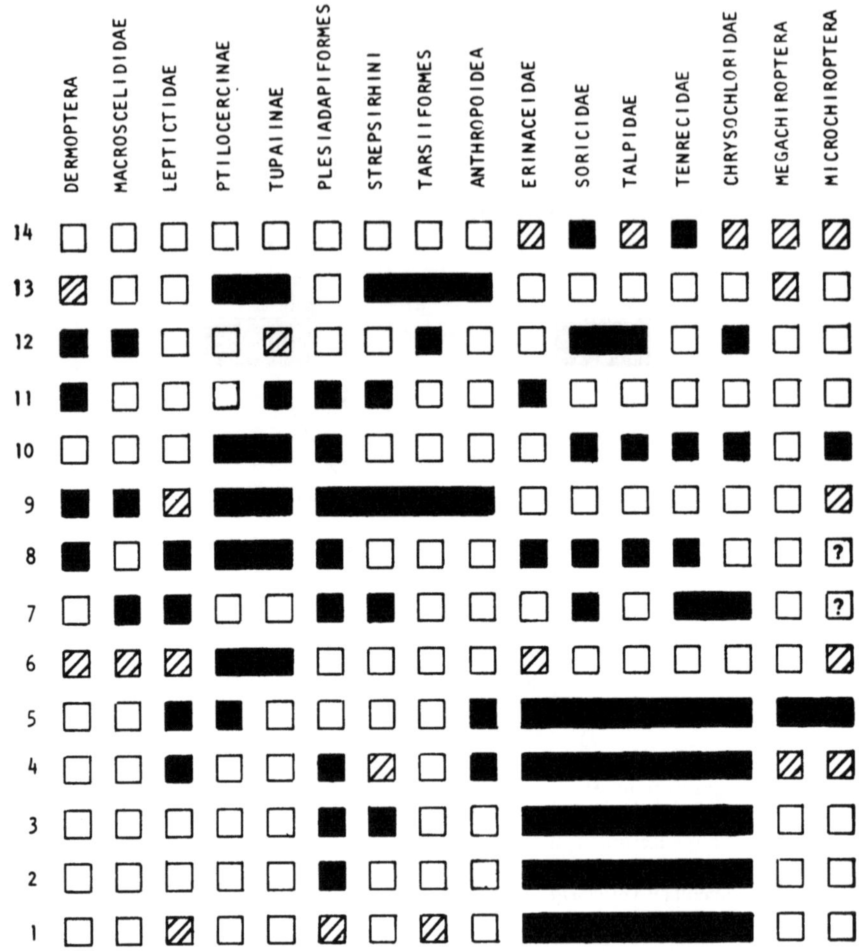

Fig. 4. Character state distributions for orbital features in tupaiids and several eutherian taxa. Solid squares represent derived character states; diagonal lines, intermediately derived; open squares, primitive states. Solid bars represent derived characters shared between selected groups which have been recruited as evidence for relationships. No distinction is made here between true synapomorphy and specialized similarity due to convergence. Numbers refer to characters listed in Table 1.

On the other hand, enlargement of the pars facialis could be effected by snout elongation (*Orycteropus, Megalohyrax, Phacochoerus,* and many artiodactyls), and horn development (bovids). Most interesting was Gregory's suggestion that enlargement or protrusion of eyes may crowd the lacrimal almost entirely out of the orbit, and the latter element is subsequently replaced by the os planum of the ethmoid or by the palatine, as in lorisids and *Tarsius*. These interpretations complicate matters. A large eye crowding the lacrimal out of the orbit renders a prominent facial extension of this bone a secondary rather than primitive trait.

Most other features of the orbital wall cited in Table 1 and Fig. 4 have received very superficial treatment, and their alleged polarities are tenuous. Muller (1934) postulated that the primitive eutherian orbitosphenoid was large and broadly contacted the frontal, excluding the maxillary-frontal and palatine-frontal contact. He attributed reduction of the orbitosphenoid to enlargement of adjacent bones resulting from expansion of the cerebral cavity, but expressed caution in applying this interpretation to the variety of mammalian conditions observed. More certain is the argument for enlargement of the dorsal moiety of the alisphenoid due to expansion of the cerebral cavity (Muller, 1934, p. 171). These hypotheses are primarily based on out-group comparisons, particularly with conditions in therapsids and other reptiles, as well as with prototherians and marsupials.

Plausible arguments for polarity of the orbital foramina are likewise elusive. Whereas the enlargement of the optic foramen is reasonably cited as a derived character linked with increased development of the visual apparatus, the small size of this foramen in some insectivores might represent a secondary specialization. Least secure are the advocated polarities for the presence, ab-

Table 1. Character states for selected orbital features (see Fig. 4)

Primitive	Derived
1. Maxilla orbital wing small or absent	1. Maxilla orbital wing moderately (int.)[1] or extensively developed
2. Fronto-maxillary contact absent	2. Fronto-maxillary contact restricted (int.) or extensive
3. Palatine orbital wing extensive	3. Palatine orbital wing small or absent
4. Lacrimal-palatine contact extensive	4. Lacrimal-palatine contact restricted (int.) or absent
5. Lacrimal facial wing large	5. Lacrimal facial wing small or absent
6. Single lacrimal foramen, exits pars facialis	6. Single foramen exits orbital wing of lacrimal (int.) or double openings are present
7. Orbitosphenoid large	7. Orbitosphenoid small
8. Alisphenoid orbital wing small	8. Alisphenoid orbital wing large
9. Optic foramen small	9. Optic foramen moderately (int.) or distinctly enlarged
10. Suboptic foramen present	10. Suboptic foramen absent
11. Foramen rotundum confluent with sphenorbital foramen	11. Foramen rotundum separated from sphenorbital foramen
12. Anterior opening of alisphenoid canal confluent with foramen rotundum	12. Anterior opening of alisphenoid canal separate from foramen rotundum (int.) or alisphenoid canal absent
13. Postorbital process of frontal absent	13. Postorbital process of frontal small (int.) or large, forming complete postorbital bar
14. Zygomatic arch complete, with large jugal	14. Zygomatic arch complete, with reduced jugal (int.) or incomplete jugal vestigial or absent

[1]int=intermediately derived character state.

sence, or positions of certain foramina in the orbitosphenoid and alisphenoid (Table 1, characters 10, 11, 12). These decisions are largely based on character distributions and interpretations of Gregory (1910) and Butler (1956).

Fortunately, not all characters of the orbital region are so enigmatic, and there seems little doubt that an enlarged postorbital process of the frontal and the loss or marked reduction of the jugal in the zygomatic arch are derived eutherian characters (Butler, 1956).

From Figure 4 it is evident that several derived orbital characters are shared by tupaiines and ptilocercines, but few are common to tree shrews and other groups. The most notable block of apomorphs comprises similar features in maxilla-palatine complex (characters 1–5) and zygomatic arch (character 14) for the Insectivora. These contribute significantly to the diagnosis of the order (Butler, 1972).

C. Auditory Region

Perhaps more than any other aspect of the skeleton, the otic region has received heavy emphasis in considerations of tupaiid, primate, and insectivore relationships. This area of the cranium provides a wealth of interesting characters and is accessible in both living and fossil specimens. For purpose of discussion, it is convenient to subdivide the otic region into three interrelated character complexes: 1) the carotid circulation (Fig. 5); 2) construction of the auditory bulla (Figs. 6–8); and 3) detailed osseous morphology of the tympanic chamber (Figs. 9–11).

1. Carotid Circulation

Considerable study of the carotid circulation has revealed several basic patterns among mammals. It appears that Mathew (1909) was correct in suggesting that the primitive eutherian carotid circulation consisted of both a median and a lateral branch. The medial internal carotid artery passed between the petrosal and the basioccipital and basisphenoid bones. The lateral internal carotid artery entered the auditory chamber and divided into a small promontory and a larger stapedial branch. More detailed accounts of this pattern and its modifications in various taxa are provided by Archibald (1977), Bugge (1974), Cartmill and MacPhee (this volume), Hunt (1974), McDowell (1958), MacIntyre (1972), McKenna (1966), Szalay (1972, 1975), and Van Valen (1966).

The carotid circulation in the middle ear regions of several eutherians and that postulated for the eutherian morphotype are diagrammatically illustrated in Figure 5. It is clear that tupaiids are notably removed from the primitive eutherians by several features (see also Fig. 12, Table 2). Unlike the latter, tree shrews lack a medial internal carotid artery, possess ossified tubes which enclose the carotid vessels, and shield the fenestra rotundum from ventral view through enlargement and bony enclosure of the lateral internal carotid artery. Bugge (1974) concluded that the carotid circulation in tupaiids is more specialized than that in all living insectivores in part because the supply to the orbit "exhibits an incipient but clear developmental tendency in the primate direction." Bugge accordingly linked tupaiids with primates based on 1) connection of the internal opthalmic artery with the distal end of the promontory, and 2) the proximal obliteration of the inferior ramus of the stapedial. Van Valen (1965, p. 27, 28) has, however, argued that the carotid similarities between primates and tupaiids are weak. The ramus inferior of the stapedial is present in *Ptilocercus,* and, unlike primates, the stapedial and promontory arteries are of subequal girth. There is certainly clear evidence of an inferior ramus in *Ptilocercus,* from osseous morphology alone (see Szalay, 1975, Fig. 2). It is noteworthy that Bugge's conclusions on the affinities of tupaiids suggested by their carotid circulation were based solely on his observations of the condition in *Tupaia*. A more detailed consideration of the carotid circulation in tupaiids (particularly *Tupaia*) is provided by Cartmill and MacPhee (this volume).

The arterial circulation in leptictid ear regions (Fig. 5) can be inferred through careful examination of grooves preserved in fossil basicrania. Such features are best known for the advanced Oligocene member of the family, *Leptictis,* but a few specimens representing earlier Paleocene and Eocene taxa show no significant difference from the plan in *Leptictis.* The carotid circulation for this group is clearly primitive, and it is worthwhile to consider its features with reference to the polarity of various character states. A small groove on the medial side of the tympanic cavity between the petrosal and the basioccipital (Fig. 5, 10) probably conveyed a medial internal carotid artery in addition to the inferior petrosal vein. Most workers would agree that the presence of this vessel is plesiomorphous, although its (inferred) reduced size in leptictids is logically derived over the primitive eutherian condition. The medial internal carotid is lost in primates, tupaiids, lipotyphlous insectivores, and many other eutherian groups. Many carnivores retain a large medial internal carotid artery but drastically reduce or lose the stapedial and promontory branches of the lateral internal carotid.

The branching sequence and course of the lateral internal carotid in leptictids are also primitive[2]. Anterior to its divergence from the smaller medial internal carotid artery, the lateral branch passes through a posterior carotid foramen in the wall of the bulla. The position of this foramen is posteromedial, an allegedly primitive condition (Archibald, 1977; MacIntyre, 1972). Well excavated sulci are present on the surface of the promontorium, showing that the promontory and stapedial branches of the lateral internal carotid diverged just anteromedial to the fenestra rotundum. The sulcus for the promontory artery shows that this vessel courses on the ventromedial face of the promontorium. Archibald (1977) and MacIntyre (1972) in comparative studies inter-

[2]Reconstruction of detailed branching patterns in fossil mammals is inferential. MacPhee (personal communication) has evidence that the vidian ramus of the promontory artery was probably a later development within eutherians.

pret this condition to be specialized. The primitive eutherian trait is basically that of the "ferungulate" ear region discussed by MacIntyre (1972, p. 287), where the groove for the promontory passes lateral to the fenestra rotundum and turns anteriorly just medial to the fenestra ovalis. In leptictids, the promontory sulcus is somewhat smaller than that for the stapedial, suggesting a corresponding disparity in vessel size. The polarity of this character in part depends on theories for the condition in the common ancestor of eutherians and metatherians. Archibald (1977) argues that the lack of evidence for a promontory artery in the early marsupials (see Clemens, 1966) suggests that if this vessel was ever present in the group, it was lost by the late Cretaceous. Based on this and other arguments, Archibald speculates that the promontory artery was an acquisition in early eutherians. Assuming this interpretation proves correct, a relatively small promontory might be recognized as an early stage in the evolution of a unique eutherian vessel; its enlargement as progressive. Under this line of argument the small promontory artery in leptictids would be interpreted as a primitive eutherian character.

The large sulcus for the stapedial artery in leptictids passes anterior and slightly medial to the fenestra rotundum, and the stapedial artery probably did not "shield" the round window ventrally. Szalay (1975, p. 97) interpreted the ventral shielding of the fenestra rotundum as primitive for primates and as a taxonomically significant similarity with erinaceoid insectivores. Besides these groups, the shielding is also present in the carnivore *Viverravus* (Matthew, 1909) and in MacIntyre's (1972, Fig. 5) "ferungulate" petrosal specimen. Archibald (1977) has accordingly argued that this feature represents a primitive eutherian character, but complicating this interpretation is the variety of conditions attributable to the "shielding." If the ventral shielding is due to the posteromedial entry of the lateral internal carotid into the tympanic cavity or the caudal tympanic process of the petrosal, then the shielding might be considered primitive. But in many groups this shielding might result from such specialized modifications as the marked enlargement of the stapedial artery or its enclosure in bony tubes or both. Therefore the contrasting states of this character by themselves are of uncertain evolutionary significance without reference to correlative features.

The pathways of the stapedial branches in leptictids are difficult to establish with certainty. The superior ramus of the stapedial either exited the tympanic chamber through a small foramen anterior to the epitympanic recess at the base of the glaserian canal (Butler, 1956) or possibly through a small foramen posterior to the epitympanic recess at the base of the tympanohyal (Novacek, 1977b). The inferior ramus of the stapedial probably exited the cavity through a small foramen at the anterior apex of the facial canal just behind the alisphenoid-petrosal suture and moved dorsad into the cerebral cavity. The direct evidence for this pathway is, however, somewhat ambiguous (Novacek, 1977b).

Virtually all primates are specialized in the loss of a medial internal carotid artery, the presence of bony tubes for the branches of the lateral internal carotid, and the loss or proximal obliteration of the ramus inferior of the stapedial artery (Fig. 5). Van Valen (1965) argued that a medial internal carotid artery was present in the earliest primates, because rodents have a medial internal carotid and rodents were probably derived from primates. He suggested that this vessel might also be present in lemurids (*fide* Saban, 1963) and lorisids (*fide* Simons, 1962). Szalay (1975) and Archibald (1977) have, however, shown that there is no evidence of a medial internal carotid in any living or fossil primate and reason that this vessel was probably absent in the earliest members of the order. Van Valen (1965) also maintained that the development of bony tubes for the carotids was a later innovation in primates, as such elements were absent in Paleocene skulls of *Plesiadapis*. A small ossified tube, however, may have been present at least proximal to the stapedial-promontory bifurcation in plesiadapids (see Szalay, 1975). However, the current available evidence suggests that ossified tubes for the carotids were independently derived in primates and tupaiids.

The carotid system is poorly known for Dermoptera and Chiroptera. The former group shows many primitive traits (Fig. 10, 12, Table 2), but specializations are also evident, such as the peculiar drainage for vessels in the petrosquamous region. These autapomorphies set dermopterans apart from tupaiids, primates, and insectivores (see Cartmill and MacPhee, this volume). The morphotypical pattern for cephalic circulation in bats is a matter of speculation, because virtually no studies involve comparisons of injected vessels (e.g., the acrylic-resin techniques employed by Bugge, 1974) in a reasonable variety of taxa. Information on megachiropterans is particularly poor. A small groove medial to the tympanic chamber is observed in basicrania of *Pteropus* (Fig. 11) and *Hypsignathus*, but I have not been able to locate arteries occupying these grooves in injected specimens. The relative sizes of the stapedial and promontory arteries for the primitive chiropteran condition are also difficult to ascertain. However, generalization is possible in respect to several features. Ossified tubes do not normally enclose the major branches of the lateral internal carotid artery; the stapedial artery is relatively large; the ramus inferior of the stapedial is present and exits the tympanic cavity through a fissure at the border of the glaserian canal; and the fenestra rotundum is not ventrally shielded by ossified tubes or enlarged arteries. From the foregoing it is apparent that the majority of features of the carotid circulation in bats are plesiomorphous and provide no essential information on their affinities with other eutherian orders.

Macroscelidids show a general resemblance to leptictids and to the primitive eutherian condition in carotid features. However, the medial internal carotid artery, if present, is probably extremely reduced in primitive elephant shrews, and some of the carotid vessels exit the tympanic chamber in bony tubes (Fig. 5, 9, 12).

The functional significance of the evolution of the carotid pattern in placentals has been considered by several workers. Hunt (1974) described changes in the cephalic arterial circulation in different carnivores as resulting

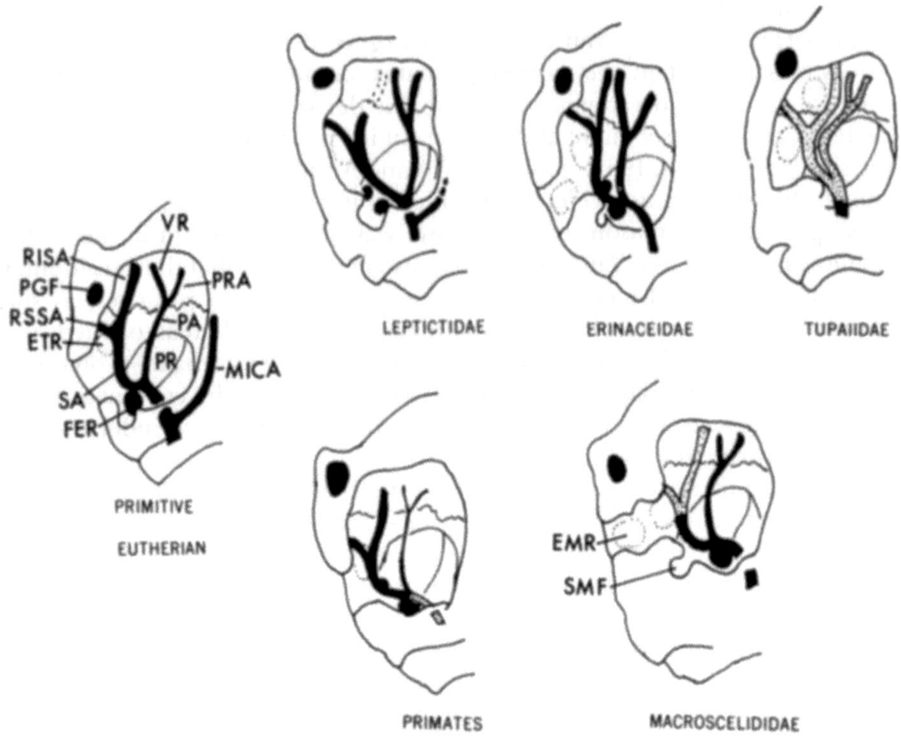

Fig. 5. Diagrams showing morphotypical reconstructions of the carotid arterial circulation and tympanic regions for several eutherian taxa. Solid black lines represent arterial branches; stippled areas, osseous tubes enclosing arteries; circles bordered by broken lines represent depressions on the roof of the tympanic chamber. For abbreviations see pp. 88–89.

from a need for shorter, more direct, and volumetrically more accommodating routes to the brain and target sensory and motor organs. This explanation has also been cited by Llorca (1934) for the embryonic obliteration of the stapedial artery in humans. Echeverra (1960), however, correlated loss of the proximal stapedial with increased growth rate of the cochlea. In his review of the eutherian cephalic arterial pattern, Bugge (1974, p. 85) emphasized the changes in relative size of the stapedial artery, its loss, or its anastomoses with adjacent vessels as of primary importance in the evolution of many placental groups. Bugge (1974) reasoned that a stapedial artery is well developed in the primitive condition where it is not encroached upon by intimate contact of the petrous part of the ear region and the osseous bulla. Close contact of the latter elements is usually correlated with reduction of the stapedial artery and stronger developments of the masticatory apparatus. Bugge cited various trends among rodents as evidence for such a relationship. It is not difficult to find exceptions to this pattern in mammals. For instance, in certain lorisids the stapedial is greatly reduced or absent, yet the bulla is expanded, leaving sufficient space below the promontorium to accommodate a stapedial, and the masticatory apparatus is not particularly specialized. Instead we find that the areas supplied by the stapedial in most mammals are primarily the responsibility of a greatly enlarged "anterior carotid" (= ascending pharyngeal) in lorisids (see Cartmill, 1975, for a recent discussion). Despite these attendant uncertainties it seems justifiable to comment on the functional significance of the primitive eutherian carotid pattern. Primitively, all three branches of the internal carotid are retained, and there is no marked diminution or loss of vessels and concomitant enlargement of others to indicate increased efficiency in blood flow. Thus, the blood supply to the brain remains primitively complex. The stapedial artery is not greatly enlarged, suggesting that there was little need for increased blood flow to the eyeball. Moreover, the stapedial was not encroached upon by neighboring bones as the result of increased development of the cochlea, ectotympanic, bullar septa, jaw apparatus, or other specializations. Finally, the promontory artery functions as an important source of blood to the brain.

2. *Auditory Bulla*

Figures 6–8 compare the bullar condition in tupaiids with other mammals. This evidence is elsewhere (Novacek, 1977c) reviewed at length, and it is postulated that: 1) The primitive condition in eutherians was probably one similar to that in monotremes, where a bulla is absent and the ectotympanic is inclined at a very low angle to the horizontal plane of the basicranium, nearly contacting the medial border of the tympanic cavity. 2) The widespread distribution of the entotympanic among eutherians suggests that it was an early innovation for these mammals, but its questionable presence in marsupials suggests that it was absent in the common ancestor of metatherians and eutherians, and it is likely that the entotympanic subsequently evolved several times within Eutheria. 3) The partial bulla of leptictids represents an early stage in the modification of the monotreme-like condition. This trait is also present in some didelphine marsupials and edentates. 4) The petrosal bulla of primates, the ectotympanic bulla of rodents, rabbits, and other eutherian orders, the composite bulla of macroscelidids, the cartilaginous bulla of some carnivores and bats, and the basisphenoid-petrosal bulla of erinaceid and tenrecid insectivores all represent derived eutherian conditions. Pathways for these modifications are uncertain in many cases. The essentials of this hypothesis are shown in Figure 8 and discussed in the accompanying legend.

The tupaiid entotympanic, though a derived eutherian trait, is phylogenetically unrevealing. This element seems to derive fully from the rostral entotympanic (Van Valen, 1965; Spatz, 1966; Cartmill and MacPhee, this volume) whereas both a rostral and caudal entotympanic are present in many eutherian taxa (see lists in Cartmill and MacPhee, this volume; Novacek, 1977c). The present story told by the entotympanic is complex and it is likely that this bone evolved several times within eutherians (Novacek, 1977c; MacPhee, 1979). Accordingly, the importance attributed to the entotympanic in considerations of tupaiid affinities seems ill-deserved.

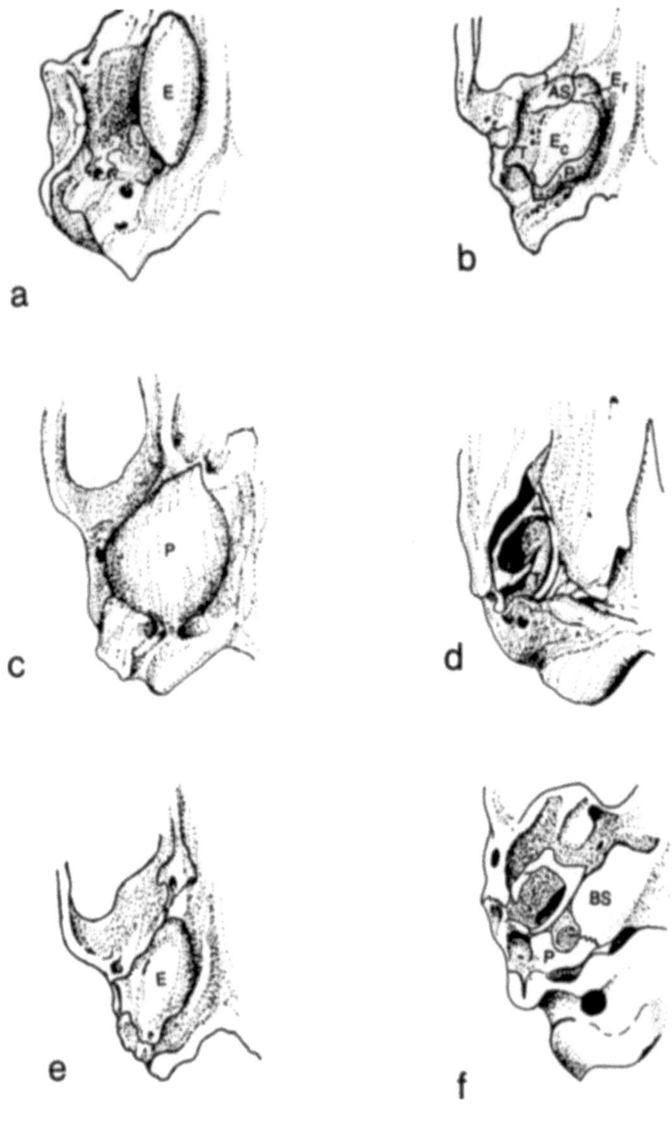

Fig. 6. Diagrams of auditory regions in several mammals showing construction of the bulla and ectotympanic annulus. (a) *Leptictis;* (b) *Macroscelides;* (c) *Lemur;* (d) *Tachyglossus;* (e) *Tupaia;* (f) *Limnogale.* Not to scale. For abbreviations, see pp. 88–89.

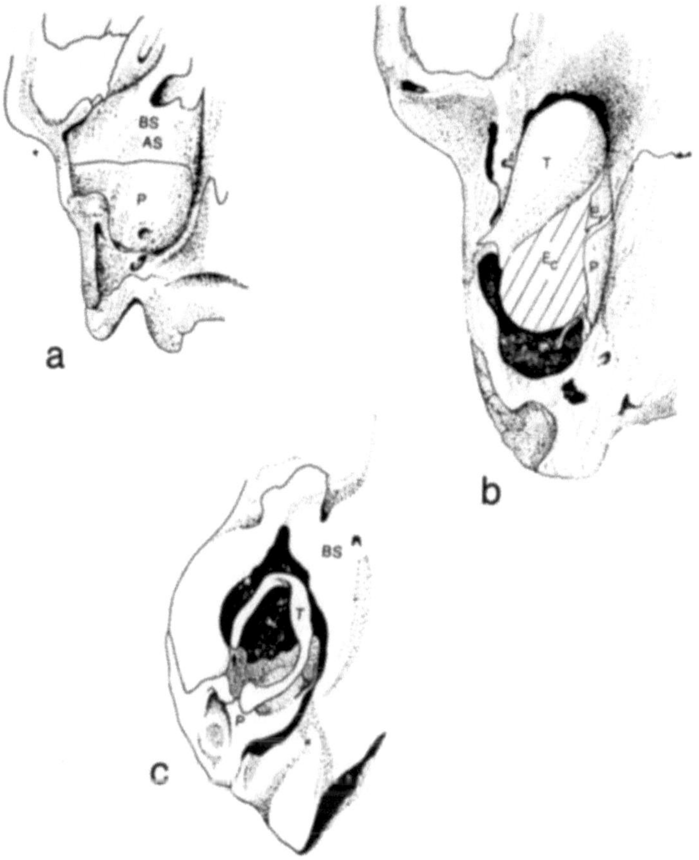

Fig. 7. Diagrams of eutherian auditory regions in three eutherian taxa showing construction of the bulla and ectotympanic ring. (a) *Brachyerix;* (b) *Nandinia;* (c) *Sorex.* Diagonal stripping for *Nandinia* indicates a cartilaginous caudal entotympanic. Not to scale. For abbreviations, see pp. 88–89.

3. Tympanic Region: Miscellaneous Features

Table 2 and Figure 12 display auditory characters for several taxa which represent a distillation of a much larger set considered elsewhere (Novacek, 1977b). A plethora of details of the tympanic roof include the shape of the promontorium, the presence or absence of a glaserian fissure, grooves for the palatine branch of nerve VII, excavation of the epitympanic recess, presence of a vidian foramen, and many others. Although these show interesting patterns of variation, ambiguity concerning their morphocline polarities precludes their effective application to studies of tupaiid affinities. Of the remaining auditory features, the structure and orientation of the ectotympanic

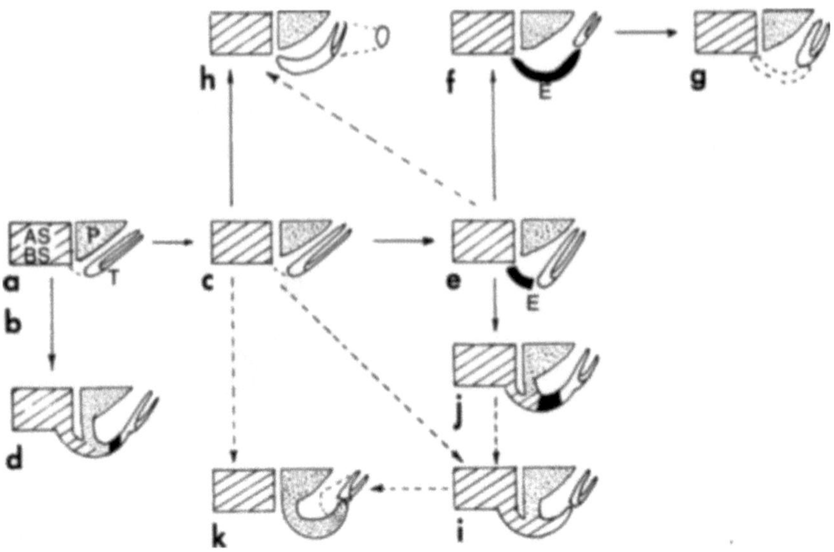

Fig. 8. Hypothesis for evolution of the auditory bulla in mammals. (a) primitive mammalian condition where the bulla is absent and the ectotympanic ring (T) is inclined at a low angle to the horizontal plane of the skull; (b), (c) primitive marsupial and eutherian conditions are similar to a; (d) marsupial bulla comprised of alisphenoid (diagonal lines; AS), petrosal (stipple, P), entotympanic (black bar), and process of the ectotympanic (open ring) (Note: MacPhee, 1979, maintains that the presence of an entotympanic in marsupials is dubious); (e) leptictid condition, with small entotympanic (black bar) bulla; (f) eutherians with well developed entotympanic bullae (e.g. many carnivores, edentates, ungulates); (g) bulla retained as a cartilaginous (entotympanic) structure (e.g. *Nandinia*, some Chiroptera); (h) eutherians with ectotympanic bulla (e.g. rodents, some insectivores, lagomorphs); (i) lipotyphlan insectivore bulla with alisphenoid, basisphenoid, and petrosal components; (j) composite bulla of macroscelidids; (k) petrosal bulla of primates. Dashed lines with arrows indicate alternative pathways of derivation. It is uncertain whether the conditions shown in h, i, and k evolved *de novo* from the primitive mammal condition or evolved through a stage where the entotympanic was present (from Novacek, 1977c).

seem most notable. Much controversy envelops the evolutionary history of this bone in various eutherian groups, particularly in reference to primate systematics (see Szalay, 1972; Gingerich, 1973, 1976; Archibald, 1977). Most workers agree, however, that primitively the eutherian tympanic was a simple ring-shaped structure which opened dorsally, lacked a recessus meatus, and was inclined at a low angle to the horizontal plane of the basicranium. These primitive features are present in tupaiids. Various modifications of the ectotympanic seem closely related to bullar development and construction (see Szalay, 1975; Hunt, 1974; Archibald, 1977; Novacek, 1977c).

In leptictids the tympanohyal bone is small and cup-shaped in ventral view and does not isolate the stylomastoid foramen from the tympanic cavity (Fig. 10). A small tympanohyal is likewise present in erinaceid insectivores and macroscelidines (Fig. 9), but this bone is enlarged and excludes the stylomas-

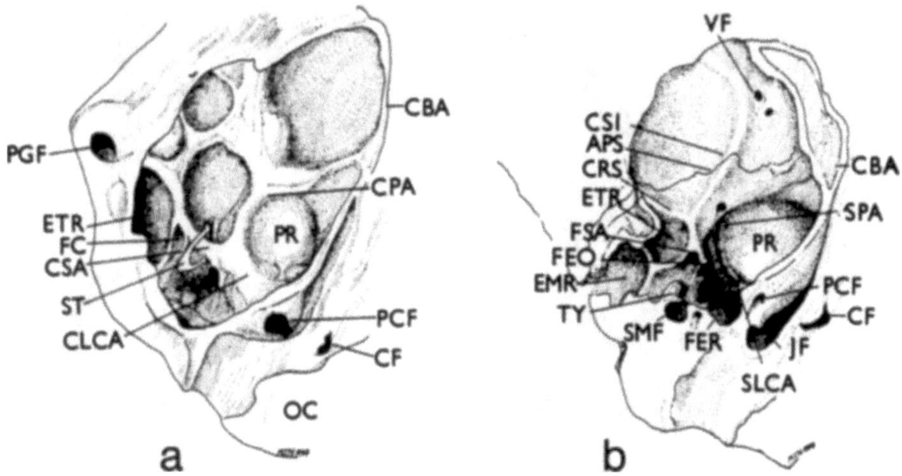

Fig. 9. Ventral views of right auditory regions in (a) *Tupaia glis* (UCMP 126–37, element collection); and (b) *Petrodromus sp.* (LACM 51157) with bullae removed (From Novacek, 1977b). Not to scale. For abbreviations, see pp. 88–89.

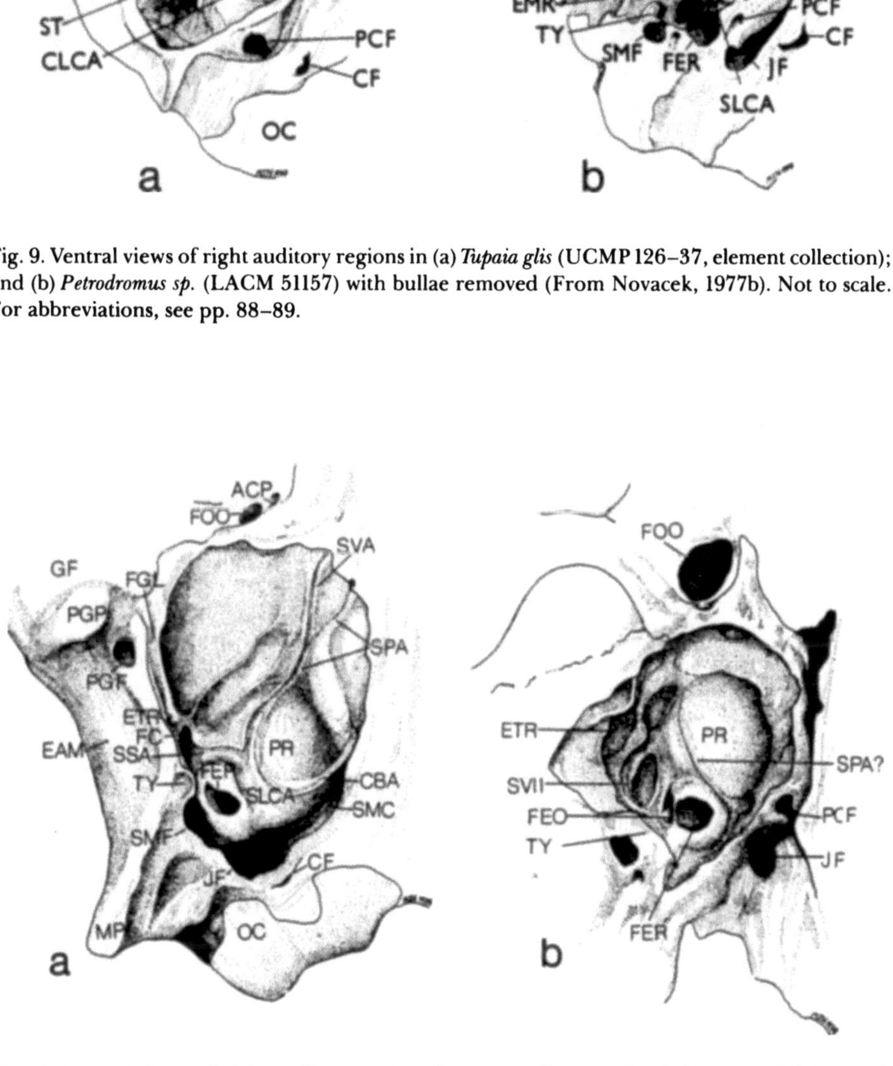

Fig. 10. Ventral views of right auditory regions in (a) an Oligocene leptictid (F:AM 94762); and (b) *Cynocephalus variegatus* (UC-MVZ 141297) with bullae removed. (From Novacek, 1977b). Not to scale. For abbreviations, see pp. 88–89.

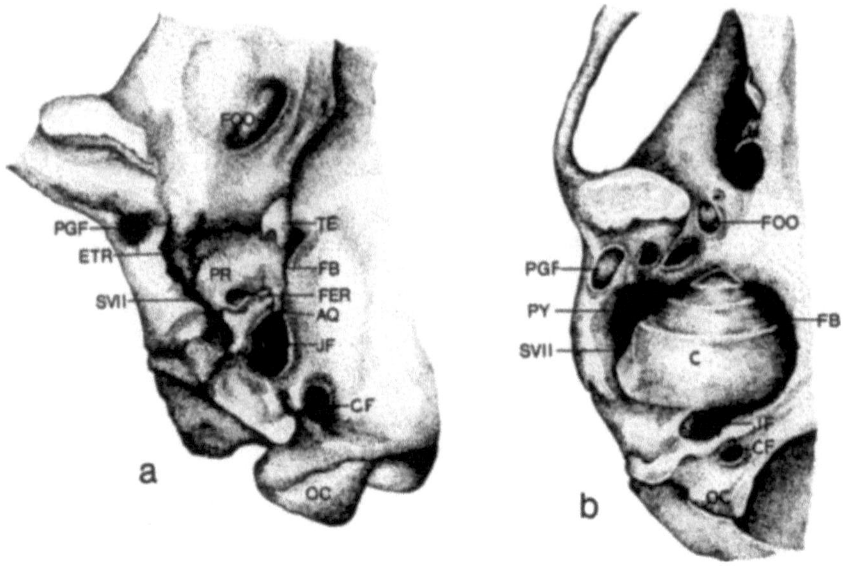

Fig. 11. Ventral views of right auditory regions in (a) *Pteropus poliocephalus* (SDSU 1152); and (b) *Rhinolophus sp.* (SDSU 1215) with bullae removed. Not to scale. For abbreviations, see pp. 88–89.

Table 2. Character states for selected auditory features (see Fig. 12)

Primitive	Derived
1. Rostral entotympanic absent	1. Rostral entotympanic present
2. Caudal entotympanic absent	2. Caudal entotympanic present
3. Ectotympanic annular	3. Ectotympanic moderately expanded ventrally (int.)[1] or markedly expanded into external auditory tube or medial auditory bulla
4. Ectotympanic inclined at acute angle to horizontal plane of cranium	4. Ectotympanic relatively vertical
5. Tympanic process of basisphenoid absent	5. Tympanic process of basisphenoid small (int.) or well developed
6. Tympanic process of petrosal lacking in auditory bulla	6. Tympanic process of petrosal partial (int.) or primary bullar element
7. Ossified tubes for intrabullar vessels are absent	7. Ossified tubes enclosing some (int.) or most intrabullar vessels
8. Medial internal carotid artery present	8. Medial internal carotid artery absent
9. Promontory artery small or absent	9. Promontory artery well developed
10. Stapedial artery large	10. Stapedial artery small (int.) or absent
11. Ramus inferior of stapedial artery present, exits via a fissure in anterior wall of tympanic cavity	11. Ramus inferior exits endocranially posterior to anterior wall of tympanic cavity (int.) or ramus inferior absent
12. Fenestra rotundum not shielded ventrally	12. Fenestra rotundum partially (int.) or completely shielded by enlarged carotid artery or bony tubes or flanges
13. Tympanohyal small process	13. Tympanohyal large, isolates stylomastoid foramen from tympanic chamber

[1]Int. =intermediately derived character states.

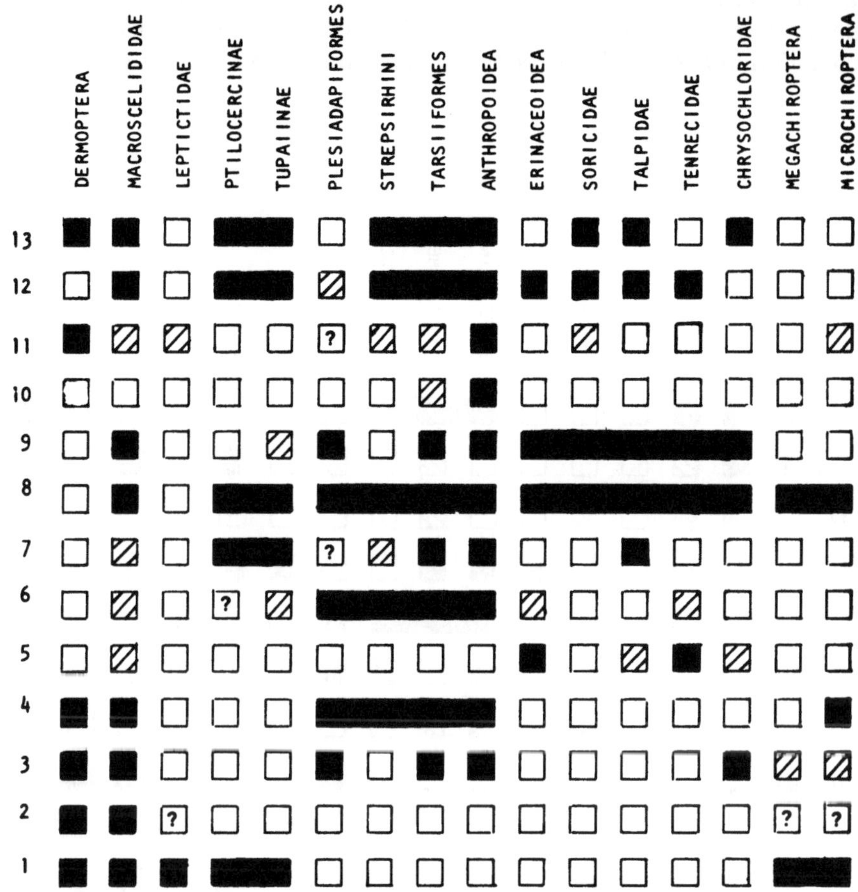

Fig. 12. Character state distributions for auditory features in tupaiids and several eutherian taxa. Numbers refer to characters listed in Table 2. For symbols, see Fig. 4.

toid foramen in tupaiids (Fig. 9) and primitive primates. Following Butler (1956), I am recognizing the leptictid condition as primitive, although admittedly the evidence for this is unclear.

D. Appendicular Skeleton

Although Gregory (1910), Matthew (1937), Winge (1941) and others have considered postcranial skeletal evidence for mammalian interordinal relationships, such features have generally received less attention in recent studies. Fossil skeletal remains are rare and available material is often unfairly ignored. This lack of historical information has impeded studies of evolutionary trends in postcranial characters. It is widely acknowledged that cases of shared specialization in postcranial features often result from convergence rather than true synapomorphy, due to the obvious similarities in adaptive needs involving locomotion and related functions. On the other hand, details of basicranial

Fig. 13. Character state distributions of hindlimb and pelvic features in tupaiids and several eutherian taxa. Numbers refer to characters listed in Table 3. For symbols, see Fig. 4.

and auditory morphology have been regarded as less subject to homoplasy, and perhaps more meaningful in systematic analysis. It is difficult to ascertain the accuracy of such views, given the incipient level of comparative studies of mammalian postcranial skeletons. However, it is anticipated that refined considerations of function and structure (Jenkins, 1971, 1974; Roberts and Davidson, 1975) and new information on fossil skeletal features (Szalay and Decker, 1974; Kielan-Jaworowska, 1977) will allow useful application of postcranial information in phylogenetic studies of higher eutherian categories.

Table 3 and Figure 13 display a small set of skeletal characters mainly confined to the appendicular anatomy. The confidence for advocated polarities varies markedly here, being notably low for scapular and forelimb characters but more secure for features of the sternum, pelvis, and distal hindlimb (see Novacek, 1977b). Ambiguity of certain character states listed in Table 3

Table 3. Character states for selected postcranial skeletal features (see Fig. 13)

Primitive	Derived
1. Vertebral border of scapula short and curved	1. Vertebral border of scapula longer, less curved
2. Scapular fossae narrow, shallow, subequal in area	2. Scapular fossae broad, deeply excavated, disparate in area
3. Metacromion process pronounced	3. Metacromion process weak or absent
4. Manubrium of sternum not enlarged	4. Manubrium enlarged, often with ventral keel
5. Deltopectoral crest on humerus strong	5. Deltopectoral crest weak or very strong[1]
6. Entepicondylar foramen present	6. Entepicondylar foramen absent
7. Supracapitular fossa shallow	7. Supracapitular fossa deep
8. Olecranon fossa shallow	8. Olecranon fossa deep
9. Capitulum spindle shaped	9. Capitulum spherical
10. Ulna robust, with sigmoid lateral curvature	10. Ulna slender, only slightly curved or strongly sigmoid with enormous olecranon process[1]
11. Iliosacral angle obtuse, greater than 30°	11. Iliosacral angle acute, less than 30°
12. Pelvic-sacral fusion limited	12. Pelvic-sacral fusion extensive
13. Anterior inferior iliac spine weak	13. Anterior inferior iliac spine pronounced
14. Greater trochanter prominent, exceeding height of femoral head	14. Greater trochanter shorter than femoral head, or weak, or absent
15. Lesser trochanter large, lamelliform	15. Lesser trochanter reduced or absent
16. Gluteal tuberosity weak or absent	16. Gluteal tuberosity prominent
17. Femoral trochlea broad and shallow	17. Femoral trochlea elongate, deeply excavated
18. Tibia-fibula broadly separated	18. Tibia-fibula distally fused (int)[2] or extensively fused
19. Distal elements of fore- and/or hindlimbs not markedly elongate	19. Marked elongation of distal elements of fore- and/or hindlimbs[3]

[1]Derived states represented by strongly divergent conditions.
[2]int=intermediately derived character states.
[3]Based on various indices, see text for discussion.

might also be noted. Although chiropterans show many obvious specializations in both fore- and hindlimb features, the derived characters states given in the table only provide a crude approximation of the actual modifications in bats. Thus, it would be misleading to ally bats and leptictids based on shared specializations of the manubrium (Table 3) without more careful consideration of the nature of these similarities.

The advocated polarities of scapular features are also suspect without a better understanding of scapular function and variation. One might postulate that the primitive eutherian scapula was somewhat "leptictid-like" with a short, curved vertebral border, small supraspinous and infraspinous fossae, and well developed metacromion process. Such features are also common to tenrecids, erinaceids, didelphid marsupials, macroscelidines, and a few lemuriform primates. However, the evolutionary and functional implications of these features are manifold. For example, Frey (1923) suggested that the range of circumduction is determined by the width of the scapular fossae: the wider and broader the fossae the greater the range of potential circumduction. Roberts and Davidson (1975), however, emphasized that the scapula is extremely mobile on the dorsum, and in running quadrapeds it rocks back and forth as the limbs are flexed and extended. Thus in some fast running animals, such as horses and canids, the scapula is narrow and seems to function as a lever that can be pulled rapidly across the relatively broad dorsum. Roberts and Davidson (1975, Fig. 2) showed, however, that the fossae are generally small in principally quadruped prosimians (e.g. *Microcebus*), while either the infraspinous or supraspinous or both fossae are broad relative to the base length of the scapula in vertical clingers and leapers (e.g. *Indri* and *Propithecus*) and slow climbers (e.g. *Nycticebus*). Since the sizes of the fossae correspond to the masses of muscle attached to them, an animal whose forelimb functions require a strong security of the shoulder joint (as in a hanger, swinger, flyer, or digger) has well developed scapular fossae (Roberts and Davidson, 1975, p. 129). Accounting for these factors, one might infer that the primitive eutherian scapula was suited for ambulatory locomotion and required few modifications for cursorial movement, but was unspecialized in any obvious way for arboreal or fossorial adaptions.

Evolutionary trends suggested in Table 3 for features of the humerus and elbow joint stem mainly from distributional information, particularly outgroup comparisons. Functional arguments for most of these trends are poorly developed. A prominent deltopectoral crest, especially pronounced in early leptictids (Novacek 1977b), is widespread among mammals and probably represents a primitive eutherian character. Based on his studies of cynodont postcranial skeletons, Gregory (1910, p. 433) suggested that the most primitive mammals had very strong epicondylar crests, broad and massive entocondyles, and a large entepicondylar foramen. It is noteworthy, however, that such features are also present in fossorial mammals where they may represent secondary specializations. Gregory (1910, p. 436) maintained that a spheroidal-shaped capitulum was the primitive state, noting this condition in a wide diversity of mammalian groups (monotremes; myrmecophagid, dasypodid, and bradypodid edentates; many marsupial families; some primitive condylarths; and *Solenodon*). Szalay *et al.* (1975) argued convincingly for the opposite trend on functional grounds: the spindle-shaped capitulum restricts pronation and supination of the antebrachium, implying that the degree of mobility of the hand seen in primates and certain other mammals was not possible in the earliest eutherians. Restriction of the antebrachium is also influenced by shallow olecranon and supracapitular fossae, and a robust, sigmoidally curved ulna, features recognized here as primitive for mammals. A caveat is introduced with consideration of polarity in characters of the ulna, because a divergence in trends for fossorial types as opposed to more cursorial forms is clearly evident (Table 3).

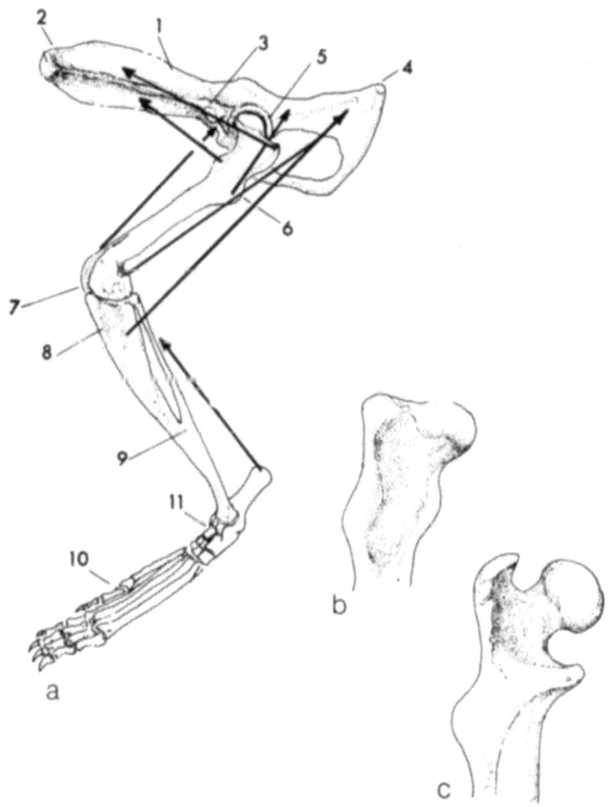

Fig. 14. (a) Diagram of hindlimb and pelvis of an Oligocene leptictid. Numbers refer to the following characters: 1) iliosacral angle, 2) gluteal surface of the iliac wing, 3) anterior inferior iliac spine, 4) superior ischial border, 5) acetabular rim, 6) femoral head and trochanter structure, 7) trochlear groove of distal femur, 8) tibial length and outline, 9) tibia-fibula fusion, 10) length of metatarsals, 11) structure of astragalocalcaneal complex. Arrows represent effort vectors for some of the major hindlimb muscles. (b) Diagram of proximal femur with derived character states. (c) Proximal femur of basically primitive construction (see Table 3).

Hindlimb and pelvic characters displayed in Table 3 comprise a subset of what I believe to be functionally related features (Novacek, 1977b). Specialized states for these are included in Fig. 14 and described in the accompanying caption. Unlike the above cited appendicular features, theories on polarities for hindlimb characters can be cast in light of some important functional reviews (Badoux, 1974; Walker, 1974; Lessertisseur and Saban, 1967; Kruger, 1958; Howell, 1932, 1944; Davis, 1964; Hildebrand, 1974; Hatt, 1932; Hall-Craggs, 1965; Jouffray, 1975; Jenkins, 1971, 1974; Gamberian and Oganesian, 1970). For example, the orientation of the pelvis at the iliosacral joint (character 12) is accountable to forces acting on the joint which vary with speed and mode of locomotion. The iliosacral joint is synostotic in older individuals of most mammalian species, and it is difficult to predict whether its stability is affected by properties of interarticular structures or by action of surrounding ligaments or muscles. Nevertheless, Badoux (1974) has nicely related some of the possible factors influencing the construction of this joint, particularly in respect to the comparative orientations of the long axis of the innominate and the long axis of the sacrum (the iliosacral angle). The femur exerts a force M on the acetabular rim whose vector moves vertical and craniad as the animal is approaching full stride (equivalent to phase III in the locomotory cycle described by Jenkins, 1971). The moment τ_m will tend to rotate the pelvis in a clockwise direction about the iliosacral axis (Fig. 15).

As Badoux (1974) notes, the ideal value of the iliosacral angle is 90° only if the force generated against the roof of the acetabulum is vertical, because the working line would then pass directly through the joint and the moment would be zero. These factors apply only under static conditions; during movement, contraction of the biceps femoris, semimembranosus, semitendinosus, and other powerful flexors of the crus and thigh generates a force acting as an increasingly horizontal component as the femur moves through its anteroposterior arc. This component is particularly emphasized during galloping or leaping where the caudo-ventral sweep of the femur is great (Davis, 1964, p. 10). The moment of $M(\tau_m)$ about the iliosacral joint (J) is the product of the force exerted on the anterior rim of the acetabulum and the perpendicular distance (l_m) between its working line and the joint:

$$\tau_m = Ml_m$$

From this relationship it is clear that a smaller iliosacral angle will result in less torque during the propulsive phase of movement. In saltatorial rodents the strong thrust generated by the hindlimb results in a large horizontal component of force M, but l_m is relatively small because the long axis of the pelvis is nearly aligned with the sacral axis, and the iliosacral angle is acute in saltatorial rodents. This angle is 13° in *Dipodomys deserti,* comparable to other small to medium-sized mammals emphasizing saltatorial or some other mode of specialized locomotion. The less acute angle for this joint in tupaiids, some primates, erinaceids, and tenrecids (Fig. 13) is therefore regarded as the primitive eutherian condition.

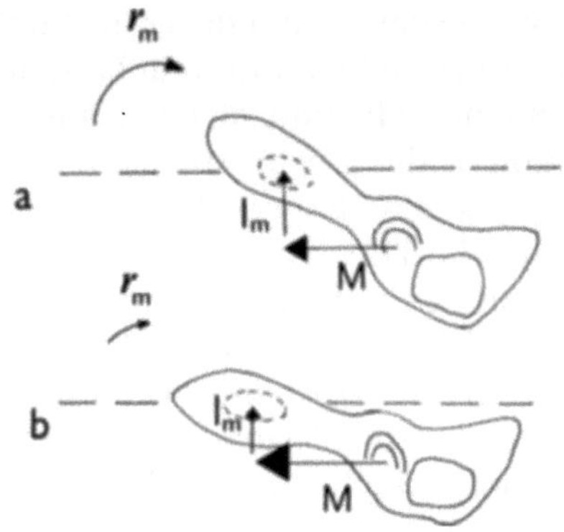

Fig. 15. Diagrams depicting forces acting on the iliosacral joint (stipple) where the iliosacral angle is obtuse (a), or acute (b). Force vectors are inferred. Symbols are explained in text.

Another means of contending with forces acting on the iliosacral joint involves the increase in the number of sacral vertebrae fused to the pelvis. In the majority of mammals, the pelvis shows some mobility, and fusion is limited to three or slightly more vertebrae. Such a condition is common to tupaiids, leptictids, didelphids, most primates and lipotyphlan insectivore groups, and some macroscelidids (Fig. 13). Stabilization of the pelvis is more crucial, however, in animals where limb excursion is pronounced. Thus, a marked degree of pelvic sacral fusion is observed in Chiroptera and Dermoptera as well as various saltatorial mammals.

Because the innominate serves as a site of attachment for muscles moving various parts of the hindlimb, its structure is an indication of relative muscle mass, power, and orientation. Of particular interest in this consideration is the relative development of the anterior inferior iliac spine (Fig. 14). This bony process is the site of origin for the rectus femoris, a long fusiform muscle that inserts on the proximal border of the patella and extends the crus and flexes the thigh. In many eutherians this muscle simply originates on a scar at the anterior lip of the acetabulum. Such a condition is also known in therapsids, monotremes, and didelphid marsupials, and is probably primitive for mammals. The prominent anterior inferior iliac spine is principally a feature of certain leaping and hopping forms (Lessertisseur and Saban, 1967, p. 830) and this process is very well developed in saltatorial and richochetal rodents, macroscelidids, lemuroid primates, and advanced leptictids (Fig. 13).

Related to limb excursion is the variation in structure of the femoral trochanters among mammals. It is postulated that in primitive eutherians the greater trochanter was prominent and served as a site of insertion for the gluteus medius and gluteus profundus, muscles which function in extending and abducting the thigh. The lesser trochanter was likewise probably well developed, and the contraction of the iliopsoas muscles inserting on this process results in flexion and lateral rotation of the thigh (Fig. 14c). Antagonistic to the deeper gluteal muscles is the action of the mm. gluteus superficialis which inserts on the third trochanter and flexes the thigh and rotates it mediad. The greater trochanter is markedly reduced in bats, dermopterans, and some primates, with concomitant reduction of deep and medial gluteal muscles (Figs. 13, 14b). Reduction of the greater trochanter is also an indication that the site of insertion of these muscles was shifted anteriorly, as in some carnivores (see Davis, 1964, p. 113–115). The medial and deep gluteals would thus remain effective in extending the femur, but their advantage in abduction is reduced. It should be noted, however, that simple correlation of leaping ability with development of the greater trochanter is not possible; even though this feature is generally characteristic of cursorial and saltatorial mammals, it is also found in mammals of strikingly different locomotor adaptations (Howell, 1944, p. 12). The lesser trochanter is small in macroscelidids, leptictids, dermopterans, and some primates (Fig. 13), suggesting that flexion and rotation of the thigh by action of the iliopsoas muscles were reduced. This suggests a priority of function for mammals where the arc of movement of the limb is pronounced in the plane parallel to the long axis of the body, but where a need arises for limitation of movement in other planes. Oddly, expansion, rather than reduction, of the third trochanter (gluteal tuberosity) is usually characteristic of cursorial or saltatorial mammals.

Despite the preliminary status of my comparisons, there seems little doubt that the polarities advocated for distal hindlimb characters listed in Table 3 are correct. In most ambulatory mammals the tibia and fibula are broadly separate, but show progressive fusion in mammals with more specialized locomotion such as dermopterans, macroscelidids, advanced leptictids, tarsiiform primates, some talpids, chrysochlorids, and leporids (but note exceptional condition in erinaceids), where there is an obvious adaptive requirement for reduction in weight of the distal limb and de-emphasis in rotational movement of the shank. These taxa also typically show an elongate and deep trochlear groove on the distal femur to allow a wide arc of extension and flexion of the crus.

Of final interest here are various quantifications of limb proportions with alleged functional significance, most notably the brachial, crural, and intermembral indices. The phyletic and evolutionary implications of these indices have been overemphasized in many studies, but such ratios seem useful if applied with reservations. The measure of the disproportionate lengths of the forelimb and hindlimb commonly used is the intermembral index

$$\left(\text{Lengths } \frac{\text{Humerus} + \text{Radius}}{\text{Femur} + \text{Tibia}} + 100\right).$$

In his classic study of mammalian locomotory adaptations, Howell (1944, p. 205) remarked: "Scrutiny of this index indicates that it is entirely worthless for indicating either speed or the lack of it, except that a relatively long hind leg often points to leaping ability." Howell calculated a value of 75 for the generalized condition in mammals, noting that the index is low (32–50) in bipedal "jumpers." The intermembral index in leptictids with referable postcranial elements is consistently around 68, a considerably higher figure than the

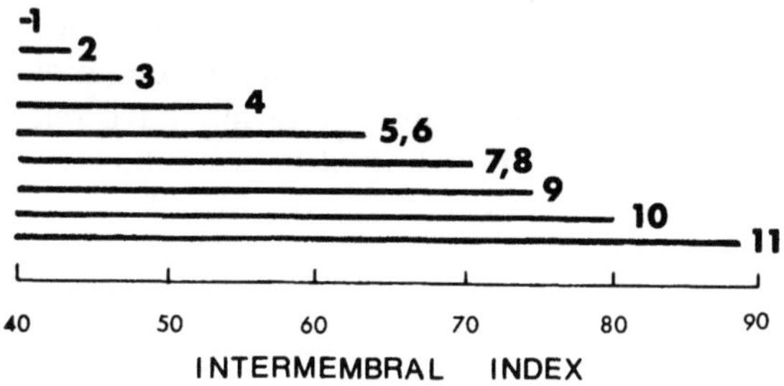

Fig. 16. Plot of several mammalian genera based on intermembral indices. Taxa are: 1, *Pedetes;* 2, *Macropus;* 3, *Dipodomys;* 4, *Avahi;* 5, *Palaeictops;* 6, *Rhynchocyon;* 7, *Leptictis;* 8, *Macroscelides;* 9, *Tupaia;* 10, *Petrodromus;* 11, *Solenodon.*

32–50 range of values seen in richochetal, bipedal rodents like *Pedetes* and *Dipodomys,* but lower than the value for *Tupaia glis* (75)[3], *Didelphis marsupialis* (78), *Echinosorex* (79), or *Solendodon paradoxus* (81). Lemuriform primates are jumpers which fall into two categories; the indriids and *Lepilemur* with indices between 56 and 65, are almost exclusively bipedal and vertical, while the remaining lemurs, with indices of 65–76, are quadrupedal and horizontal in their thrust locomotion (Jouffray, 1975). *Sciurus carolinensis,* an "arboreal scrambler" (Gingerich, 1976, p. 84), shows an identical intermembral index (68) to some leptictids and plesiadapids. With the exception of *Petrodromus,* elephant shrews show low values for intermembral indices, and were thus favorably compared by Evans (1942) to richochetal rodents. It is generally believed, however, that macroscelidids do not show bipedal hopping but are basically quadrupedal, using their hind legs for kicking back leaves or thumping the ground like rabbits (see Evans, 1942). Some of these comparisons are graphically portrayed in Fig. 16.

Figure 17 shows the brachial (BI) and crural (CI) indices of some small

$$\mathrm{BI} = \text{Length}\,\frac{\text{Radius}}{\text{Humerus}} \times 100 \qquad \mathrm{CI} = \text{Length}\,\frac{\text{Tibia}}{\text{Femur}} \times 100$$

mammåls. These ratios are only crude indications of locomotory adaptations, and creatures of wide ranging habitus may be artificially lumped by them. Nevertheless, some interesting patterns are revealed. For instance, macroscelidids show marked variation in the relative size of the antebrachium but are

[3]Jenkins (1974) in a recent study has emphasized the considerable amount of variation in tupaiid locomotion, ranging from the predominantly terrestrial *Urogale everetti* and *Tupaia (Lyonogale) tana* to the arboreal-terrestrial habits of *Tupaia glis longicauda* and *T. javanica*. Field and laboratory observations of *Dendrogale* and *Anathana* are virtually lacking. Jenkins' analysis leads him to suggest that although *Tupaia glis* is both arboreal and terrestrial, this species is (p. 92) ". . . particularly adept at rapid locomotion in an environment that necessitates abrupt changes in direction or elevation, and is capable of upward leaps of approximately one meter." Following the categories established by Gambarian and Oganesian (1970), Jenkins characterized *Tupaia glis* as a mammal using a "primitive rebounding jump" where the hind limbs provide most of the propulsive force and the forelimbs act primarily as shock absorbers.

generally in the upper range of values for the crural indices (120–140), surpassing even some richochetal rodents in relative elongation of the crus. Lemuriform primates, on the other hand, generally have relatively long antebrachii (obviously correlated with arboreal habitus) but show low values for crural indices. The crural index for the scansorial *Tupaia glis* is also low. Terrestrial and plantigrade insectivores not showing leaping or hopping in their locomotion receive low scores on the crural scale, as do ground squirrels and marmots. Leptictids and *Sciurus* are of intermediate position, approaching some bipedal rodents in the degree of crus elongation.

The functional advantage of a relatively elongated lower leg in running and leaping forms is obvious. Such animals require a "higher gear ratio" lever system, where muscles can move the joints through wider angles at higher speeds because of the increased ratio of out-lever arm (l_o) to the in-lever arm (l_i). A general discussion of these principles is provided by Hildebrand (1974). A high value for the crural index coupled with a low intermembral index suggest locomotion where an elongate crus functions in leaping, but where the propulsive action of the front limb has been de-emphasized. With these basic principles in mind, it is possible to draw a cautious inference from the above cited data on indices: primitive eutherians probably showed conservative values for intermembral, brachial, and crural indices, in keeping with a basically unspecialized, ambulatory mode of locomotion.

The above reviewed appendicular characters do not provide strong evi-

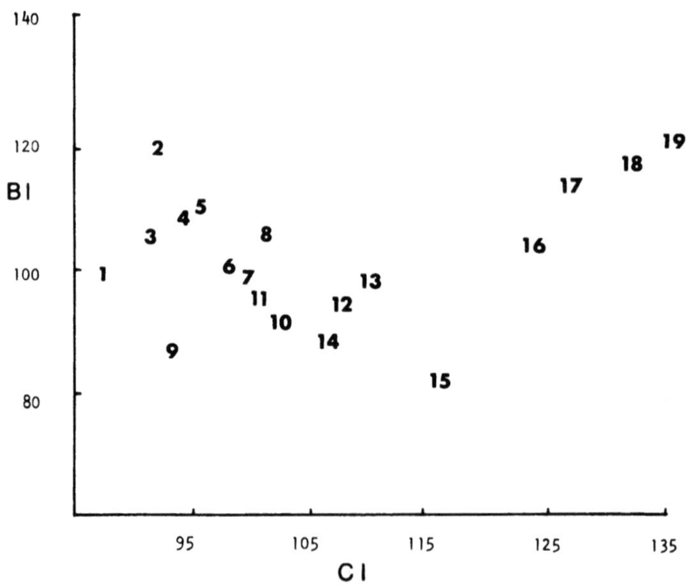

Fig. 17. Bivariate plot of mammalian genera based on brachial (BI) and crural (CI) indices. Taxa are: 1, *Solenodon;* 2, *Avahi;* 3, *Propithecus;* 4, *Hapalemur;* 5, *Lemur;* 6, *Daubentonia;* 7, *Cheirogaleus;* 8, *Microcebus;* 9, *Plesiadapis;* 10, *Tupaia;* 11, *Citellus;* 12, *Frictops;* 13, *Leptictis;* 14, "Menat leptictid"; 15, *Pedetes;* 16, *Rhynchocyon;* 17, *Nasilio;* 18, *Dipodomys;* 19, *Macroscelides.*

dence for eutherian sub-categories, with the possible exception of a macroscelidid-leptictid grouping and the expected linkage of mega- and microchiropterans (Fig. 13). Tupaiids notably lack derived character states among the features considered here. However, this information is useful when applied to phyletic analysis incorporating a larger set of osteological features (see below).

E. Proximal Tarsus

Of particular interest to considerations of tupaiid affinities are features of the calcaneum and astragalus; an interest augmented by Szalay's (1977b) study of tarsal structure in tree shrews, primates, and dermopterans. Gregory (1910), Matthew (1937), and other authors recruited evidence from tarsal morphology for support of certain interordinal relationships, but Szalay (1977b) has notably extended this work in his comprehensive and useful survey of Eutheria. The character states listed in Table 4 and plotted in Figure 18 generally follow those defined by Szalay and Decker (1974), Szalay *et al.* (1975), and Szalay (1977b). The calcaneal and astragalar morphology of the late Cretaceous genus *Protungulatum* was taken by Szalay and Decker (1974, p. 231) to represent ". . . an astragalocalcaneal complex more primitive than that of any known palaeoryctoids or other known Eutheria." Many of the primitive character states listed in Table 4 are present in this genus.

Relevant to reconstructions of the morphotypical tarsus in eutherians is a recent description by Kielan-Jaworowska (1977) of the Late Cretaceous palaeoryctoid *Asioryctes*. The tarsus of this genus is the oldest known for any eutherian, and it shows some of the features recognized as primitive by Szalay and Decker (1974): the calcaneal cuboid facet is concave and oriented obliquely to the long axis of the calcaneum; the peroneal tubercle is pronounced; the calcaneal fibular facet is present; the astragalar tibial trochlea is broad and short with a low, lateral trochlear border; and the distal astragalar facets are moderately convex. Kielan-Jaworowska (1977, p.75) concludes that: "The tarsus of *Asioryctes* represents the most primitive type found in therian mammals." Indeed, some of the tarsal features in this animal seem markedly more archaic than those in *Protungulatum* or the late Cretaceous North American palaeoryctoid, *Procerberus*. Unlike the latter, the astragalus is situated medial to the calcaneum and its contact with that element is limited to the sustentacular facet. The tibial trochlea of the astragalus is also very poorly developed, resembling a condition in mammal-like reptiles, particularly bauriamorph therapsids (Kielan-Jaworowska, 1977). Thus, it might be inferred that the primitive eutherian calcaneal–astragalar complex was more conservative in structure than that predicted by Szalay and Decker (1974). However, the polarities advocated by these authors seem basically correct because the *Protungulatum* condition is plesiomorphous compared with the tarsii in other eutherians they consider.

It is possible to achieve a superordinal hierarchy based on synapomorphic features of the mammalian tarsus, and Szalay (1977b) has utilized such char-

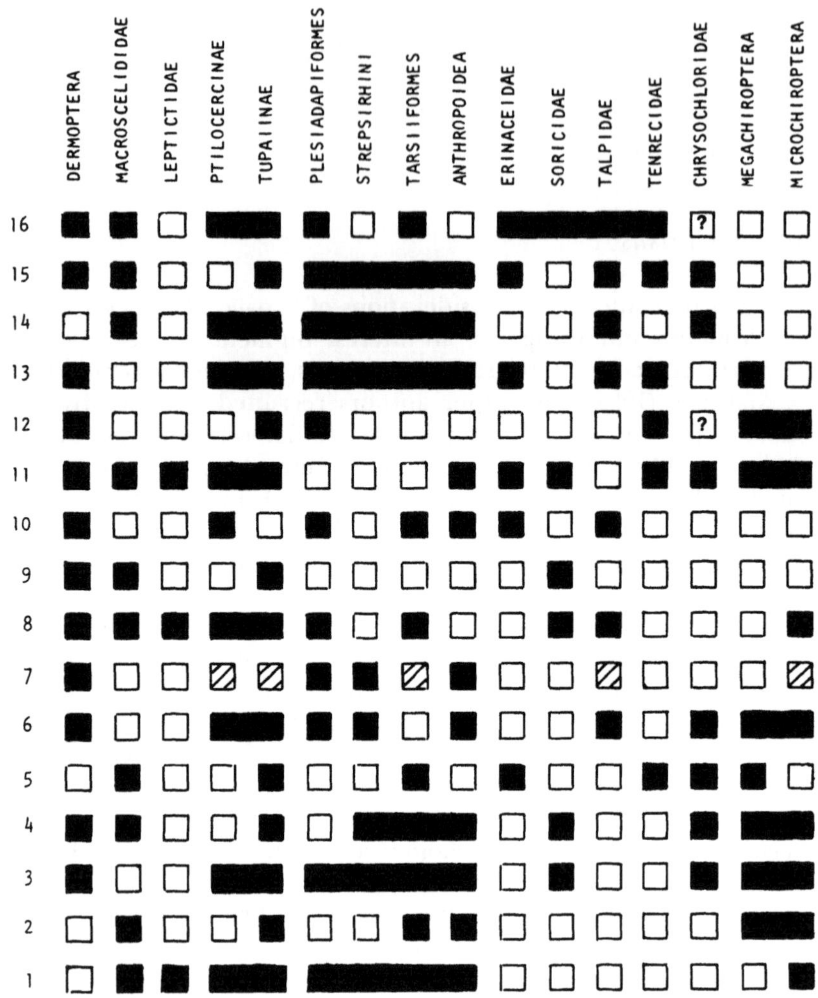

Fig. 18. Character state distributions for astragalar and calcaneal features in tupaiids and several eutherian taxa. Numbers refer to characters listed in Table 4. For symbols, see Fig. 4.

acters in a phylogenetic analysis of eutherians. I have elsewhere (Novacek, 1977b) commented on his conclusions regarding relationships of leptictids, lagomorphs, insectivores, zalambdalestids, and macroscelidids (Fig. 19). Here I confine attention to Szalay's advocated monophyly for the Archonta, to which he relegates tupaiids, primates, dermopterans, and chiropterans. Szalay (1977b) cites the following tarsal features as diagnostic for the Archonta: 1) the tibial trochlea of the astragalus is proximo-distally elongate; 2) traces of the superior astragalar foramen persist; 3) the lateral border of the astragalar body is high and sharply crested, while the medial border is low and rounded; 4) the astragalar sustentacular facet tends to be confluent with the naviculo-astragalar facet medially; 5) the astragalocalcaneal facet (of the calcaneum) is

Table 4. Character states for selected calcaneal and astragalar features[1] (see Fig.18)

Primitive	Derived
1. Calcaneal fibular facet pronounced	1. Fibular facet weak or absent
2. Calcaneal peroneal tubercle strong	2. Peroneal tubercle weak or absent
3. Calcaneal astragalar-calcaneal facet not strongly arched, with ellipsoidal outline in dorsal view	3. Astragalar-calcaneal facet curvature strong, with small radius; sigmoidal or cylindrical outline in dorsal view
4. Calcaneal astragalocalcaneal facet oriented obliquely to long axis of calcaneum	4. Astragalocalcaneal facet acutely angled, nearly parallel or nearly transverse to long axis of calcaneum
5. Calcaneal cuboid facet deeply concave	5. Cuboid facet with shallow concavity
6. Calcaneal cuboid facet oriented obliquely to long axis of calcaneum	6. Calcaneal cuboid facet oriented transversely to long axis of calcaneum
7. Astragalar tibial trochlea broad and short	7. Tibial trochlea moderately (int)[2] or extremely long and narrow
8. Lateral astragalar trochlear border low and rounded	8. Lateral trochlear border high and often crestiform[3]
9. Medial astragalar trochlear border low and rounded	9. Medial trochlear border high and often crestiform[3]
10. Astragalar trochlea limited to body	10. Astragalar trochlea extending to neck
11. Superior astragalar foramen present	11. Superior astragalar foramen absent
12. Astragalar fibular shelf prominent	12. Fibular shelf weak or absent
13. Astragalar sustentacular facet separate from distal astragalar facets	13. Sustentacular facet confluent with distal astragalar facets
14. Calcaneal-astragalar facet (on astragalus) slightly concave	14. Calcaneal-astragalar facet strongly concave
15. Naviculoastragalar and spring ligament facets moderately convex	15. Naviculoastragalar and spring ligament facets strongly convex or flattened
16. Grooves on calcaneum and astragalus for tendon of mm. flexor digitorum fibularis poorly defined	16. "Flexor fibularis" grooves or troughs well defined.

[1]Tarsal characters and advocated polarities generally follow those discussed by Szalay and Decker (1974) and Szalay *et. al.* (1975).

[2]int = intermediately derived character states.

[3]Trochlear crests may form the borders of distinctive troughs on the proximal trochlea (see Fig. 20).

aligned proximo-distally and has a sigmoid outline, allowing for both rotational and translational movement; 6) the fibular facet is almost completely eliminated; and 7) the calcaneal cuboidal facet is slightly convex and aligned transversely, rather than obliquely to the long axis of the calcaneum.

Although Szalay (1977b) provides important characterizations of tarsal variation in these eutherians, I find problematic his application of some of the above-cited features to the Archonta problem. While the tibial trochlea in these forms is elongate (character 1 above), the elongation is much more pronounced in tupaiids and dermopterans (Figs. 20 and 21; and Szalay 1977b; Fig. 6) and there is concomitant distal flaring of the trochlea. Moreover, the trochlea in *Notharctus* is much shorter and broader and virtually confined to the astragalar

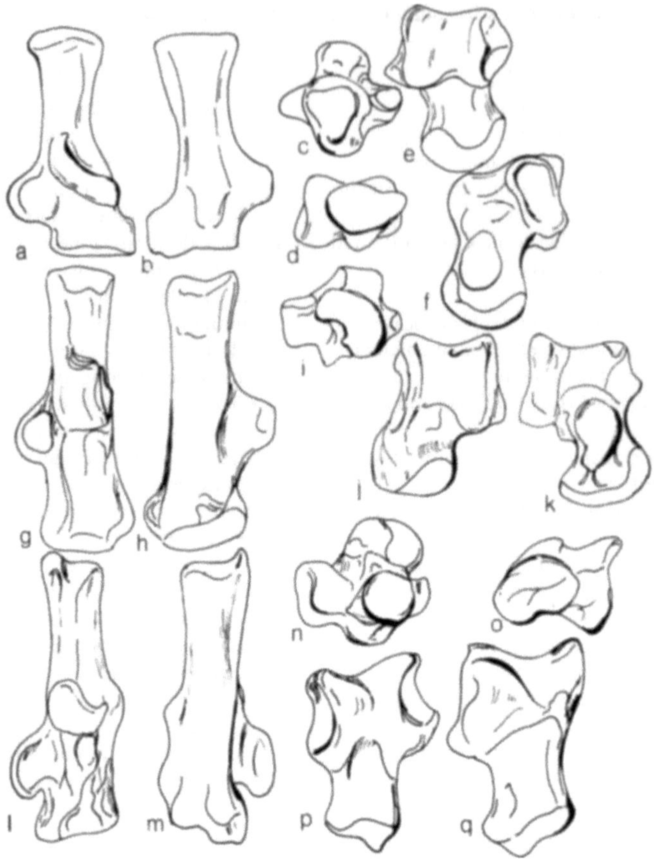

Fig. 19. Astragali and calcanea of leptictids (a-f), lagomorphs (g-k), and macroscelidids (l-o). a, dorsal; b, ventral; c, distal views of left calcaneum, and d, distal; e, dorsal; f, ventral views of right astragalus of *Prodiacodon*. g, dorsal; h, ventral; i, distal views of left calcaneum of *Paleolagus*, and j, dorsal; k, ventral views of left astragalus of *Pseudictops*. l, dorsal; m, ventral; n, distal views of left calcaneum, and p, dorsal; q, ventral; o, distal views of left astragalus of *Petrodromus*. Not to scale.

body, thus approaching the condition in *Protungulatum*. If such is the morphotypical condition for primates, primitive members of this order lacked an important archontan trait. The persistence of the superior astragalar foramen (character 2) is a plesiomorphic eutherian trait (according to Szalay and Decker, 1974), and therefore serves little purpose in adjoining various archontans. It is noteworthy that traces of this foramen are not apparent in extant dermopterans (Fig. 21). A high, sharp lateral trochlear crest is present in plesiadapids, other primates, dermopterans, and tupaiids, but this apomorphic feature is also common to lagomorphs, leptictids, macroscelidids, and several other eutherian groups (Szalay, 1977b; Novacek, 1977b). Its diagnostic role as a uniquely derived archontan trait is therefore diminished. Furthermore, both the lateral and medial trochlear crests are sharp and prominent in *Ptilocercus*

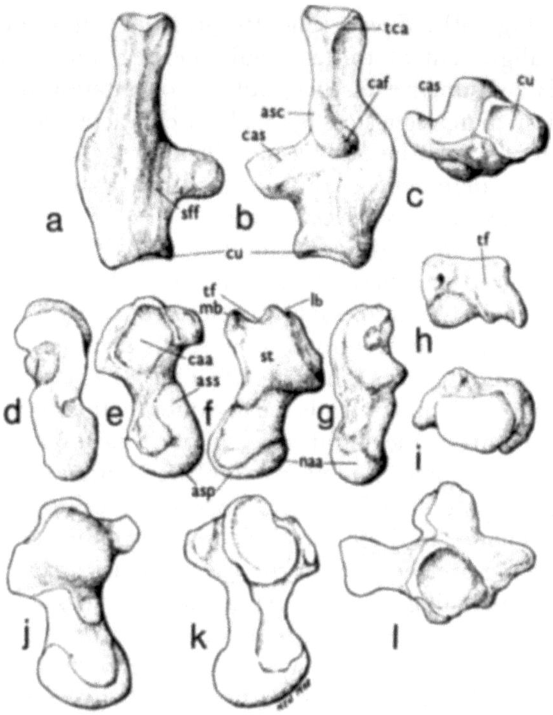

Fig. 20. Astragali and calcanea of *Tupaia montana,* UC-MVZ 120926 (a-i); and *Urogale everetti,* UC-MVZ 124126 (j-l). a, ventral; b, dorsal; c, distal views of left calcaneum. d, medial; e, ventral; f, dorsal, g, lateral, h, proximal; i, distal views of left astragalus. j, dorsal; k, ventral views of right astragalus; l, distal view of left calcaneum. X4. For abbreviations, see pp. 88-89.

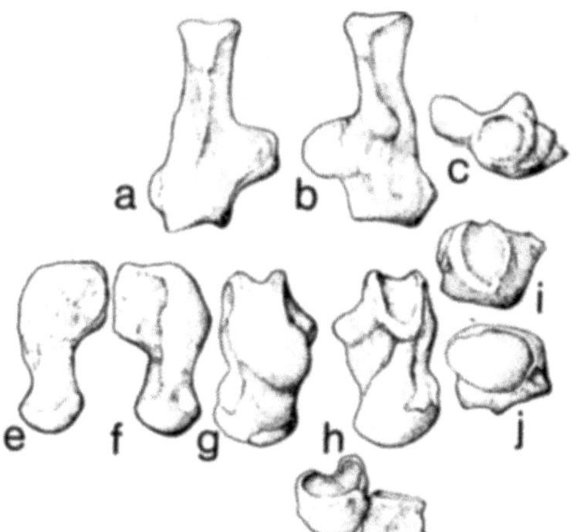

Fig. 21. Proximal tarsal elements of *Cynocephalus variegatus,* UC-MVZ 141297. a, ventral; b, dorsal; c, distal views of left calcaneum. e, lateral, f, medial; g, dorsal; h, ventral; i, proximal; j, distal views of the left astragalus. k, dorsal view of navicular (left) and cuboid (right). X2.

(Szalay 1977b, Fig. 6D), *Tupaia* (Fig.20), and dermopterans (Fig. 21). The proximodistal alignment of the astragalocalcaneal facet on the calcaneum (character 5) is a distinctive feature, but the orientation of this facet shows greater variation than suggested by Szalay's characterization. For example, *Tupaia* shows an oblique orientation (Fig. 20), approaching the primitive eutherian condition; its alignment certainly differs from that in dermopterans (Fig. 21), *Ptilocercus, Plesiadapis,* or *Phenacolemur.*

From the above we might conclude that at least four of the seven tarsal characters recognized by Szalay (1977b) for the Archonta are of dubious phylogenetic significance. An additional problem concerns Szalay's inclusion of the Chiroptera within the Archonta, because his detailed tarsal comparisons did not extend to that group. The true affinities of bats are far from understood, but preliminary analysis leads me to doubt their archontan membership based on tarsal features. One might expect foot characteristics to be anomalous in the only mammals that employ hindlimbs as hangers in roosting and as stabilizers for a flapping wing. The megachiropteran calcaneum and astragalus bear little resemblance to these tarsal elements in other alleged archontans. Particularly striking are the shift in the calcaneal astragalar articulation and the weak development of guiding ridges and articular surface area of the tibial trochlea (Fig. 22). Even more divergent is the structure of microchiropteran tarsus (Fig. 22f). Although bats may share some derived tarsal features of other "archontans" (Fig. 18), the basic differences between these complexes in respect to form and function are so great as to render any similarities highly suspect examples of synapomorphy.

The functional implications of the above comparisons were considered at length by Szalay and Decker (1974) and Szalay (1977b), who emphasized that the derived tarsal characteristics of dermopterans, Paleocene primates, and *Ptilocercus* suggested an arboreal-scansorial habitus for early archontans. Jenkins (1974) and Kay and Cartmill (1974), however, argue that early primates

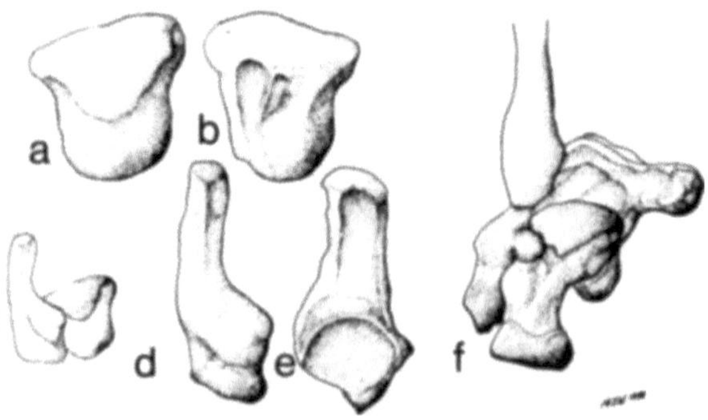

Fig. 22. Astragali and calcanea of *Pteropus poliocephalus.* (a-e) and *Macroderma gigas* (f). a, dorsal; b, ventral views of right astragalus (X6). c, articulation of right astragalus and calcaneum (X3). d, dorsal; e, medial views of right calcaneum (X6). f, dorsomedial view of articulated right calcaneum and astragalus (X5).

and tupaiids show a mode of scansorial-terrestrial locomotion harking back to the primitive eutherian condition. On the basis of my preliminary survey of foot structure, I agree with Szalay (1977b) that some primates, dermopterans, and tupaiids share derived features in tarsal structure which suggest arboreal adaptations, but these are neither abundant nor pronounced. Szalay's argument (anticipated by many earlier views), that the glissant dermopteran and volant chiropteran roles resulted from arboreal adaptations like those of primates and tupaiids, is clearly reasonable, but sorely needed are more detailed studies of the postcranial skeleton in "archontans" and other eutherians (see Szalay and Drawhorn, this volume). These regions of the skeletons are so much more intensively studied in primates and (to a lesser extent) tupaiids than in most other mammalian groups, as to defy an objective basis for comparison. Ignorance is particularly evident in the case of chiropterans, but the foregoing preliminary analysis suggests few, if any, tarsal features linking bats with other alleged archontans.

In the context of this discussion, the recent contribution of Szalay and Drawhorn (this volume) should be considered. These authors further argue for the monophyly of the Archonta (but now express uncertainty regarding the archontan affinities of bats) based on morphology and inferred adaptations of the tarsus. Their objections to some of the above remarks stem from what they recognize as fundamental differences in our application of the morphotype concept and hypotheses for character transformation[4]. More specifically, they feel that I overlook the "gestalt" of the tarsal complex and place too much emphasis on the independent appearance of isolated features in divergent mammalian groups as a basis for doubting the tarsal evidence for archontan monophyly. It should be clear from the forgoing discussion that I certainly do not deny the phylogenetic value of an "entire homologous character complex." Nor do I feel, as Szalay and Drawhorn suggest many paleontologists do, that tarsal characters are any more susceptible *a priori* to convergence than other cranioskeletal features (see my statements under "Methodology" and "Appendicular Skeleton"). However, the recognition of the derived tarsal complex for the archontan morphotype is a working hypothesis, not an empirical observation. Szalay and Drawhorn have provided original insights on character adaptation and integration, but certainly there is room for further scrutiny of such theories through detailed functional and developmental studies. Pending such work, the distributions of isolated characters cannot easily be dismissed if they show discrepancy with a particular phylogenetic conclusion based on recognition of a "gestalt" character complex. My own assertions on the relationships of the orbital bones (see above) are also vulnerable from this perspective.

The practical problems in application of the form-function complex might

[4]Contrary to my statements, Szalay and Drawhorn feel that the assessment of character transformation is not essential to cladistic methodology, but the distinction of apomorphy, synapomorphy, and symplesiomorphy is. Regardless of traditional views and semantic debates in this area, I see no difference between the two lines of inquiry.

also be considered. Verbal dissection of a "gestalt" character complex may seem artificial, but it is usually necessary for the simple communication of ideas and examination of hypotheses. Although many characters are recognized and described with respect to their functional roles, contrasting views on the evolution of a character complex may ultimately reside in arguments concerning one or a few of its components. Likewise, an isolated character often plays a pivotal role in phylogenetic assessment. For example, Szalay and Drawhorn (this volume) distinguish the tarsal complex of living and fossil primates on a single derived character—the deep groove on the calcaneum for the tendon m. flexor fibularis. Although this feature has interesting functional implications and relationships to other moving parts, much of the latter is purely inferential (particularly in fossils), being based on a single item of empirical evidence.

Finally, Szalay and Drawhorn criticize my practice of assessing features in advanced members of superspecific taxa to falsify hypotheses bases on shared derived similarity of the ancestor of that group to another taxon. They make particular reference to my statements concerning tarsal features in *Notharctus*, but I readily acknowledge that this genus shows many features (including

Table 5. Character states for miscellaneous cranial and postcranial features (see Fig. 23)

Primitive	Derived
1. Infraorbital canal long	1. Infraorbital canal short
2. Posterior nasals broad, expanded	2. Posterior nasals narrow
3. Palatal fenestrae absent	3. Palatal fenestrae present (int.)[1] and numerous
4. Occiput not posteriorly expanded	4. Occiput posteriorly expanded, visible in dorsal view
5. Anterior border of the coronoid process vertical or slightly inclined	5. Anterior border of the coronoid process markedly inclined
6. Jaw condyle distinctly lower than coronoid process	6. Jaw condyle higher or subequal to height of coronoid process
7. Postglenoid process present, moderately developed	7. Postglenoid process weak or absent or very pronounced[2]
8. Supraorbital foramen absent	8. Supraorbital foramen present[3]
9. Origin of internal pterygoid muscle confined to ectopterygoid process	9. Origin of internal pterygoid muscle extending to medial wall of orbit
10. Origin of temporalis muscles extending anteriorly over frontals	10. Origin of temporalis muscles posteriorly confined, barely reaching frontals
11. Scaphoid and lunate unfused	11. Scaphoid and lunate fused
12. Os centrale present	12. Os centrale absent or fused with other elements
13. Metatarsals not greatly elongated	13. Some (int.) or all metatarsals markedly elongated
14. No metatarsals significantly reduced	14. Some metatarsals reduced(int.) or lost or fused with other elements

[1]int = intermediately derived character states.
[2]Derived states represented by strongly divergent conditions.
[3]Advocated polarity of this character is highly tentative.

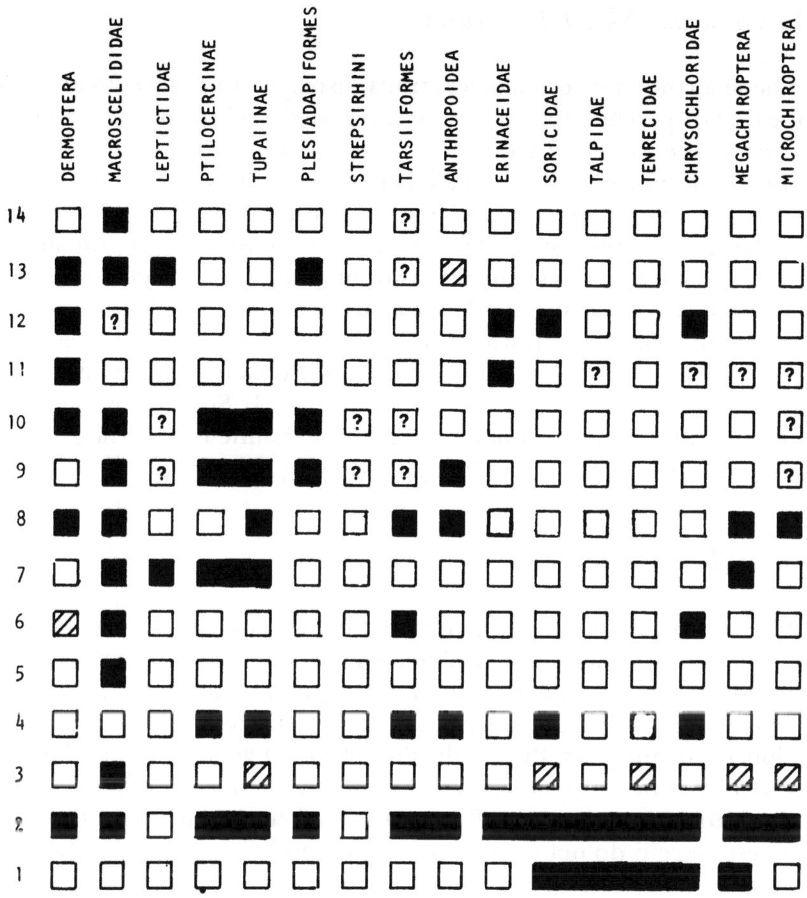

Fig. 23. Character state distributions for miscellaneous skeletal features in tupaiids and several eutherian taxa. Numbers refer to characters listed in Table 5. For symbols, see Fig. 4.

tarsal ones) "advanced" for primates. I do, however, note that the broad, short tibial trochlea of the astragalus in *Notharctus* is a condition present in the eutherian morphotype but not (*viz* Szalay, 1977b) in the primate morphotype. Szalay and Drawhorn naturally assume their reconstruction of the ancestral primate tarsus to be valid, and ascribe little significance to trochlear morphology in *Notharctus.* The character might, of course, represent a secondary reversal, but an alternative interpretation is clearly possible. Such a character might indeed be a primitive eutherian trait retained in a taxon with an otherwise "advanced" tarsal complex. Hence, I do not question Szalay and Drawhorn for applying morphotype reconstruction in phylogenetics; I question the nature of the morphotype (in respect to one or a few details) they reconstruct. The recognition of mosaicism as an important aspect of evolutionary processes certainly allows for such an inquiry.

F. Miscellaneous Skeletal Features

The assortment of osteological traits displayed in Figure 23 and Table 5 is not easily dispatched to one or two integrated regions of the skeleton, but they are considered because they have been cited in various studies of tupaiid affinities. I will not provide a detailed character analysis here; arguments for the polarities advocated in Table 5 may be found in Butler (1956), McDowell (1958), Gregory (1910), and Novacek (1977b) among others. Tupaiines and ptilocercines share derived features in jaw musculature, postglenoid structure, occiput, and nasal morphology. These characters provide little information on special tupaiid relationships with other groups. For example, the reduction in width of the posterior region of the nasals in probably derived, but this character is widely distributed among eutherians (Fig. 23). Some important features of the jaw musculature cannot be directly examined in fossil groups and require more detailed study in living forms (Fig. 23, characters 9, 10).

4. Tupaiid Affinities

It is obvious from the foregoing that interpretations of evolutionary polarities for many eutherian skeletal features are speculative. Nevertheless, it is of interest to forge on with an inquiry into tupaiid relationships, accepting for the moment that relevant morphoclines are true as given. Even with such abandoned assumptions, the results do not prove very revealing. My conclusions are conservative, generally corresponding to those of other contributers to this volume (e.g. Luckett, Butler, Cartmill and MacPhee); the features considered herein do not provide strong support for any of the traditionally recognized alliances between tupaiids and other eutherian groups. Therefore, the ordinal distinction of tree shrews is presently appropriate, but the superordinal affinities of this group remain elusive.

Such a statement hardly deserves the scrutiny of large numbers of cranioskeletal features, but I believe that this evidence at least provides useful and instructive comment on various hypotheses suggested for tupaiid affinities. Few of these published interpretations show close agreement, although they can be categorized very generally as theories which advocate 1) macroscelidid-tupaiid relationships as Menotyphla, or 2) membership within the Insectivora without implications of special affinity to a particular insectivore group, or 3) close ties with primates, particularly non-anthropoids, or 4) allocation to the superordinal category "Archonta". The first of these suggestions, that advocating a menotyphlan grouping, has been adequately laid to rest by recent systematic work (Patterson, 1965; Butler, 1972; McKenna, 1975, and others). In an earlier consideration, Butler (1956) cited 16 skull features in common between macroscelidids and tupaiids as evidence for the Menotyphla. Many of these characters are also present in dermopterans and primates, five are primitive, at least two are of dubious polarity, and six are shared with leptictids (Novacek, 1977b). Butler rightly emphasized the problem with these "defining"

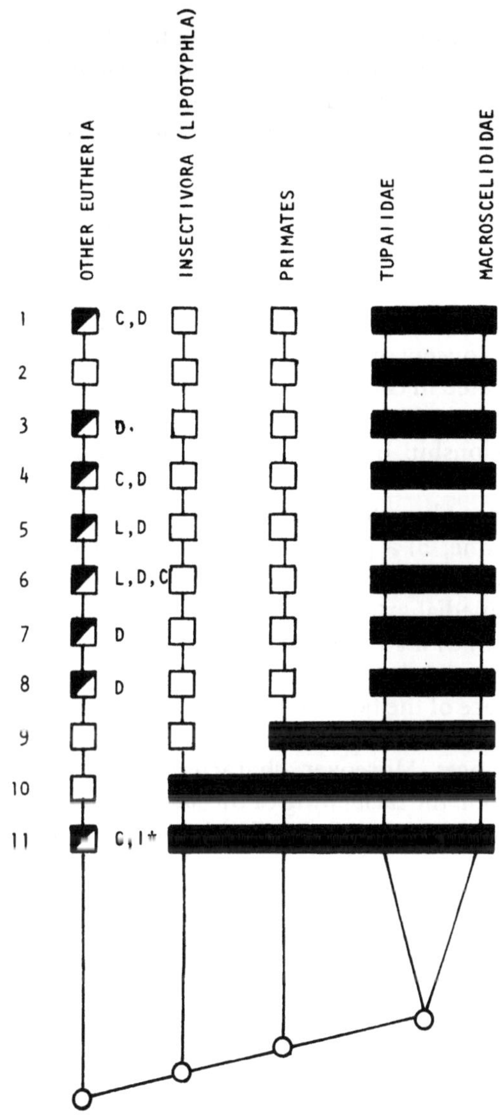

Fig. 24. Cladogram displaying a sister group relationship between tupaiids and macroscelidids (Menotyphla). Solid bars represent shared, derived characters accepted as synapomorphic to support this hypothesis. Open squares represent primitive characters. Semi-shaded squares represent derived character states in other Eutheria (C, Chiroptera; D, Dermoptera; L, Leptictidae; I*, Insectivora and several other groups) which must be accepted as independent acquisitions under this hypothesis. Note that few derived features are unique to "Menotyphla." Numbers refer to the following derived characters; (1) rostral entotympanic present, (2) ossified tubes for intrabullar circulation, (3) tympanohyal well developed, isolating stylomastoid foramen from the tympanic cavity, (4) ulna slender and elongate, (5) lateral trochlear crest of astragalus prominent, (6) superior astragalar foramen of trochlea absent, (7) origin of temporalis muscles posterior to frontals, (8) distinct grooves on astragalus and calcaneum for tendon of the flexor fibularis muscles, (9) calcaneal fibular facet weak or absent, (10) fenestra rotundum shielded ventrally by enlarged carotid artery or osseous tubes or flanges, (11) medial internal carotid artery absent.

characteristics and subsequently (1972) recognized elephant shrews and tree shrews as separate and not closely related mammalian orders, Macroscelidea and Scandentia respectively. This view was generally correspondent to Patterson's (1965) regarding the isolated position of elephant shrews. The shared derived characters of macroscelidids and tupaiids distilled from the distributions surveyed in this paper are depicted in Figure 24. It is evident from the figure that these characters are largely present in other eutherian groups; Menotyphla cannot claim a sufficiently large set of uniquely derived osteological features.

Any hypothesis allying tupaiids closely with the Insectivora is strongly influenced by our concept of the latter taxon. Acceptance of a broadly defined Insectivora, advocated in earlier studies of this group, leads to the inclusion of the macroscelidids and a plethora of archaic eutherians, such as the Leptictidae. A close relationship between leptictids and tupaiids was suggested by McDowell (1958), who emphasized the following similarities: 1) an entotympanic bulla; 2) an annular ectotympanic; 3) a distinct facial wing of the lacrimal; 4) a large jugal bone; 5) a palatine with an orbital wing which contacts the lacrimal; 6) a true postglenoid process; 7) the lack of a bony canal for ramus inferior of the stapedial artery, 8) alisphenoid canal separated by bone from the gasserian fossa, 9) a strong dorsum sellae with a clinoid process, 10) an extensive pubic symphysis, and 11) an astragalar trochlea which curves over the proximal surface of the body of the astragalus. If one accepts the polarities suggested in this review, characters 2, 3, 4, 5, 6, 7, and 10 are merely plesiomorphous resemblances. Moreover, characters 3 and 5 were inaccurately described by McDowell: the facial wing of the lacrimal in leptictids (where known) is markedly reduced compared with that in certain tupaiids, and the palatine is separated from contact with the lacrimal by a very narrow process of the maxilla. The derived features shared by tupaiids and leptictids are far from unique (Fig. 26). Consequently, a case for strong affinity between these groups is greatly weakened when the "supportive" characters are scrutinized. Lack of resemblance is even more obvious between tupaiids and lipotyphlan insectivores (Fig. 26), and such an association is surely discrepant with the features reviewed above.

Of all the theories on tupaiid relationships, arguments for close primate ties have received the greatest attention. Such views were originally espoused by Carlsson (1922) and elaborated upon by Le Gros Clark (1959, 1971), but have been criticized at length by several authors (Van Valen, 1965; Campbell, 1974; Luckett, 1969; and others). As Cartmill and MacPhee (this volume) effectively demonstrate, many of the alleged resemblances between tupaiids and primates in basicranial, orbital, and carotid features are erroneous, nonhomologous, or plesiomorphous. Derived features shared by these groups (large optic foramen, absence of a medial internal carotid artery, shielding of the fenestra rotundum, and a few tarsal characters) are certainly not restricted to these taxa (Fig. 25). Other noted similarities show marked variation in primates (Fig. 13). My analysis accordingly underscores the views of those authors skeptical of close tupaiid-primate affinities.

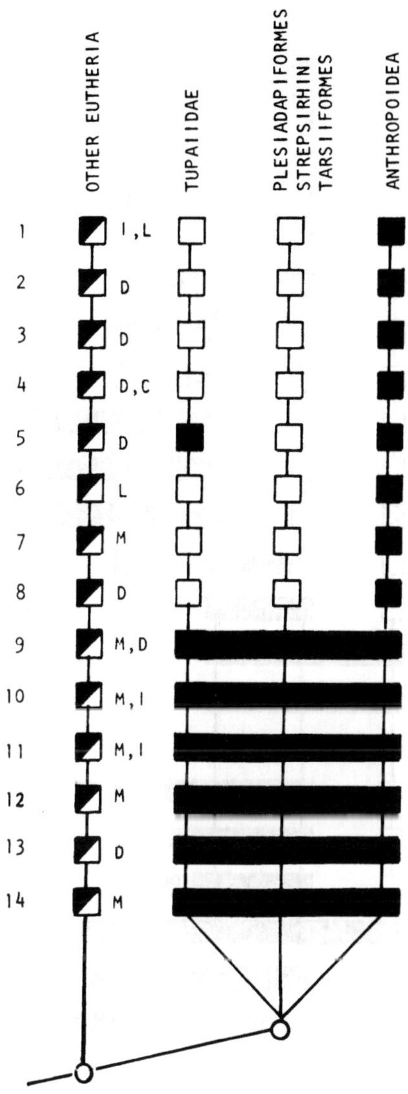

Fig. 25. Cladogram depicting evidence for Carlsson's (1922) hypothesis of close relationship between tupaiids and primates. Letters indicate derived characters present in other eutherian groups (C, Chiroptera, D, Dermoptera, I, Insectivora, L, Leptictidae, M, Macroscelididae). Other symbols are explained in the caption for Fig. 24. Numbers refer to the following derived characters: (1) facial wing of lacrimal reduced or absent, (2) stapedial artery absent, (3) ramus inferior of the stapedial artery absent, (4) superior scapular border relatively long and notably curved, (5) deltopectoral crest on humerus weak, (6) iliosacral angle acute, (7) peroneal tubercle on calcaneum weak or absent, (8) trochlea extending to astragalar neck, (9) optic foramen large, (10) medial internal carotid artery absent, (11) fenestra rotundum ventrally shielded by carotid arteries or osseous tubes or flanges, (12) calcaneal fibular facet weak or absent, (13) astragalar sustentacular facet confluent with distal astragalar facets, (14) calcaneal astragalar facet strongly concave.

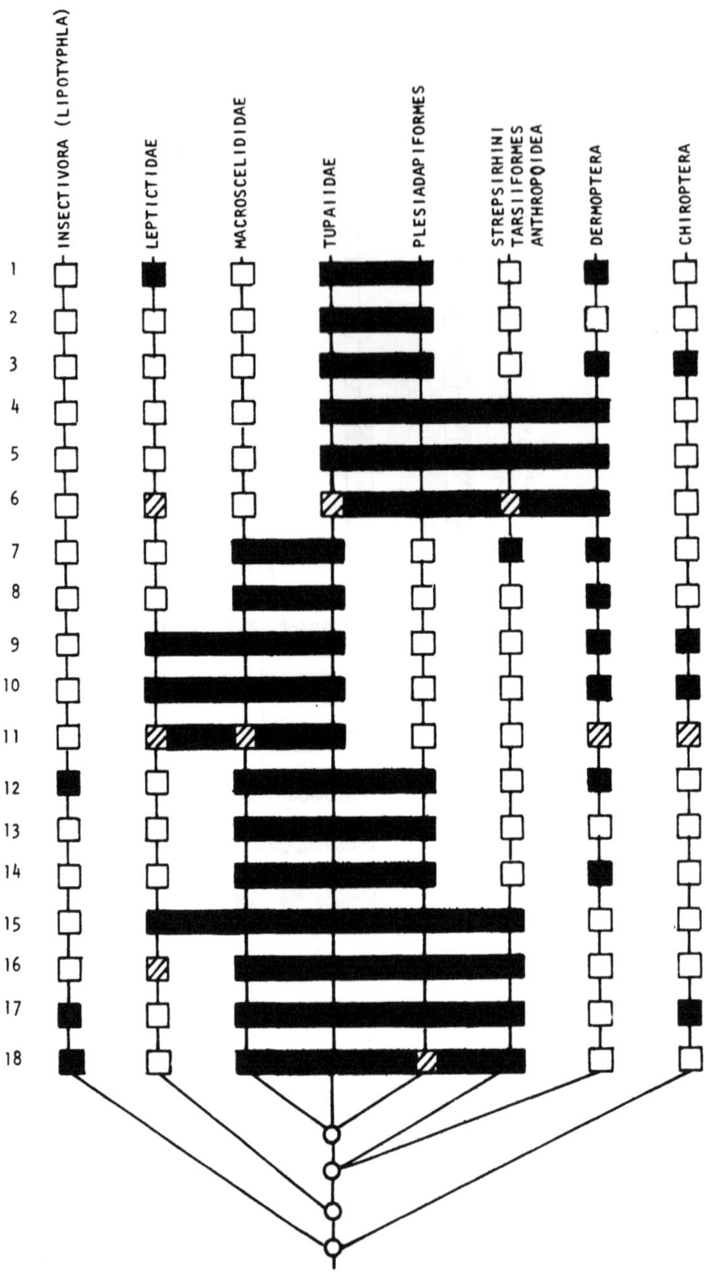

Fig. 26. Cladogram depicting alternative relationships for tupaiids. Solid squares represent derived eutherian features; squares with diagonal lines, intermediately derived features; open squares, primitive eutherian features; solid horizontal bars, derived characters shared by groups as indicated. Branching sequence represents the maximum resolution permitted by the character distribution. See text for further explanation. Numbers refer to the following derived characters: (1) alisphenoid orbital wing well developed, (2) suboptic foramen absent, (3) calcaneal cuboid facet

In view of the suspected antiquity of tree shrews it is not surprising that search for its nearest sister group has proven extremely difficult. At a less refined level, however, it would be advantageous to discover associations of several orders whose distinctive and unifying traits are evident. A diversity of such superordinal groupings has been proposed (e.g., Simpson's, 1945, cohorts), but response to these concepts has oscillated between enthusiasm and rejection. Gregory's (1910) Archonta did not rest favorably with many subsequent treatments (Simpson, 1945), but this concept, though modified through the exclusion of macroscelidids, has been revived by McKenna (1975), Goodman (1975), and Szalay (1977b). The most explicit support for the Archonta is provided by the latter author; although Gregory (1910) is credited with detailed documentation and phylogenetic consideration of this group, there is a virtual lack of discussion in Gregory's monograph concerning the morphological evidence for the Archonta (see also comments of Luckett, this volume). I have expressed some doubt in forgoing sections of this paper concerning the validity of Szalay's recognition of the Archonta, but the best means of scrutinizing this hypothesis is to examine the likelihood of alternative arrangements. The problem can be represented graphically in a single diagram (Fig. 26) for character distribution wherein tupaiids occupy a central position. To the right of tupaiids are taxa comprising the Archonta (*sensu* Szalay, 1977b), while groups to the left are generally excluded by this concept. The degree of shared, derived resemblance decreases towards the edges of the diagram. Accordingly the shared apomorphs are most abundant between tupaiids, plesiadapiforms, and macroscelidids. The diagram includes only one-fourth the total number of features considered in the foregoing character analysis, because derived states for the omitted characters are rare in occurrence at both sub- and superordinal levels or are not present in tupaiids.

The supportive evidence for the Archonta consists of characters 1 through 6. Although these features are primarily restricted to alleged archontans, few of them are typical of all members of that group. For example, the transverse orientation of the cuboidal facet on the calcaneum (Fig. 26, character 3) excludes some of the primates. The pronounced curvature of the astragalar calcaneal facet (4) is not characteristic of chiropterans. Derived non-tarsal features are few (characters 1, 2). Only two derived characters seem common enough to provide some meaningful basis for the Archonta, but even these fail to account for the variations observed in Chiroptera. The confluence of

←

oriented transversely to long axis of calcaneum, (4) astragalo-calcaneal facet curvature with small radius, (5) astragalar sustentacular facets confluent with distal astragalar facts, (6) astragalar trochlea elongate, (7) tympanohyal well developed, (8) ulna slender and elongate, (9) rostral entotympanic present, (10) superior astragalar foramen absent, (11) opening of lacrimal foramen confined to orbit or double openings are present, (12) distinct grooves are present on the astragalus for "flexor fibularis" tendons, (13) origin of internal pterygoid muscle extending to medial wall of orbit, (14) origin of temporalis muscles posterior to frontals, (15) calcaneal fibular facet weak or absent, (16) optic foramen large, 17) medial internal carotid artery absent, 18) fenestra rotundum shielded ventrally by carotid arteries or bony tubes or flanges.

the sustentacular and distal astragalar facets (5) is not shared by microchiropterans, and the elongation of the astragalar trochlea (6) is not even remotely approached in megachiropterans. Accordingly, the application of these tarsal features either mandates a stricter definition of the Archonta which excludes bats (as proposed by Dene *et al.*, this volume), or suggests that the chiropteran condition is grossly modified from the archontan morphotype.

Characters 7 through 18 in Figure 26 have derived states shared by tupaiids and non-archontan groups. However, such distributions do not tip the scales in favor of alternative relationships because virtually all these features are present in at least some archontans. For example, if one recognizes a rostral element of the entotympanic (character 9), the lack of a superior astragalar foramen (10), and modifications of the number and positions of the lacrimal foramina (11) as synapomorphic for tupaiids, macroscelidids and leptictids, one must accept the independent acquisition of such features in dermopterans and chiropterans. Characters 12–18 show more widely distributed specializations, and merely support some remote affinity between tupaiids, macroscelidids, and primates (particularly Plesiadapiformes). The last auditory characters (17–18) show apomorphs which extend to lipotyphlan insectivores. Naturally I see no clear pattern for resolution in Figure 26, and I can only issue the following conservative remarks. The divergence of lipotyphlan insectivores, on one hand, and the chiropterans, on the other, seems more remote than that of other clades shown in Figure 26. Only three tarsal characters support a close interrelationship between tupaiids, primates, and dermopterans. There is virtually no convincing evidence to support a concept of the Archonta which includes chiropterans. Lastly, the acceptance of a close relationship between tupaiids, macroscelidids, and leptictids requires independent evolution of the supportive apomorphs in dermopterans and bats.

No *a posteriori* means of weighting characters has been applied in this analysis, but the utility of such a procedure might be entertained. For example, one might juxtapose two alternative hypotheses relevant to the distributional patterns of characters 4–11. According to Szalay and Decker (1974), tarsal features (Fig. 26, characters 4, 5, 6) supporting a relationship between tupaiids, primates, and dermopterans are integrated features of a form-function complex. These characters would therefore be assigned a relatively high ranking in the system proposed by Hecht and Edwards (1976). An even higher value would be bestowed upon the innovative acquisition of a rostral element in the entotympanic (character 9, Fig. 26), a feature suggesting the alternative tupaiid-macroscelidid-leptictid relationship. Nevertheless, the strong possibility of independent evolution of this element and problems in interpretations of homology (Novacek, 1977b) obviate the application of this character for the analysis of tupaiid relationships. Other characters (7, 8, 10, 11) suggesting close leptictid-tupaiid-macroscelidid relationships are either simple loss or gradational modifications and thus receive low weights under the Hecht and Edwards scheme. Application of such a weighting procedure therefore very tentatively favors a cladistic grouping of tupaiids, primates, and dermopterans over other arrangements, but I believe that a phylogenetic decision stemming from this application is unwarranted. To reiterate, emphasis should be placed on a careful search for additional characters where *a posteriori* character weighting does little to elucidate the presently available evidence.

From the foregoing analysis it is clear that many characters relevant to various problems in eutherian phylogeny have little direct bearing on the question of tupaiid affinities. This is largely due to supposed retention of primitive states for such characters in tree shrews, a condition which in itself suggests the remote divergence of this group in eutherian history. A search for special similarities between tupaiids and other eutherian groups leads to the conclusion that a satisfactory hypothesis for tupaiid relationships is not currently available. While the osteological evidence reviewed herein proves useful for testing several well known interpretations, a novel theory which will stand the scrutiny of this evidence is not clear to me. Accordingly it should be emphasized that proposed classifications of tupaiids within the Eutheria have little, if any, phylogenetic basis from skeletal data. While I favor the ordinal distinction of tupaiids (*viz.* Butler, 1972), this is admittedly a matter of convenience spawned by ignorance, and it provides little phylogenetic information, other than serving to underscore the divergent nature of this group. The zeal to discover the nearest relatives of primates or insectivores is understandable from many biological perspectives; yet it appears that the traditionally popular implication of tupaiids in this role is not clearly supported by the skeletal evidence so frequently cited.

ACKNOWLEDGMENTS

I am grateful to various colleagues, particularly Drs. M. Cartmill, R. MacPhee, W. P. Luckett, F. S. Szalay, and M. C. McKenna for their useful comments on this paper, without implicating them as co-conspirators for any erroneous descriptions, tabulations, syntheses, or judgments. Completion of this study was aided by support from the National Science Foundation and the San Diego State University Grant-in-Aid and Faculty Fellowship Programs.

ABBREVIATIONS

Anatomical

ACP	posterior opening of the alisphenoid canal
AL	alisphenoid (orbital process)
APS	alisphenoid-petrosal suture
AQ	aquaductus cochlea
AS	alisphenoid (tympanic process)
BS	basisphenoid
C	cochlea
CBA	bony contact with auditory bulla
CF	condyloid foramen
CLCA	canal for the lateral internal carotid artery
CPA	canal for the promontory artery
CRS	canal for the ramus superior of the stapedial artery
CSA	canal for the stapedial artery
CSI	canal for the ramus inferior of the stapedial artery
E	entotympanic
E_c	caudal entotympanic
E_r	rostral entotympanic
EAM	external auditory meatus
EMR	recessus meatus of the external auditory canal
ETR	epitympanic recess
F	frontal
FB	basicochlear fissure
FC	facial canal
FER	fenestra rotundum
FEO	fenestra ovalis
FGL	glaserian fissure
FOO	foramen ovale
FSA	foramen for the stapedial artery
GF	glenoid fossa
H	hamular process
JF	jugular foramen
L	lacrimal
M	maxilla
MICA	medial internal carotid artery
MP	mastoid process
O	orbitosphenoid
OC	occipital condyle
P	petrosal
PA	palatine
PAR	promontory artery
PCF	posterior carotid foramen
PF	promontory foramen
PGF	postglenoid foramen
PGP	postglenoid process
PR	promontorium
POF	postorbital process of frontal
PRA	promontory ramus of the promontory artery
PY	pyriform fenestra
RISA	inferior ramus of the stapedial artery
RSSA	superior ramus of the stapedial artery
SA	stapedial artery
SLCA	sulcus for the lateral internal carotid artery
SMC	sulcus for the medial internal carotid artery
SMF	stylomastoid foramen
SPA	sulcus for the promontory artery
SSA	sulcus for the stapedial artery
ST	stapes
SVII	sulcus for the facial nerve
SVA	sulcus for the vidian ramus of the promontory artery
T	ectotympanic
TE	flange for the eustachian tube
TY	tympanohyal
VF	vidian foramen
VR	vidian ramus of the promontory artery
ZY	zygomatic arch
asc	astragalocalcaneal facet of the calcaneum
asp	astragalar "spring" ligament facet
ass	astragalar sustentacular facet
caa	calcaneoastragalar facet
caf	calcaneal fibular facet
cas	calcaneal sustentacular facet
cu	cuboid facet
lb	lateral border of the trochlea
mb	medial border of the trochlea
naa	naviculoastragalar facet
sff	sustentacular groove for tendon of m. flexor digitorum fibularis

st	superior tibial facet of the trochlea	LACM	Los Angeles County Museum of Natural History
tca	tuber of the calcaneum	SDSU	San Diego State University Vertebrate Collections
tf	trochlear groove for tendon m. flexor digitorum fibularis	UCMP	University of California Museum of Paleontology
Institutions		UC-MVZ	University of California Museum of Vertebrate Zoology
F:AM	Frick Collections: American Museum of Natural History		

5. References

Archibald, J. D. 1977. The ectotympanic bone and internal carotid circulation of eutherians in reference to anthropoid origins. *Jour. Human Evolution* 6: 609–622.

Badoux, D. M. 1974. An introduction to biomechanical principles in primate locomotion and posture, pp. 1–43. *In* F. A. Jenkins, Jr. (ed.). *Primate Locomotion*. Academic Press, New York.

Bock, W. J. 1973. Philosphical foundations of classical evolutionary classification. *Syst. Zool.* 21: 364–374.

Bock, W. J. 1977. Adaptation and the comparative method, pp. 57–82. *In* M. K. Hecht, P. C. Goody, and B. M. Hecht (eds). *Major Patterns in Vertebrate Evolution*. Plenum Press, New York.

Bonde, N. 1974. Origin of higher groups: Viewpoints of phylogenetic systematics. *Colloque International C.N.R.S.* 218: 293–305.

Brundin, L. 1966. Transantarctic relationships and their significance as evidenced by chironomid midges. With a monograph of the subfamilies Podonominae and Aphrotemiinae and the Austral Heptagyinae. *K. Svenska Vetensk. Acad. Handl.* 4th Ser. 11 (1): 1–472.

Bugge, J. A. 1974. The cephalic arterial system in insectivores, primates, rodents, and lagomorphs, with special reference to the systemic circulation. *Acta Anat.* 87 (suppl. 62): 1–160.

Butler, P. M. 1956. The skull of *Ictops* and a classification of the Insectivora. *Proc. Zool. Soc. London* 126: 453–481.

Butler, P. M. 1972. The problem of insectivore classification, pp. 253–265. *In* K. A. Joysey and T. S. Kemp (eds.). *Studies in Vertebrate Evolution*. Winchester Press, New York.

Butler, P. M. and A. T. Hopwood. 1957. Insectivora and Chiroptera from Miocene rocks of Kenya Colony. *British Mus. (Nat. Hist.) Fossil Mammals of Africa* 13: 1–35.

Campbell, C. B. G. 1974. On the phyletic relationships of tree shrews. *Mammal Rev.* 4: 125–143.

Carlsson, A. 1922. Uber die Tupaiidae und ihre Beziehungen zu den Insectivora und den Prosimiae. *Acta Zool.* 3: 227–270.

Cartmill, M. 1975. Strepsirhine basicranial structures and affinities of Cheirogaleidae, pp. 313–354. *In* W. P. Luckett and F. S. Szalay (eds.) *Phylogeny of the Primates*. Plenum Press, New York.

Clemens, W. A. 1966. Fossil mammals of the type Lance Formation, Wyoming. Part II. Marsupialia. *Univ. Calif. Pub. Geol. Sci.* 62: 1–122.

Darlington, P. J. 1970. A practical criticism of Hennig-Brundin "Phylogenetic Systematics" and Antarctic biogeography. *Syst. Zool.* 19:1–18.

Davis, D. D. 1964. The giant panda. A morphological study in evolutionary mechanisms. *Fieldiana: Zool. Mem. Chicago Nat. Hist. Mus.* 3: 1–339.

Dawson, M. R. 1967. Fossil history of the families of recent mammals, pp. 12–53. *In* S. Anderson and J. K. Jones, Jr. (eds.). *Recent Mammals of the World*. Ronald Press, New York.

Echeverra, M. 1960. Note sur l'arterè stapédienne. *C. R. Assoc. Anat.* 46: 243–246.

Eldredge, N. 1977. Cladism and common sense. *North American Paleontological Convention II. Abstracts of Papers, Jour. Paleont.* 51 (suppl. 2, part III): 10.

Englemann, G. F. and E. O. Wiley. 1977. The place of ancestor-descendant relationships in phylogeny reconstruction. *Syst. Zool.* 26: 1–12.
Evans, F. G. 1942. Osteology and relationships of the elephant shrews (Macroscelididae). *Bull. Amer. Mus. Nat. Hist.* 80: 85–125.
Frey. H. 1923. Untersuchungen über den Scapula, speziell über äussere Form und Abhangigkeit von Funktion. *Zeitschr. Gesamt. Anat.* 68: 276–324.
Fuchs, H. 1910. Über das Pterygoid, Palatinum, und Parasphenoid der Quadrupeden, insbesondere der Reptilia und Säugetiere. *Anat. Anz.* 36:33–95.
Gamberian, P. P. and R. O. Oganesian. 1970. Biomechanics of the gallop and the primitive rebounding jump in small mammals. *Proc. Acad. Sci. U.S.S.R. Biol. Ser.* 3: 441–447.
Gingerich, P. D. 1973. Anatomy of the temporal bone in the Oligocene anthropoid *Apidium* and the origin of the Anthropoidea. *Folia Primat.* 19:329–337.
Gingerich, P. D. 1974. Stratigraphic record of early Eocene *Hyopsodus* and the geometry of mammalian phylogeny. *Nature* 248: 107–109.
Gingerich, P. D. 1976. Cranial anatomy and evolution of the early Tertiary Plesiadapidae (Mammalia, Primates). *Mus. Paleont. Univ. Michigan Papers on Paleontology* 15: 1–117.
Goodman, M. 1975. Protein sequence and immunological specificity: Their role in phylogenetic studies of primates, pp. 219–248. *In* W. P. Luckett and F. S. Szalay (eds.). *Phylogeny of the Primates*. Plenum Press, New York.
Gregory, W. K. 1910. The orders of mammals. *Bull. Amer. Mus. Nat. Hist.* 27: 1–524.
Gregory, W. K. 1920. Studies of the comparative myology and osteology; no. IV. A review of the evolution of the lacrimal in vertebrates with special reference to that of mammals. *Bull. Amer. Mus. Nat. Hist.* 42: 95–263.
Haeckel, E. 1866. *Generelle Morphologie der Organismen*. G. Reimer, Berlin.
Haines, R. W. 1950. The interorbital septum in mammals. *Jour. Linnean Soc. London (Zool.)* 41: 585–607.
Hall-Craggs, E. C. B. 1965. An analysis of the jump of the lesser galago *(Galago senegalensis)*. *Proc. Zool. Soc. London* 147: 20–29.
Hatt, R. T. 1932. The vertebral column of richochetal rodents. *Bull. Amer. Mus. Nat. Hist.* 63: 599–738.
Hecht, M. K. 1976. Phylogenetic inference and methodology as applied to the vertebrate record. *Evol. Biol.* 9: 335–363.
Hecht, M. K. and J. Edwards. 1976. The determination of parallel or monophyletic relationships: The proteid salamanders—a test case. *Am. Nat.* 110: 653–677.
Hennig, W. 1950. *Grundzuge einer Theorie der Phylogenetischen Systematik*. Deutscher Zentral Verlag, Berlin.
Hennig, W. 1966. *Phylogenetic Systematics*. University of Illinois Press, Urbana.
Hennig, W. 1975. "Cladistic analysis or cladistic classification?": A reply to Ernst Mayr. *Syst. Zool.* 24: 244–256.
Hildebrand, M. 1974. *Analysis of Vertebrate Structure*. John Wiley and Sons, New York.
Howell, A. B. 1932. The saltatorial rodent *Dipodomys*, the functional and comparative anatomy of its muscular and osseous systems. *Proc. Am. Acad. Arts Sci.* 67: 377–536.
Howell, A. B. 1944. *Speed in Animals, Their Specializations for Running and Leaping*. Hafner, New York.
Hull, D. L. 1970. Contemporary systematic philosophies. *Ann. Rev. Ecol. Syst.* 1: 19–54.
Hunt, R. M., Jr. 1974. The auditory bulla in Carnivora: An anatomical basis for reappraisal of carnivore evolution. *Jour. Morph.* 143: 21–76.
Jenkins, F. A., Jr. 1971. Limb posture and locomotion in the Virginia opossum (*Didelphis marsupialis*) and other non-cursorial mammals. *Jour. Zool.*. 165: 303–315.
Jenkins, F. A., Jr. 1974. Tree shrew locomotion and the origins of primate arborealism, pp. 85–115. *In* F. A. Jenkins, Jr. (ed.) *Primate Locomotion*. Academic Press, New York.
Jepsen, G. L. 1970. Bat origins and evolution, pp. 1–64. *In* W. A. Wimsatt (ed.) *Biology of the Bats*. Volume 1. Academic Press, New York.
Jouffray, F. K. 1975. Osteology and myology of the lemuriform postcranial skeleton, pp. 149–192. *In* I. Tattersall and R.W. Sussman (eds.). *Lemur Biology*. Plenum Press, New York.

Kavanaugh, D. H. 1972. Hennig's principles and methods of phylogenetic systematics. *The Biologist* 54: 115–127.

Kay, R. F. and M. Cartmill. 1974. The skull of *Palaechthon nacimienti*. *Nature* 252: 37–38.

Kermack, K. A. and Z. Kielan-Jaworowska. 1971. Therian and non-therian mammals. *Zool. Jour. Linnean Soc. London (suppl. 1)*: 103–115.

Kielan-Jaworowska, Z. 1977. Evolution of therian mammals in the Late Cretaceous of Asia. Part II. Postcranial skeleton in *Kennalestes* and *Asioryctes*. *Palaeont. Polonica* 37: 65–83.

Krishtalka, L. 1976. Early Tertiary Adapisoricidae and Erinaceidae (Mammalia, Insectivora) of North America. *Bull. Carnegie Mus. Nat. Hist.* 1: 1–40.

Krüger, W. 1958. Der Bewegungsapparat, pp. 1–176. *In* W. Kükenthal (ed.). *Handbuch der Zoologie*, Vol. 8 (13–14). Walter de Gruyter and Co., Berlin.

Le Gros Clark, W. E. 1934. On the skull structure of *Pronycticebus gaudryi*. *Proc. Zool. Soc. London* 1934: 19–27.

Le Gros Clark, W. E. 1959. *The Antecedents of Man*. 1st ed. Edinburgh University Press, Edinburgh.

Le Gros Clark, W. E. 1971. *The Antecedents of Man*. 3rd ed. Quadrangle Books Inc., Chicago.

Lessertisseur, J. and R. Saban. 1967. Squelette appendicular. *In* P. P. Grassé (ed.) *Traité de Zoologie; Anatomie, Systématique Biologie. Mammifères, Téguments Squelette*. Tome XVI, Fasc. 1. Masson et Cie, Paris.

Llorca, F. O. 1934. L'artère stapédienne chez l'embryon humain et la cause mécanique probable de son atrophie. *Arch. Anat. Antrop. Lisboa* 16: 199–207.

Luckett, W. P. 1969. Evidence for the phylogenetic relationships of tree shrews (family Tupaiidae) based on the placenta and foetal membranes. *J. Reprod. Fert., Suppl.* 6: 419–433.

MacIntyre, G. T. 1972. The trisulcate petrosal pattern of mammals. *Evolut. Biol.*, 6: 275–303.

MacPhee, R. D. E. 1979. Entotympanics, ontogeny and primates. *Folia Primat.* 31:23–47.

McDowell, S. B., Jr. 1958. The Greater Antillean insectivores. *Bull. Am. Mus. Nat. Hist.* 115: 113–214.

McKenna, M. C. 1966. Paleontology and the origin of the primates. *Folia Primat.* 4: 1–25.

McKenna, M. C. 1975. Toward a phylogenetic classification of the Mammalia, pp. 21–46. *In* W.P. Luckett and F. S. Szalay (eds.). *Phylogeny of the Primates*. Plenum Press, New York.

Matthew, W. D. 1909. The Carnivora and Insectivora of the Bridger Basin, middle Eocene. *Mem. Am. Mus. Nat. Hist.* 9: 291–567.

Matthew, W. D. 1937. Paleocene faunas of San Juan Basin, New Mexico. *Trans. Am. Philos. Soc. N. S.* 30: 1–510.

Mayr, E. 1974. Cladistic analysis or cladistic classification? *Z. Zool. Syst. Evol.-Forsch.* 12: 94–128.

Miller, G. S., Jr. 1907. The families and genera of bats. *Bull. U. S. Nat. Mus.* 57: 1–282.

Muller, J. 1934. The orbitotemporal region in the skull of the Mammalia. *Archiv. Neerl. Zool.* 1: 118–259.

Nelson, G. J. 1973. The higher level phylogeny of the vertebrates. *Syst. Zool.* 22: 87–92.

Novacek, M. J. 1976. Insectivora and Proteutheria of the later Eocene (Uintan) of San Diego County, California. *Nat. Hist. Mus. Los Angeles County Contr. Sci.* 283: 1–52.

Novacek, M. J. 1977a. A review of Paleocene and Eocene Leptictidae (Eutheria: Mammalia) from North America. *PaleoBios* 24: 1–42.

Novacek, M. J. 1977b. Evolution and relationships of the Leptictidae (Eutheria: Mammalia). Ph. D. Thesis, University of California, Berkeley.

Novacek, M. J. 1977c. Aspects of the problem of variation, origin, and evolution of the eutherian auditory bulla. *Mammal Rev.* 7: 131–149.

Patterson, B. 1965. The fossil elephant shrews (family Macroscelididae). *Bull. Mus. Comp. Zool.* 133: 295–385.

Piveteau, J. 1957. *Traité de Paléontologie*, Vol. 7. Masson et Cie Paris.

Platnick, N. I. 1977. Parallelism in phylogeny reconstruction. *Syst. Zool.* 26: 93–96.

Roberts, D. and I. Davidson. 1975. The lemur scapula, pp. 125–147. *In* I. Tattersall and R. W. Sussman (eds.). *Lemur Biology*. Plenum Press, New York.

Rose, K. D. and E. L. Simons. 1977. Dental function in the Plagiomenidae: Origin and relationships of the mammalian order Dermoptera. *Contr. Mus.Paleon. Univ. Michigan* 24: 221–236.

Saban, R. 1963. Contribution a l'étude de l'os temporal des primates. *Mem. Mus. Natl. d'Hist. Nat.* (nouv. ser., ser. A.) 29: 1–378.
Salomon, M. I. 1930. Considérations sur l'homologie de l'os lachrymal chez les Vertébrés supérieurs. *Acta Zool.* 11: 151–183.
Schaeffer, B., M. K. Hecht, and N. Eldredge. 1972. Phylogeny and paleontology. *Evol. Biol.* 6: 31–46.
Sigé. B. 1974. *Pseudorhynchocyon cayluxi* Filhol, 1892, insectivore géant des phosphorites du Quercy. *Palaeovertebrata* 6: 33–46.
Simons, E. L. 1962. A new Eocene primate genus, *Cantius*, and a revision of some allied European lemuroids. *Bull. Brit. Mus. Nat. Hist. Geol.* 7:1–36.
Simons, E. L. 1972. *Primate Evolution, An Introduction to Man's Place in Nature*. MacMillan, New York.
Simpson, G. G. 1945. The principles of classification and a classification of mammals. *Bull. Am. Mus. Nat. Hist.* 85: 1–350.
Simpson, G. G. 1975. Recent advances in methods of phylogenetic inference, pp. 3–19. *In* W. P. Luckett and F. S. Szalay (eds.). *Phylogeny of the Primates.* Plenum Press, New York.
Sokal, R. R. 1975. Mayr on cladism—and his critics. *Syst. Zool.* 24: 257–262.
Spatz, W. B. 1966. Zur Ontogenese der Bulla Tympanica von *Tupaia glis* Diard 1820 (Prosimiae, Tupaiiformes). *Folia Primat.* 4: 26–50.
Szalay, F. S. 1968. The beginnings of primates. *Evolution* 22: 19–36.
Szalay, F. S. 1969. Mixodectidae, Microsyopidae and the insectivore-primate transition. *Bull. Am. Mus. Nat. Hist.* 140: 197–330.
Szalay, F. S. 1972. Cranial morphology of the early Tertiary *Phenacolemur* and its bearing on primate phylogeny. *Am. J. Phys. Anthrop.* 36: 59–76.
Szalay, F. S. 1975. Phylogeny of primate higher taxa: The basicranial evidence, pp. 91–125. *In* W. P. Luckett and F. S. Szalay (eds.). *Phylogeny of the Primates*. Plenum Press, New York.
Szalay, F. S. 1977a. Ancestors, descendants, sister groups and testing of phylogenetic hypotheses. *Syst. Zool.* 26: 12–19.
Szalay, F. S. 1977b. Phylogenetic relationships and a classification of the eutherian Mammalia, pp. 315–374. *In* M. K. Hecht, P. C. Goody, and B. M. Hecht (eds.). *Major Patterns in Vertebrate Evolution*. Plenum Press, New York.
Szalay, F. S. and R. L. Decker. 1974. Origins, evolution, and function of the tarsus in Late Cretaceous eutherians and Paleocene Primates, pp. 223–259. *In* F. A. Jenkins, Jr. (ed.). *Primate Locomotion*. Academic Press, New York.
Szalay, F. S., I. Tattersall, and R. L. Decker. 1975. Phylogenetic relationships of *Plesiadapis*—postcranial evidence, pp. 135–166. *In* F. S. Szalay (ed.) *Approaches to Primate Paleobiology*. S. Karger, Basel.
Van Valen, L. 1965. Treeshrews, primates, and fossils. *Evolution* 19: 137–151.
Van Valen, L. 1966. Deltatheridia, a new order of mammals. *Bull. Am. Mus. Nat. Hist.* 132: 1–126.
Van Valen, L. 1967. New Paleocene insectivores and insectivore classification. *Bull. Am. Mus. Nat. Hist.* 135: 217–284.
Walker, A. 1974. Locomotor adaptations in past and present prosimian primates, pp. 349–381. *In* F. A. Jenkins, Jr. (ed.) *Primate Locomotion*. Academic Press, New York.
Williston, W. W. 1925. *The Osteology of the Reptilia*. Cambridge, Massachusetts.
Winge, H. 1941. *The Interrelationships of the Mammalian Genera*. Vol. 1. C. A. Reitzels Forlag, Copenhagen.

6. *Note Added in Proof*

Two recent publications bear significantly upon theories for the evolution of the internal carotid arterial system in mammals. Presley's (1979; *Acta Anat.* 103: 238–244) detailed embryological studies led him to question the generally accepted view that primitive mammals had developmentally equivalent medial and promontory internal carotid arteries. In all mammals studied by Presley, the internal carotid is derived as a single vessel from the embryonic dorsal aorta. This vessel occupies either a "medial" or a "promontory" position in the adult according to differential growth in the fetal auditory region. If Presley's observations are found to apply for a larger sample of mammalian groups, they would render doubtful certain reconstructions of the internal carotid system in fossil mammals. For example, it would be unreasonable to hold the interpretation that leptictids had both a promontory and medial internal carotid artery, if one accepts Presley's arguments. Preservation in the leptictid basicranial region suggests that a promontory artery coursed through the tympanic cavity, but offers only ambiguous evidence for a medial internal carotid artery (see discussion above).

The validity of inferring "soft" anatomy from the osseous evidence is further questioned by Conroy and Wible (1978; *Folia Primatol.* 29: 81–85), who show that the canal alleged to convey the promontory artery in *Lemur variegatus* contains instead the internal carotid nerve. Hence, the various tubes, grooves, and foramina cited by this author and many others as evidence for arterial patterns in fossil mammals must be regarded as ambiguous, and sometimes even misleading, clues to the unpreserved anatomy.

Tupaiid Affinities: The Evidence of the Carotid Arteries and Cranial Skeleton

3

M. CARTMILL and R.D.E. MacPHEE

1. Introduction

The order Insectivora as classically conceived is a paraphyletic or "wastebasket" taxon, to which systematists since Cuvier have relegated various early eutherian mammals and small extant groups that would not fit anywhere else. Other systematists, offended by the use of taxa that lack unique defining specializations (autapomorphies), have labored to empty this wastebasket as fast as it could be filled. There are two ways to do this. One is to fragment the Insectivora into a series of narrowly-defined ordinal groupings which appear to be strictly monophyletic: Dermoptera, Macroscelidea, Erinaceomorpha, Soricomorpha, and so forth. This strategy involves emptying the wastebasket by heaping its contents around it, rather than within it. The result is perhaps tidier from some standpoints, but not from others. The second way is to seek evidence for special affinities of certain insectivoran groups to other eutherian orders, and to empty the Insectivora by assigning its members elsewhere—e.g., by associating colugos with bats or elephant shrews with rabbits and hares. This strategy involves an increased risk of creating unnatural (i.e., polyphyletic) groupings. Whether or not this is worth the risk depends on whether one

---

M. CARTMILL AND R.D.E. MacPHEE • Departments of Anatomy and Anthropology, and Department of Anatomy, Duke University, Durham, North Carolina 27710.

finds insubstantial but potentially refutable speculation more offensive than an untestable confession of ignorance.

Tree shrews are among the taxa that have been repeatedly expelled from Insectivora by these procedures. Various systematists (see Luckett, this volume) have attempted to deal with tupaiids by combining them with elephant shrews to form an order Menotyphla, lumping them with Primates, or separating them from all other eutherians as a distinct order Tupaioidea (or Scandentia). Gregory's (1910) radical alternative was to unite Menotyphla, Primates, Dermoptera, and Chiroptera within a single supraordinal taxon, Archonta. However, all recent supporters of Archonta exclude elephant shrews, and some also omit the bats (McKenna, 1975; Szalay, 1977; Dene *et al.*, this volume; cf. Fig. 14)

We will only touch in passing on the possible affinities of tree shrews to bats, elephant shrews, and colugos, because these hypothetical relationships have not been supported in detail by many authorities. A great deal more evidence has been amassed bearing on the possible affinities of tree shrews to primates. Virtually every part of the mammalian body has been investigated by researchers seeking special resemblances between tree shrews and primates, and various authors have claimed to detect such resemblances in the musculature, brain, viscera, postcranial skeleton, dentition, chondrocranium, adult skull, circulatory system, genitalia, pattern of placentation, and biochemistry of the two groups. Our assessment of these claims will be restricted to features of cranial morphology. Some of the cranial features that have been claimed as special resemblances linking tree shrews to primates in general (and Malagasy lemurs in particular) are not in fact shared by both groups—e.g., the formation of the bulla "from the inflation of the entotympanic" (Gregory, 1910). Others are probably retained primitive eutherian traits, like the absence of a tympanic process of the basisphenoid (Saban, 1956/57) or the position of the carotid foramen "at the posterior margin of the bulla," which Jones (1929) invoked to prove tupaiid-strepsirhine affinities. Yet other supposed cranial resemblances between tree shrews and primates are simply irrelevant—for instance, the fact that both groups have some members with long faces and others with shorter faces, which Gregory (1910, p. 322) lists as a reason for thinking that primitive primates resembled tree shrews. These, and many other cranial features that have been adduced to demonstrate primate affinities of tree shrews, can be dismissed out of hand. But other tupaiid-primate resemblances might well be synapomorphies. These can be divided into the following two groups:

(1) *Resemblances to primates in general.* These include reduction of the alisphenoid's tympanic process (Saban, 1956/57); reduction of the anterior process of the malleus (ibid.; Segall, 1970); bony tubes enclosing the major arteries in the walls of the tympanic cavity (ibid.); loss of the medial branch of the internal carotid (McKenna, 1963, 1966); loss of the stapedial artery's inferior ramus and a corresponding reduction of the Glaserian fissure (Le Gros Clark, 1925; Saban, 1956/57; Bugge, 1974); origin of the central retinal artery from the internal carotid (Bugge, 1974); loss of the suboptic foramen (Butler, 1956); separation of the foramen rotundum from the superior orbital fissure (Greg-

ory, 1910; Carlsson, 1922); a complete postorbital bar (Gregory, 1910; Carlsson, 1922; Saban, 1956/57); retreat of the parietal and squamosal from the orbital wall (Saban, 1956/57); and reduction of the number of ectoturbinals from 3 to 2 (Le Gros Clark, 1925; Saban, 1956/57).

(2) *Special resemblances to Malagasy lemurs.* These include the "free" and "intrabullar" ring-shaped ectotympanic (Jones, 1929; Saban, 1956/57; Le Gros Clark, 1959); a lacrimal bone which touches the zygomatic and palatine (Le Gros Clark, 1959); and a large zygomatic (or "malar") foramen (Gregory, 1910; Carlsson, 1922). These and other specifically lemur-like traits of tupaiids led Simpson (1945) to classify tree shrews with the Malagasy lemurs, not simply as Prosimii *incertae sedis.*

The significance of these similarities was discounted by Van Valen (1965), whose studies of macerated skulls and fossil material persuaded him that the lemur-like features of tree shrews are either convergences or primitive retentions (tupaiid-lemur symplesiomorphies), and that the tupaiids should accordingly be excluded from Primates. Van Valen noted that dissection of preserved material and study of early developmental stages would be needed to render unequivocal judgments on certain features of cranial anatomy that have been invoked to justify inclusion of tupaiids in the primate order. We agree, and the present study incorporates both sorts of information. Most of Van Valen's conclusions are fully borne out thereby. We provide here a detailed reanalysis of the supposed resemblances between tupaiids and primates in features of the carotid circulation and ear region, which seem to us to be somewhat more important than the others.

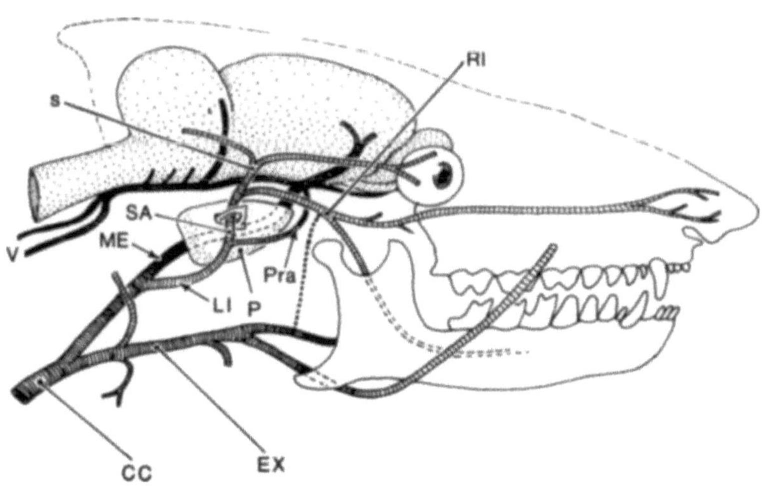

Fig. 1. Schematic drawing of major cephalic arteries in a primitive eutherian mammal. Stipple indicates the petrosal (P) and brain. The route of the maxillary artery, through which the external carotid (EX) annexes the branches of the ramus inferior (RI) of the stapedial artery (SA) in tupaiids and primates, is indicated by a line of dots. The medial internal carotid (ME) is shown as an entity separate from the promontory artery (Pra). Abbreviations as in Table 2.

2. The Evidence of the Carotid System

It is generally accepted that the common carotid divided into three major branches in ancient placental mammals (Fig. 1). The external carotid supplied only the upper neck, occiput, tongue, and lateral parts of the face. The medial entocarotid (medial internal carotid) entered the braincase via a canal between the petrosal and basioccipital, and emptied into the cerebral arterial circle. The internal carotid proper (lateral internal carotid) travelled to the posterior end of the promontorium of the petrosal, where it divided into a small promontory artery (to the cerebral arterial circle) and a larger stapedial artery. This last supplied those parts of the head that the maxillary artery supplies in *Homo,* plus most of the orbital contents.

In the following descriptions we accept this framework for identifying carotid arteries and their branches. However, we wish to record our feeling that current concepts of the homologies of different carotid vessels are based on a number of unproven assumptions. In particular, we doubt that there is any method which permits one to distinguish consistently between the medial internal carotid and the promontory branch of the lateral internal carotid. Most authors simply assume that any large vessel which travels along the basicranium outside the confines of the tympanic cavity is a medial internal carotid, and any which travels inside this chamber is a lateral internal carotid. Rigid application of this topological argument forces the conclusion that megachiropterans, for example, possess a medial internal carotid while microchiropterans have only the lateral vessel (Tandler, 1899; Grosser, 1910). Medial internal carotids and promontory arteries follow essentially the same route to the cerebral arterial circle, and appear to have identical relationships with the internal carotid nerve (cf. Story, 1951). The internal carotid in feloid Carnivora has a somewhat intermediate position, being partly intrabullar and partly extrabullar (Davis and Story, 1943; Hunt, 1974). No mammal, whether extinct or extant, fetal or adult, has been unequivocally shown to have both a medial internal carotid and a promontory artery. (Questionable counter-examples are offered by Matthew, 1909, and Hunt, 1974.) This may mean that the two arteries are homologous. However, in the absence of convincing evidence for this, we acquiesce to the conventional view that they are separate entities.

The carotid arrangement of most lipotyphlans is highly conservative, although they lack any vessel that could be described as a medial internal carotid. Figure 2 illustrates the major branches of the lateral internal carotid in the erinaceoid *Echinosorex gymnurus.* The promontory artery is relatively small; most of the blood traversing the internal carotid stem goes into the stapedial artery. After passing between the crura of the stapes, the stapedial artery divides into a superior and an inferior ramus.[1] The ramus superior

[1] In some fetal erinaceoids (e.g., *Erinaceus europaeus*), the stapedial artery gives off a third branch, the ramus posterior, just before passing through the stapes (MacPhee, 1977b). This vessel travels out of the middle ear via the stapedius fossa. It has not been investigated systematically, and its distribution in mammals is unknown.

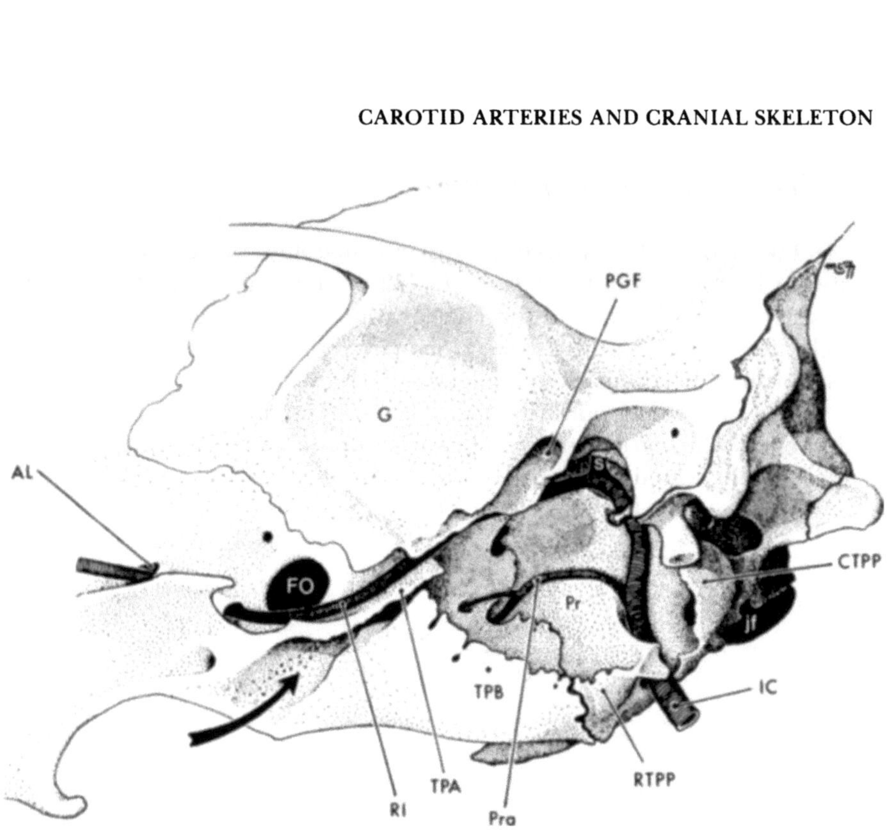

Fig. 2. Left ear region of *Echinosorex gymnurus*, illustrating primitive arrangement of the branches of the lateral internal carotid (IC). Based principally on USNM 487894; arteries reconstructed from that and other USNM specimens in the same collecting series which preserve the arteries in the ear region. The ectotympanic, malleus, and incus have been removed. The smaller (superior) terminal branch of the promontory artery (Pra) is the artery of the pterygoid canal ("Vidian" artery). Arrow indicates position of auditory tube in life. Abbreviations as in Table 2.

passes upward through the tegmen tympani and emerges endocranially, medial to the postglenoid foramen. Here it enters the sinus canal and runs forward (emitting meningeal branches en route), passing into the orbit at the front edge of the alisphenoid. The ramus inferior of the stapedial curves forward along the roof of the tympanic cavity, from which it emerges via a foramen (the Glaserian fissure) between the tympanic process of the alisphenoid and the anteriorly-directed styliform process (Henson, 1974) on the crus anterior of the ectotympanic. From there, it runs forward below foramen ovale and enters the alisphenoid canal. Its branches presumably supply the infraorbital and mandibular regions, as in *Erinaceus europaeus* (Bugge, 1974), and correspond to those of the human maxillary artery—minus the meningeal vessels, represented here by branches from the ramus superior.

The stapedial artery of primates is considerably simplified. In anthropoids, it does not normally persist into adult life. In typical Malagasy lemurs, the ramus inferior is absent in the adult, but the ramus superior persists in approximately its primitive form. In both groups, the external carotid artery gives off an anastomotic vessel, the maxillary artery (Fig. 1), which annexes the infraorbital and mandibular branches of the stapedial ramus inferior.

A. The Carotid Arteries of Tupaia

Like most other aspects of tupaiid anatomy, the carotid circulation is known in detail only for *Tupaia glis*. The following description is based on dissection of a specimen of this species which was injected with latex immediately after its death at the Duke University Primate Center. In most respects, its morphology conforms to Steuerwald's (1969) description.

The external carotid of *Tupaia* (Fig.3), like that of primates, has annexed the vessels which are primitively branches of the stapedial ramus inferior. The specimen dissected differed from that described by Saban (1963) in lacking a

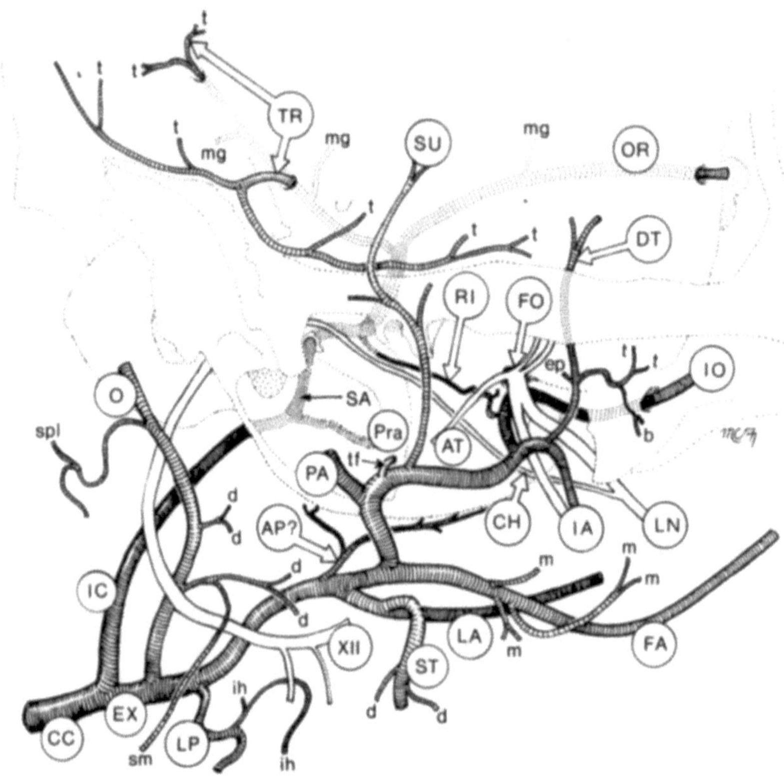

Fig. 3. Right lateral view of the principal carotid arterial branches of *Tupaia glis*, shown in relation to the skull (dotted outline). Nerves shown in white. Some of the vessels have been displaced slightly to display underlying vessels; for instance, the proximal maxillary artery (distal to the origin of PA) has been pulled ventrally forward. The putative ascending pharyngeal artery (AP?) probably represents an anastomosis that has annexed the AP's anterior branches; its posterior branches are annexed by the occipital artery (O). Abbreviations as in Table 2.

complete ascending pharyngeal artery; the more caudal branches of this vessel appear to have been annexed by the occipital. In this it resembled various nonprimates, and differed from *Lemur* and cheirogaleids (Cartmill, 1975 and unpublished observations). The specimen resembled those primates, however, in having a complex and contorted laryngeopharyngeal trunk arising at the point where the superior thyroid artery arises in anthropoids and carnivorans. We are not yet able to say whether this resemblance is a symplesiomorphy or not.

The external carotid of *Tupaia* divides terminally into a small superficial temporal artery and a much larger maxillary artery. The latter vessel runs forward deep to the neck of the mandible, sends off a large common trunk for the deep temporal and inferior alveolar arteries, and curves medially between the medial and lateral pterygoid muscles. As it nears foramen ovale, it gives off a small anterior tympanic branch that runs backward toward the Glaserian fissure. The maxillary artery continues medially forward, runs under foramen ovale deep to the branches of the mandibular nerve, traverses the alisphenoid canal, and reaches the periorbita. Here it divides into the usual branches. The course and distribution of the maxillary artery after the departure of its inferior alveolar branch are identical to those of the extrabullar part of the stapedial ramus inferior of erinaceids, with which this part of the artery is clearly homologous.

The internal carotid's primary branches are enclosed in bony tubes derived from the petrosal and entotympanic (MacPhee, 1977b). The branching pattern is precisely like that of *Echinosorex,* except that the ramus inferior of the stapedial is minute. The canal transmitting the ramus inferior runs forward in the roof of the tympanic cavity between the entotympanic and the epitympanic wing of the sphenoid (MacPhee, 1977b). The ramus inferior emerges through the medial wall of the Glaserian fissure and anastomoses directly with the anterior tympanic branch of the maxillary artery.

The stapedial ramus superior divides intracranially into two branches. The posterior branch (probably that identified as the "cerebellar artery α" by Boulay and Verity, 1973, p. 33) follows the petrosquamous sinus backward and sends off meningeal and temporal branches; the latter emerge through foramina in the parietal and supply the posterior half of the temporalis muscle. The anterior branch of the ramus superior runs forward in the sinus canal and gives off meningeal rami. Emerging in the orbit, it sends off frontal, lacrimal, and ethmoidal branches. It also supplies the eyeball via a large ciliary branch, which arches over the optic nerve and receives the small ophthalmic artery (Bugge, 1974, Pl. 2C).

The promontory artery trifurcates below the base of the brain into posterior communicating, ophthalmic, and anterior rami, whose relations and distributions are described by Steuerwald (1969). The minute ophthalmic artery passes into the orbit along the anteroinferior surface of the optic nerve and empties into the ciliary branch of the stapedial ramus superior.

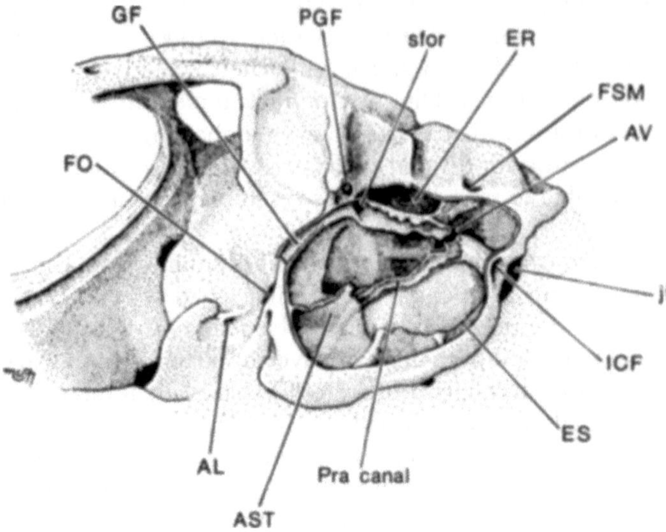

Fig. 4. Ear region of *Dendrogale murina*, USNM 320779. The floor of the bulla has been removed and the bony stapedial canal broken open to display the foramen (sfor) for the stapedial ramus superior. The ramus inferior apparently continued forward through the Glaserian fissure (GF), which also surrounds the chorda tympani and the ectotympanic's styliform process. The ossicles and ectotympanic have been removed. Note the hiatus in the stapedial canal where the artery passes through the stapes, across the apertura fenestrae vestibuli (AV). Abbreviations as in Table 2.

B. Comparisons and Discussion

Le Gros Clark (1926) discovered a ramus inferior of the stapedial artery in stained sections of *Ptilocercus*. Our dissection of a macerated skull of *Ptilocercus* (USNM 488068) in which the dried arteries remained intact in their bony tubes supports his observations in every detail. The ramus inferior of *Ptilocercus* is larger than that of *Tupaia;* it has about the same diameter as the promontory artery. The bony arterial canals in *Dendrogale,* which we have been able to expose in only a single specimen (Fig. 4), suggest that this genus retains a relatively large ramus inferior as well.

Le Gros Clark (1926), Saban (1963), and Bugge (1972) were unable to find the stapedial ramus inferior in *Tupaia,* and concluded that this genus resembles primates in lacking this vessel in the adult. However, a large ramus inferior was reported in *Tupaia* by Hyrtl in 1845 (*fide* Gregory, 1920, p. 174), and Steuerwald (1969) found a diminutive ramus inferior in four injected adult *Tupaia.* We have found a stapedial ramus inferior in two fetal, two infant, and one adult *Tupaia glis* (Fig. 3). Its minute size makes it easy to overlook in *Tupaia,* although the canal which transmits it is usually discernible in dried skulls.

Saban (1963) noted the anterior tympanic ramus of the maxillary artery in *Tupaia,* but failed to detect its anastomosis with the stapedial. We are inclined to endorse Saban's (1963, p. 89) suggestion that the anterior tympanic artery of *Homo* is homologous with the distal part of the stapedial ramus inferior, and are correspondingly puzzled by his claim (ibid., p. 138) that the ramus inferior which Le Gros Clark found in *Ptilocercus* "ne peut. . .s'agir que de l'artère tympanique antèrieure." This is a distinction without a difference.

The stapedial ramus superior of *Tupaia,* like that of macroscelidids and most extant lipotyphlans (Bugge, 1974), supplies blood to the eye as well as to the other orbital contents. The persistence in *Tupaia* of a small ophthalmic branch of the internal carotid is probably a primitive therian trait; its constant absence in some lipotyphlans seems likely to be a derived feature reflecting a secondary reduction of the eyes. Among other extant insectivorans, the ophthalmic artery occurs variably in erinaceids and macroscelidids, and is present in the fetus in at least some chrysochlorids and tenrecids (Tandler, 1899; Roux, 1947; Bugge, 1972). Extant reptiles display an ophthalmic branch of the internal carotid, which closely resembles that of *Tupaia* in its relative size and connections (Shindo, 1914). In the injected specimen of *Tupaia glis* dissected in the course of this investigation, the ophthalmic artery originated from the internal carotid stem directly opposite the posterior communicating artery, as it does in *Lemur catta* (Bugge, 1972) and as its embryonic homolog, the primitive dorsal ophthalmic artery, does in *Homo* (Padget, 1948). In *Homo, Lemur variegatus, Microcebus,* and some specimens of *Tupaia* (Tandler, 1899; Bugge, 1972; Cartmill, 1975), the ophthalmic artery originates further down the stem of the internal carotid in the adult. The variability of this feature makes us doubt that it has the phyletic significance claimed for it by Bugge (1972). In summary, neither the presence nor the connections of the ophthalmic artery afford grounds for linking the tupaiids with the primates.

The rami temporales of the stapedial ramus superior (Fig. 3) have not been remarked by previous investigators, although Grosser (1901) and Buchanan and Arata (1969) described them in Microchiroptera. Bugge's (1972, 1974) illustrations show that they exist in a number of insectivorans and rodents, and apparently homologous vessels are found in reptiles (Shindo, 1914). We have found these rami dried in place in *Ptilocercus* skulls (on one of which they were demonstrably continuous with the stapedial ramus superior), and foramina for them exist in some other tupaiids. In primates, these vessels have been annexed by the superficial temporal artery. Foramina connecting the petrosquamous sinus with the temporal fossa exist in some skulls of *Hapalemur griseus,* but dissections of newborn and adult *H. griseus* from the Duke University Primate Center reveal no temporal rami of the stapedial artery and demonstrate a superficial temporal artery of the usual primate type. Although the trait may be variable in this species, present evidence suggests that absence of the temporal rami of the stapedial ramus superior is a primate synapomorphy not seen in tupaiids.

3. The Evidence of the Ear Region

The auditory region of the skull has a rather constant morphology within individual higher taxa of mammals, but varies in a complex fashion from one such taxon to another. It has therefore been widely (and sometimes uncritically) used as a taxonomic touchstone in classifying mammals. Otic characters have played an especially important role in assessments of the affinities of tree shrews. Most previous descriptions of the tupaiid ear region have dealt only with the adult arrangement, which shows some striking similarities to those seen in Malagasy lemurs. Van Valen's (1965) examination of the skulls of young tree shrews and lemurs convinced him that the supposed similarities are convergences overlying more fundamental differences in developmental history. While Van Valen's conclusions have been widely adopted, they are based on studies of macerated skulls, which cannot afford conclusive evidence of the ontogeny and relationships of the bones and surrounding tissues of the ear region. We here present new comparative anatomical findings on the origin, development, and relationships of the structures of the auditory region in tree shrews, based on a survey of sectioned fetuses of tupaiids and other mammals (MacPhee, 1977b).

A. Ontogeny of the Tympanic Floor in Mammals

The floor or ventral wall of the middle ear separates an air-filled sac, the tympanic cavity, from the structures of the upper neck. In adult mammals, this floor may be composed of bone, cartilage, or membrane, or some combination of the three. The ontogeny of the tympanic floor is roughly analogous to that of the cranial vault. The latter is initially composed of the primitive meninges, which differentiate around the growing neural mass. These membranes provide a substrate for the growth of bones and help to direct that growth along predetermined planes. Similarly, a sheet of dense connective tissue (here termed the fibrous membrane of the tympanic cavity) differentiates beneath the expanding tympanic cavity and provides a substrate for the appearance and expansion of skeletal elements (Fig. 7).

Skeletal elements in the floor of the tympanic cavity fall into two categories: tympanic processes and entotympanics. Tympanic processes (Fig. 5) form as extensions of neighboring bones that spread into the floor of the tympanic cavity, growing directly along the inner (intratympanic) surface of the fibrous membrane (Figs. 7b, 10b). With one exception (mentioned below), tympanic processes arise as periosteal outgrowths; they are not preformed in cartilage, even if their parent bones are (Starck, 1975). Entotympanics, on the other hand, usually arise from cartilaginous rudiments, and have their own ossification centers. Entotympanics appear and grow *within* the fibrous membrane, which acts as their periosteum (Fig. 7a); tympanic processes, which merely adjoin the membrane's inner surface, are always enclosed by perios-

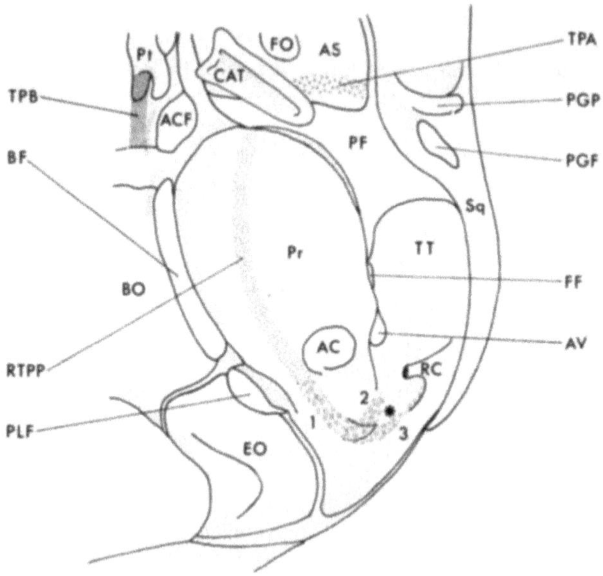

Fig. 5. Schematic representation of the left ear region of an idealized fetal eutherian (ventral aspect), illustrating sites where tympanic processes initially arise during ontogeny. The tympanic processes of the alisphenoid and basisphenoid (TPA, TPB) may fuse during development; the latter often incorporates material derived from the pterygoid bone (Pt). The rostral tympanic process of the petrosal (RTPP) always appears first in the middle of the ventral surface of the petrosal's pars cochlearis, though its longitudinal extent varies. In most eutherians, the caudal tympanic process of the petrosal (CTPP) arises along the arc identified by the numbers 1 and 2. In undoubted primates, it forms in areas 1 and 3, and thus surrounds the origin of the stapedius (asterisk). Abbreviations as in Table 2.

teum derived from that surrounding their parent bones (Fig. 7b). Tympanic processes and entotympanics often fuse secondarily with other skeletal elements during ontogeny (Klaauw, 1930), and so cannot always be correctly identified in the adult skull.

Even a cursory survey of the ear region in living and extinct mammals reveals that any of the bones adjoining the ear region can contribute tympanic processes to the formation of the tympanic cavity's floor, and that more than one entotympanic may form in that floor's fibrous membrane. Although it has been said that there are only two entotympanics, a rostral and a caudal one (Klaauw, 1922, 1931), Hunt (1974) has argued that as many as three different entotympanics may exist in Carnivora. Entotympanics are not found in all extant mammals, although they are widely distributed (Table 1). They differ greatly in their ontogenetic origin, position in the floor of the cavity, and relationship to other elements of that floor. These and other facts suggest that entotympanics have appeared independently in several different eutherian lineages, although the evidence is inconclusive (MacPhee, 1977b). In some

Table 1. Incidence of Entotympanics in Modern Mammals[a,b]

Subclass: PROTOTHERIA	
Order: Monotremata	NO
Subclass: THERIA	
Supercohort: Marsupialia	?NO
Supercohort: Eutheria	
Cohort: Edentata	
Order: Cingulata	YES R,C
Order: Pilosa	YES C
Cohort: Epitheria	
Magnorder: Ernotheria	
Grandorder: Anagalida	
Order: Macroscelidea	YES R,C
Order: Lagomorpha	NO
Magnorder: Preptotheria	
Grandorder: Ferae	
Order: Carnivora	YES R, C[c]
Grandorder: Insectivora	
Order: Erinaceomorpha	NO
Order: Soricomorpha	NO
Grandorder: Archonta	
Order: Scandentia	YES R[d]
Order: Dermoptera	?YES R,C
Order: Chiroptera	YES R,C
Order: Primates	NO
Grandorder: Ungulata	
Order: Tubulidentata	NO
Order: Artiodactyla	NO
Order: Cetacea	NO
Order: Perissodactyla	?YES
Order: Hyracoidea	YES R,C
Order: Proboscidea	?NO
Order: Sirenia	?YES
Magnorder: ?Preptotheria *incert. sed.*	
Order: Pholidota	YES C
Cohort: Epitheria *incert. sed.*	
Order: Rodentia	NO

[a]NO=entotympanic(s) absent; YES=entotympanic(s) present; ?NO=probably absent; ?YES=possibly present; R=rostral entotympanic; C=caudal entotympanic. Identification of rostral and caudal entotympanics is based on describers' categorizations; it is not certain that all elements in each category are homologous with each other.

[b]This list is provisional, since many orders have not been investigated using appropriate embryological techniques. Data are drawn from numerous sources and discussed in detail in MacPhee (1977b). The classification, categories, and taxon names are those of McKenna (1975). Our use of them does not imply endorsement.

[c]Hunt (1974) claimed that two different caudal entotympanics occur in Carnivora in addition to the rostral entotympanic.

[d]The entotympanic of adult *Ptilocercus* appears to contact only the tympanic process of the alisphenoid and not the cartilage of the auditory tube. This is not necessarily evidence that the entotympanic of *Ptilocercus* is a caudal one, but in view of the close ontogenetic relationship between the entotympanic and the tubal cartilage in *Tupaia*, its position in the pentailed tree shrew is most unusual.

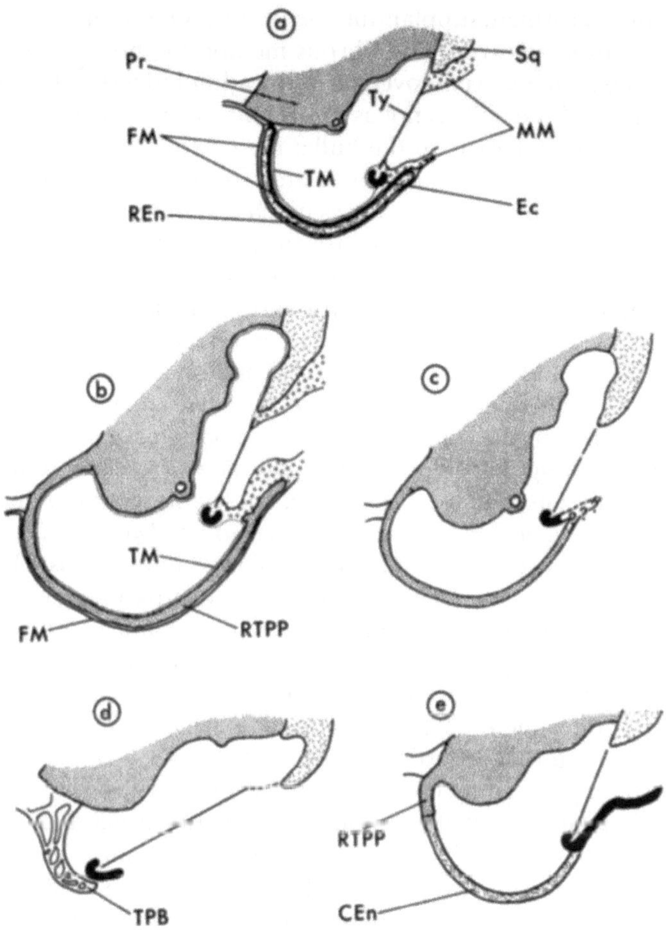

Fig. 6. Schematic coronal sections through the ear regions of some adult eutherians. (*a*), tree shrew (e.g., *Ptilocercus*); (*b*) modern Malagasy lemur (e.g., *Lemur*); (*c*), modified lemuriform condition, seen in some extinct primates (e.g., *Necrolemur*, *Megaladapis*); (*d*), erinaceoid lipotyphlan (e.g., *Erinaceus*); (*e*), elephant shrew (e g., *Elephantulus*). An aphaneric ectotympanic is seen in illustrations *a* and *b*; *d* illustrates the semiphaneric condition, and *e* the phaneric. Illustration *c* depicts a condition which may be considered aphaneric, phaneric, or semiphaneric, depending on the extent to which the ectotympanic contributes to the osseous meatus (dashed line); the point is uncertain owing to the absence of detectable sutures in relevant specimens. Soft tissues are indicated in the first two drawings; note that the entotympanic (REn) is encased within the fibrous membrane (FM) in *a*, while the rostral tympanic process of the petrosal (RTPP) in *b* is not. Abbreviations as in Table 2.

mammals, the skeletal elements in the tympanic cavity's walls grow very little during ontogeny, and do not form a complete floor (or auditory bulla) beneath the middle ear cavity in the adult. The fibrous membrane thus persists in the adult, surrounding the air-filled spaces of the middle ear. In other species, a complete bulla is formed, supplanting the fibrous membrane as the functional floor of the tympanic cavity. The fibrous membrane, however, remains even in these species as an external covering of the adult bulla (Fig. 6b). The degree of completeness of the bulla is not as important for purposes of systematics as the elements that contribute to the bulla; the osseous bulla may vary considerably in completeness among closely related species (e.g., within Didelphidae or Viverridae), but almost always varies much less in its constituents.

The periosteum that covers the inferior edge of the ectotympanic usually contacts the fibrous membrane in fetal mammals (Fig. 10b). If another bone growing along or in the fibrous membrane spreads far enough into the tympanic floor, its leading edge comes into close relation with the lower part of the ectotympanic, and in most cases sutural tissues are formed (Fig. 10d). In such instances, the element in the floor of the cavity forms a simple edge-to-edge articulation with the ectotympanic. The floor element will subsequently cease to expand along that edge, or else continue to grow and push the rim of the ectotympanic laterally. In either case, the ectotympanic remains visible from the ventral aspect, and is therefore said to be *phaneric*.[2] The formation of sutural tissues between the two bones is followed in many cases by complete bony fusion across the interface. In other cases, for unknown reasons, sutural tissue formation between floor element and ectotympanic is deficient, and the contiguous edges of the two bones are not tightly bound together (Fig. 10c). The floor element is therefore free to continue growing laterally beyond the ectotympanic, following the fibrous membrane (which extends laterally to the cartilage of the external acoustic meatus). If the floor element is a tympanic process, it will expand along the inner surface of the fibrous membrane, and will therefore delaminate the membrane from the periosteum of the ectotympanic as it grows beyond that bone. Such a subtympanic extension of any element in the floor of the tympanic cavity will occlude the ectotympanic from view, either partly (the *semiphaneric* condition) or completely (the *aphaneric* condition). Among extant eutherian mammals, the aphaneric condition is found only in tree shrews and Malagasy lemurs. This fact has always provided one of the strongest reasons for regarding tree shrews as persistently primitive lemuroids (cf. Le Gros Clark, 1959, 1971).

B. Ontogeny of the Tympanic Floor in Tupaiids

As many as three elements may be found in the tupaiid bulla: (1) an entotympanic, (2) tympanic process of the petrosal, and (3) tympanic process of the alisphenoid. The last is found only in *Ptilocercus* (Fig. 12). We will describe

[2] The traditional use of the terms "intrabullar" and "extrabullar" to describe the position of the ectotympanic is illogical (Simons, 1974; Cartmill, 1975). The terms "aphaneric" ("unseen"), "phaneric" ("seen") and "semiphaneric" (for the intermediate condition) are taken from MacPhee (1977b).

them in the order given above. The description is abstracted from the more detailed accounts of Spatz (1966) and MacPhee (1977b).

The entotympanic of tupaiids develops endochondrally within the fibrous membrane, which thus serves as its perichondrium (Fig. 7a). In *Tupaia glis,* the only tree shrew that has been adequately studied, its ontogeny is peculiar in two respects. First, it does not originate from a separate cartilaginous rudiment, but is connected from its earliest appearance with the cartilage of the auditory tube (Fig. 8a). In this respect, it resembles the rostral entotympanics found in the hyrax *Procavia* (Klaauw, 1922) and the armadillo *Dasypus* (Reinbach, 1952). We infer that the entotympanic of *Tupaia* corresponds to the rostral one of Klaauw (1931). Second, the entotympanic of *T. glis* fuses while still cartilaginous with two outgrowths of the auditory capsule; (1) the caudal tympanic process of the petrosal, and (2) the tegmen tympani (which forms part of the roof of the tympanic cavity). In a few other mammals, the unossified *caudal* entotympanic fuses with the petrosal's caudal tympanic process, but such fusion in-

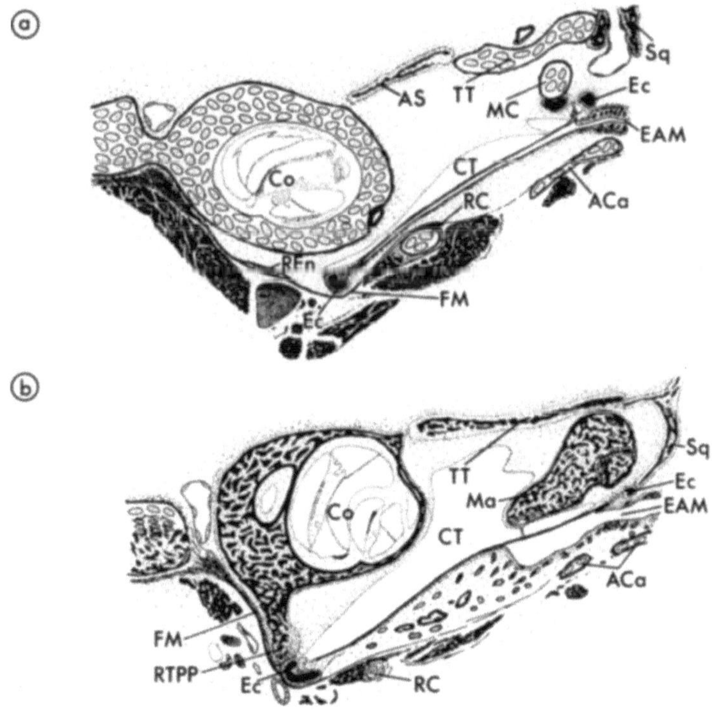

Fig. 7. Semischematic coronal sections through the auditory regions of late fetal *Tupaia glis* (a) and *Microcebus murinus* (b). The rostral entotympanic (REn) of *T. glis* arises in cartilage and grows *within* the fibrous membrane (FM) of the tympanic cavity (CT). The rostral tympanic process of the petrosal (RTPP) of *M. murinus* develops as a periosteal outgrowth of the auditory capsule and grows *along* the fibrous membrane rather than within it. Abbreviations as in Table 2.

volving a *rostral* entotympanic is apparently unique to tupaiids. Fusion between entotympanic and tegmen tympani (Fig. 8b) may not be exclusive to tree shrews, but it is certainly a rare occurrence; according to Klaauw (1931), this fusion is otherwise known only in *Rhinoceros*.

Tympanic processes of the mammalian petrosal are of two sorts: those which arise rostral to the aperture of the fossula fenestrae cochleae, and those which arise caudal to it (Figs. 2, 5). The latter, if developed prior to extensive ossification of the auditory capsule, originate in cartilage (unlike other tympanic processes). The tympanic process of the petrosal in *Tupaia* is of the caudal sort, and originates as a cartilaginous outgrowth of the mastoid region (Figs. 8a, 11a). It does not surround the origin of the stapedius muscle as it does in undoubted prosimians. This process, which had been overlooked prior to Spatz' (1966) study, remains small throughout ontogeny (Fig. 12a).

The third element of the tupaiid bulla, the tympanic process of the alisphenoid, is not seen in tupaiines but occurs in *Ptilocercus*, where it forms part of the anterior bullar wall (Fig. 12b). It is demarcated from the body of the alisphenoid by a deep groove. Klaauw (1931), misled by his erroneous interpretation of the macroscelidid alisphenoid, took this groove for a suture and regarded this part of the alisphenoid as a separate element. As far as we can judge from adult skulls, there is no suture at this point. The groove is continuous at its upper end with the Glaserian fissure which transmits the chorda tympani and the ramus inferior of the stapedial artery, and in poorly cleaned skulls of *Ptilocercus* these structures can be seen running along the groove. We

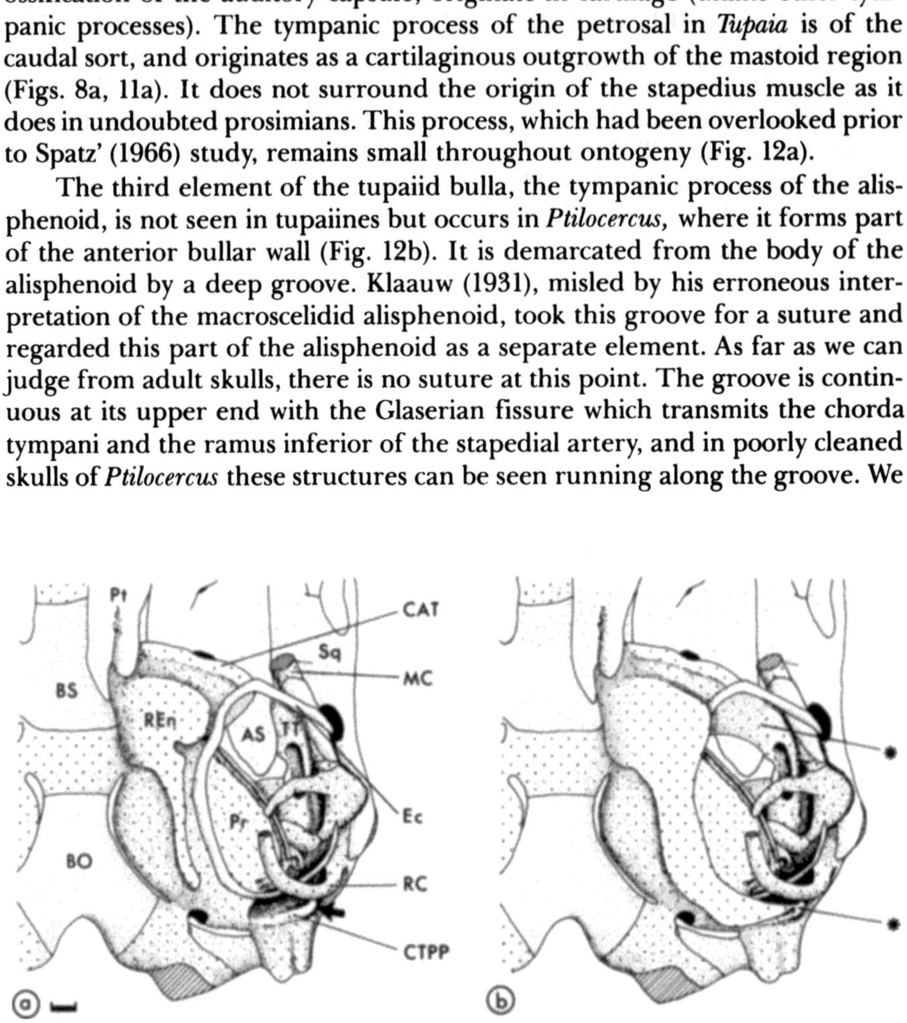

Fig. 8. Schematic reconstructions of the left auditory regions of (a) newborn and (b) eight-day-old specimens of *Tupaia glis* (modified from Spatz, 1966). The rostral entotympanic (REn) is continuous with the cartilage of the auditory tube (CAT). By the eight-day stage, it has fused with the tegmen tympani and the caudal tympanic process of the petrosal (asterisks), and has partially covered the ectotympanic (Ec). The arrow points to the position of the muscular origin of the stapedius (cf. Fig. 11a). Coarse regular stipple indicates cartilage; other structures shown are bony. The scale represents 0.5mm. Abbreviations as in Table 2.

suspect that the groove is homologous with the sulcus for the ramus inferior on the alisphenoid's tympanic process in erinaceids (Fig. 2). In tupaiines, the alisphenoidal process is vestigial or absent, and the sulcus for the ramus inferior runs along the alisphenoid-entotympanic suture. Comparison with other mammals suggests that reduction of the tympanic process of the alisphenoid is a tupaiine autapomorphy, and that Gregory (1910) was correct in regarding this process as an extremely ancient feature of the therian tympanic floor.

C. *Ectotympanic Ontogeny in Tupaiids and Lemurs*

Like that of Malagasy lemurs, the tupaiid ectotympanic retains an unmodified horseshoe shape throughout life, and is aphaneric in the adult. However, the aphaneric condition is produced in the two groups by different ontogenetic processes. In fetal Malagasy lemurs, as in fetal mammals generally, the fibrous membrane of the tympanic cavity is attached to the ectotympanic's periosteum in early developmental stages (Fig. 7b). In later stages, the developing petrosal bulla expands laterally to abut the ectotympanic, but a complete array of sutural tissues is not formed between the ectotympanic and the petrosal bulla's leading edge. The bulla continues to grow laterally, below and past the ectotympanic, along the inner surface of the fibrous membrane—thus secondarily delaminating the fibrous membrane from the ectotympanic's periosteum (MacPhee, 1977a,b). During postnatal life, the bulla becomes increasingly inflated, and the tympanic cavity and its lateral wall expand laterally

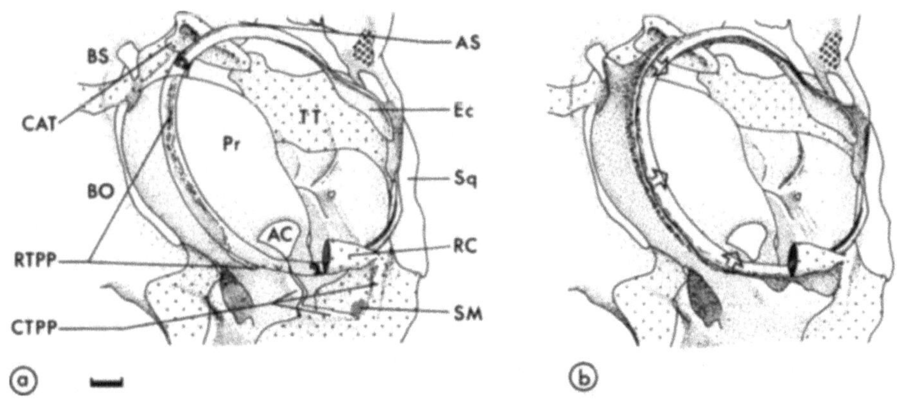

Fig. 9. Schematic reconstructions of the left auditory regions of two late fetal specimens of *Microcebus murinus*. The central part of the ectotympanic has been removed in (a) to show the position of the rostral tympanic process of the petrosal (RTPP). Later in development (b), the rostral and caudal tympanic processes have coalesced into a single broad petrosal plate, which is beginning to cover the ectotympanic (arrows). The muscular origin of the stapedius (SM) is covered by the caudal process (cf. Fig. 11b). Scale and shading conventions as in Fig. 8. Abbreviations as in Table 2.

below the medial end of the meatus; but the ectotympanic becomes fixed in space owing to the fusion of its posterior crus with the squamosal and the rearmost part of the bulla. A gap therefore appears between the lower edge of the ectotympanic and the internal face of the bulla's lateral wall (Fig. 6b). In the living animal, this gap is filled by a semitubular "anulus membrane"—the medial end of the meatus, ventrally enclosed in a wrapping of intratympanic mucosa—which stretches across from the ectotympanic to the meatal porus of the bulla (Cartmill, 1975).

The fibrous membrane of fetal *Tupaia,* unlike that of lemurs or other mammals, scarcely contacts the ectotympanic; in some fetuses of *T. glis,* there is an appreciable gap between the two (Fig. 7a). Thus, when the leading edge of the entotympanic reaches the position of the ectotympanic, neither a true articulation nor sutural tissues are formed (Fig. 10a). The entotympanic continues its unimpeded growth within the fibrous membrane, eventually passing below and beyond the ectotympanic and rendering it aphaneric. There is no postnatal expansion of the tympanic cavity below the ectotympanic (Fig. 6a), the meatal soft tissues are not separated from the bullar wall, and there is no membranous bridge or anulus membrane like that of lemurs at any stage of ontogeny.

D. *Comparisons and Discussion*

1. *Composition of the Bulla*

The evidence is now conclusive that the lemuriform bulla is formed through the coalescence and growth of only two elements, the rostral and caudal tympanic processes of the petrosal (MacPhee, 1977b, 1979). The rostral tympanic process is a periosteal outgrowth of the underside of the cochlear part of the auditory capsule, and has no equivalent in tree shrews (Figs. 8, 9). The caudal tympanic processes of tree shrews and lemurs are undoubtedly homologous, but they display different character states. In all primates, not just lemurs, the caudal tympanic process completely encloses the origin of the stapedius muscle (Figs. 9, 11b). It does not in tree shrews (Saban, 1963; cf. Figs. 8, 11a). This is a significant difference, since complete enclosure of the stapedius (as far as is now known) is unique to primates.

These facts imply that the tupaiid and lemuriform bullae are non-homologous, and suggest that the last common ancestor of the two groups lacked an osseous bullar floor. There are, however, alternative interpretations. Kampen (1905) and Spatz (1966) have posited that ancestral lemurs might have had an entotympanic which fused to the petrosal during ontogeny (as in *Tupaia*), and that this entotympanic began to fuse earlier and earlier until at some point it lost its developmental independence altogether (an example of so-called "primordial fusion"—Beer, 1937). There is no reason to believe that this has happened; but even if it had, it would not argue for tupaiid-lemuriform affinities, since the presumptive entotympanic region in lemurs ossifies as an

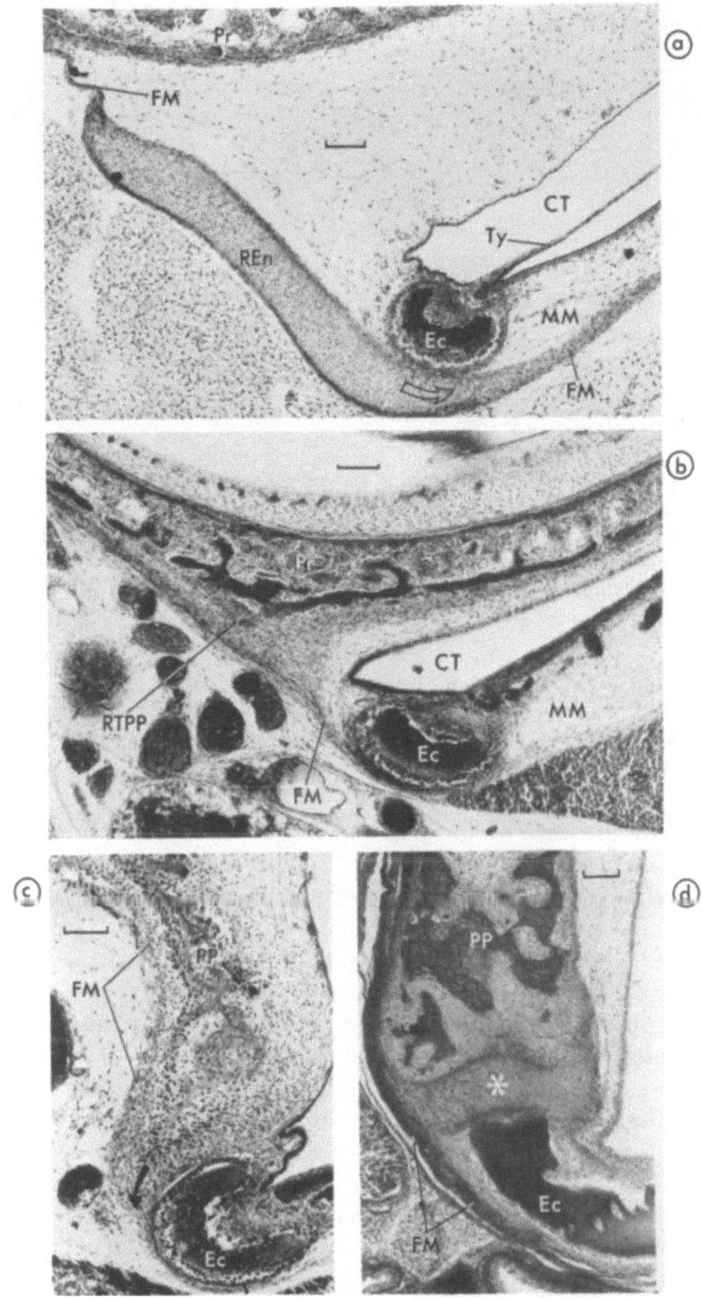

Fig. 10. Differences in bullar ontogeny between tree shrews and primates. (*a*), *Tupaia glis* (8 days old, MPIH 1960/77); (b), *Microcebus murinus* (younger fetus, MPIH 1964/41); (c), *M. murinus* (older fetus, MPIH 1964/43); (d), the lorisiform primate *Loris tardigradus* (near-term fetus, MPIH 1666). The rostral entotympanic (REn) of *T. glis* expands within the fibrous membrane (FM) and grows beneath the ectotympanic (Ec) without forming an edge-to-edge articulation with it. The petrosal bullae of the primate specimens, growing laterally adjacent to the fibrous membrane, form articulations with the ectotympanic. In *Microcebus*, the articulation is transitory and is lost as the bulla grows around the ectotympanic (*c*, arrow), but in *Loris* the articulation persists because sutural tissues (asterisk) are formed. These tissues are not represented in *M. murinus*. Scales represent 0.1 mm; abbreviations as in Table 2.

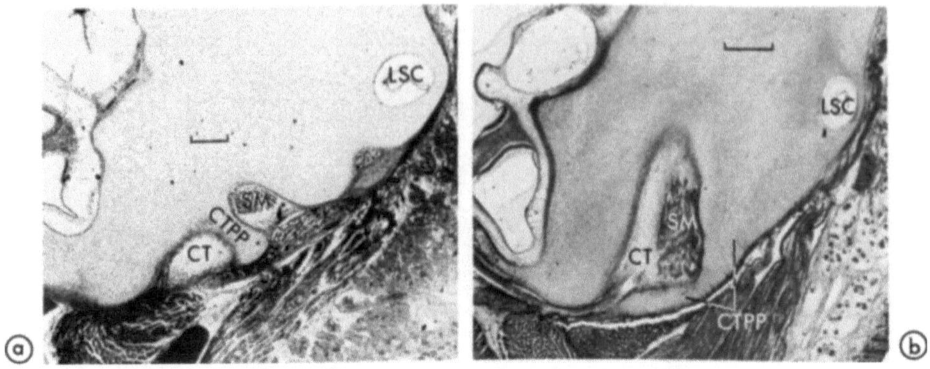

Fig. 11. Differences between tupaiids and primates in the relationship of the caudal tympanic process of the petrosal (CTPP) to the stapedius muscle (SM). In tree shrews (*a*; newborn *Tupaia glis*, MPIH 1959/4), the origin of the stapedius is exposed on the sidewall of the auditory capsule. In primates (*b*; fetal *Propithecus sp.*, AI 210/211), it is entirely enclosed and lies morphologically within the tympanic cavity (CT). Scales represent 0.25 mm (*a*) and 0.5 mm (*b*). Abbreviations as in Table 2.

extension of the petrosal's *rostral* tympanic process, whereas the cartilaginous entotympanic of *Tupaia* fuses with the *caudal* tympanic process. The tupaiid entotympanic does not fuse, even in the adult, with the rostral (promontorial) part of the petrosal (cf. *Dendrogale*, Fig. 4). There is no evidence that the rostral tympanic process of the petrosal in any extant mammal represents a suppressed entotympanic (MacPhee, 1977b).[3] This process's mode of ossification and relation to the fibrous membrane are not like those of the entotympanic, and some mammals (e.g., elephant shrews) have a large rostral tympanic process of the petrosal (Fig. 6e) *and* both entotympanics (Klaauw, 1929). The secondary fusion of the (rostral) entotympanic with the petrosal's caudal tympanic process in *Tupaia glis* does not indicate that this element is presently undergoing suppression in tree shrews, as Spatz (1966) implies. The formation of cartilaginous bridges or commissures between unrelated chondrocranial structures is exceedingly common during mammalian ontogeny (Starck, 1967), and has no particular significance.

Le Gros Clark (1959, 1971), inverting Kampen's (1905) hypothesis, suggested that the tupaiid condition may be derived from a lemur-like condition, and that the entotympanic of the tree shrew is a detached petrosal outgrowth which has gained developmental independence. The mode of ossification of the tupaiid entotympanic, and its relations to the fibrous membrane, refute this hypothesis. We conclude that the rostral tympanic process of lemurs and the entotympanic of *Tupaia* are simply non-homologous, and that neither could have been derived from the other.

[3]Starck (1975) suggests that the cartilage he found in the bullae of young *Tarsius bancanus* might represent an entotympanic, but we regard this as extremely doubtful. The vexed question of entotympanic suppression in primates is examined in detail elsewhere (MacPhee, 1979).

2. *Form and Relationships of the Ectotympanic*

In tree shrews and typical lemurs, the ectotympanic is a simple ring, obscured ventrally by the floor of the bulla. Both its anular shape and its aphaneric disposition have been cited as characters linking the two groups together. We will assess the two characters separately.

a. Anular Shape. The mammalian ectotympanic (Figs. 8, 9, 10) forms in the fetus as a simple ring of membranous bone. It frames the eardrum, which is attached to the tissue filling the tympanic sulcus. In most mammals, the outer or meatal edge of the ring (lateral to the sulcus) proliferates laterally during later fetal and postnatal life, forming a short bony collar around the medial end of the membranous meatus. This growth does not occur in tupaiids and typical lemurs, in which the anulus retains its fetal shape into adult life. Some *Microcebus murinus* specimens display a short meatal expansion of the ectotympanic, which is drawn out laterally toward a similar but medially-directed meatal expansion of the petrosal (Cartmill, 1975). In *Allocebus*, *Megaladapis*, and palaeopropithecines, the ring is linked to the bulla's meatal porus by a complete bony tube (Saban, 1963; Tattersall, 1973; Cartmill, 1975; Szalay, 1972, 1975), but it remains to be demonstrated that any part of the tube is an ectotympanic derivative (Cartmill and Kay, 1978). The bony meatal tube in all these forms presumably develops between the meatal and mucosal layers of the anulus membrane, and is often called an "ossified anulus membrane"; but this is a misnomer (MacPhee, 1977a).

Lateral expansion of the ectotympanic is, then, rare in Malagasy lemurs, and is almost certainly apomorphous within that group. The unexpanded anular ectotympanic might therefore represent a lemur-tupaiid synapomorphy, if it could be shown that this character state was not found in the ancestral eutherians. But this, in our opinion, is not demonstrable. Although most Theria have a somewhat expanded ectotympanic, a simple anulus persists into

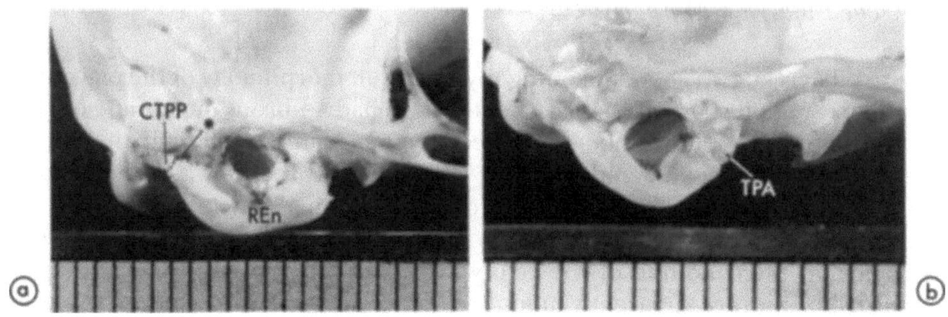

Fig. 12. Differences in the adult bullae of (*a*) *Tupaia tana* (USNM 174612) and (*b*) *Ptilocercus lowii* (USNM 112611). The bulla is formed by an entotympanic (probably rostral in both genera), a minute caudal tympanic process of the petrosal (CTPP), and (in *Ptilocercus* only) a tympanic process of the alisphenoid (TPA). The probable extent of the CTPP in *T. tana* is indicated by a dashed line. The origin of the stapedius muscle lies in a groove (asterisk) which also accommodates the stylomastoid foramen. The scales are in millimeters. Abbreviations as in Table 2.

adult life in many fossil and extant lipotyphlans (McDowell, 1958). The ectotympanics of didelphid marsupials are slightly expanded, but the expansion occurs *medial* to the tympanic sulcus (Segall, 1969, 1970). The possibility that the simple anular ectotympanic represents a lemur-tupaiid symplesiomorphy cannot be ruled out.

b. Aphanery. Since the ectotympanic is covered by different elements in tree shrews and lemurs, their last common ancestor is unlikely to have had a completely aphaneric ectotympanic. However, an incipiently aphaneric condition appears in other therian mammals, and may be primitive in this group. In young stages of the hedgehog *Erinaceus* and the tenrec *Hemicentetes*, for example, the tympanic process of the basisphenoid comes into close relation with the ectotympanic, but no sutural tissues form; so the basisphenoid continues to grow a short distance beyond the ectotympanic, and ends up covering a small part of its medial section in the adult (Fig. 6d). A much closer approach to the aphaneric condition is seen in dasyurid marsupials (Jones and Lambert, 1939). In these animals, the body of the ectotympanic is covered in the adult by a subtympanic extension of the (alisphenoid) bulla, but the ectotympanic is prolonged laterally to the lip of the bullar porus, where it projects very slightly beyond the alisphenoid. In diprotodont marsupials, the ectotympanic appears to fuse to the margin of the bullar porus. *Dasyurus* juveniles have a simple anular ectotympanic which is only partly overlapped by tympanic wings of the alisphenoid and petrosal; a similar arrangement is seen in some didelphids (Archer, 1976). Although little is known about the ontogeny of the marsupial ear region, we suspect that sutural tissues do not form between the bullar floor elements and the body of the ectotympanic in this group either. If the absence of such sutural tissues is primitive for Theria, then expansion of tympanic processes or entotympanics in any early therian lineage would be expected to lead initially to a semiphaneric or aphaneric condition.

3. The Eutherian Morphotype

What is known about the ontogeny and adult morphology of the placental auditory region suggests that basal eutherians had a fibrous membrane of the tympanic cavity, a horseshoe-shaped ectotympanic, and small tympanic processes of the alisphenoid and petrosal. The petrosal process was of the caudal type, and did not enclose the origin of the stapedius muscle. We are not so confident as some other investigators are about the presence of an entotympanic and the condition of the ectotympanic in the eutherian morphotype. Entotympanics are apparently unique to eutherians,[4] but some eutherians lack

[4]The marsupial "entotympanic" described by Kampen (1905) is evidently an enlarged caudal tympanic process of the petrosal and not an independent entity (MacPhee, 1977b). The supposed entotympanic identified by Carlsson (1926) in adult *Dasyuroides byrnei* also appears to us to be continuous with the petrosal's mastoid region. Jones (1949) regarded the petrosal component of the bulla of *Dasycercus cristicauda* "as including an entotympanic element," but he found no separate entotympanic even in pouch young. We are not aware of any good evidence for the existence of an entotympanic in a marsupial.

them and at least two non-homologous types of entotympanic occur among the other Eutheria. Klaauw (1931) suggested that the entotympanic in its most primitive state takes the form of a small pyramidal plate of bone lying behind the auditory tube and grooved by the sulcus for the medial entocarotid. However, the element bearing this sulcus has been identified as the *caudal* entotympanic in pilosan Xenarthra and the *rostral* entotympanic in Carnivora. If either entotympanic element was present in the ancestral eutherian, it was subsequently lost without trace in several different lines of descent—some of which apparently reverted to an askeletal bulla, while others evolved a second entotympanic to replace the ancestral one. This is perhaps slightly less probable than the hypothesis that entotympanics were developed independently in several different lineages. The condition of the ectotympanic in ancient eutherians is equally indeterminate. Evidence can be selectively marshalled to "prove" that it was covered or uncovered, anular or expanded. Furthermore, there is no reason to believe that any of these character states of the ectotympanic cannot be replaced by its opposite later on in the evolution of a lineage (MacPhee, 1977a). These and several other cranial traits (e.g., the presence or absence of various blood vessels, the mosaic patterns formed by membrane bones, the discreteness of certain cranial foramina) are probably more subject to reversible evolutionary change than has been appreciated by those investigators who have not taken their ontogenetic origins into account.

4. *Other Features of the Ear Region*

Most studies of the auditory ossicles of tupaiids emphasize morphological comparisons with primates, insectivorans, and elephant shrews, and rarely take sufficient account of function and variation in mammals generally. It is true that ossicular anatomy seems arrestingly similar in tree shrews and primates (e.g., Doran, 1878; Gregory, 1910; Le Gros Clark, 1926, 1959; Saban, 1956/57; Werner, 1960; Segall, 1970). However, more extensive comparison demonstrates that many of their apparent shared specializations are abundantly distributed among therians (cf. Fleischer, 1973). Because we have not undertaken original investigations of mammalian ear bones, we cannot be certain that all resemblances between tupaiids and primates can be plausibly interpreted as convergences. However, available data indicate that the worth of some ossicular features commonly employed in tupaiid systematics has been overvalued in the past. Two prominent examples are the condition of the anterior process of the malleus and the form of the stapedial crura.

The anterior process offers a difficult problem in anatomical definition. In a number of mammals it is mostly or entirely formed by the gonial, an independent bone which fuses during ontogeny with the anterosuperior part of the malleus. In many other instances, however, the gonial is welded to a continuous series of plates and shelves arising from the malleus' head. Since it is usually not possible to define morphological boundaries in these instances (at least in adults), the anterior process will be here regarded as the gonial plus associated malleolar outgrowths, if any (cf. Henson, 1961).

Fleischer (1973) has shown that the anterior process of ancient therians was relatively enormous and broadly ankylosed with the anterior arm of the ectotympanic (thus severely limiting the mobility of the ossicular chain). Among living therians, minor variations on this primitive theme are found in all bats, most lipotyphlans, some edentates, and a number of rodents and marsupials. By contrast, in tupaiids and primates the laminae and buttresses characteristic of the plesiomorphous condition are absent, and the process itself is reduced to a small spike which is partly or wholly free of the ectotympanic. Yet structural simplification of the anterior process, with consequent improvement of ossicular mobility, is an extremely common trend in therians (Fleischer, 1973). Forms as different as *Bradypus, Cynocephalus, Oryctolagus, Mustela*, and *Spalax* exhibit similarly reduced anterior processes, and it seems very likely that reductions have independently occurred in numerous lineages. We see no reason to believe that anterior-process simplification in tree shrews and primates represents a synapomorphy rather than a convergence.

Equally doubtful is the systematic value of the form of the stapedial crura. Gregory (1910) pointed out that the crura of tupaiids, macroscelidids, and lemurs are relatively straight, while in lipotyphlans they are usually highly convex. This morphological contrast has no meaning. Straight crura can be found in representatives of every major therian group, and even within primates crura can be straight, convex, or slightly concave (Fleischer, 1973; Werner, 1960). Further, there is no evident correlation between stapedial artery size and the degree of crural bowing. For example, the crura of *Homo* are no less convex than those of the soricid *Crocidura* (Wassif, 1948), despite the fact that the stapedial artery involutes embryonically in man.

Other features which do not vary in any way consistent with the usual taxonomic divisions of Theria, except at the grossest levels, include the neck length of the malleus, the axis of incudal-malleolar rotation, and the form of the stapedial footplate. However, there is some reason to believe that variations in these features is explicable in functional terms, as Webster and Webster (1975) have shown for heteromyid rodents. It is obvious that further insights into the evolutionary significance of ossicular characters, whether in tupaiids or in other mammals, will only come from detailed investigations of acoustic function.

It may also be noted that the tensor tympani muscle of tree shrews, as described by Saban (1963), is unlike that of primates (and most other mammals which have been investigated). There is some doubt about the identification of this muscle in tupaiids; although Saban (1963) claimed that it is present in *Tupaia,* the tensor tympani is bilaterally absent in the sectioned specimens of *T. glis* studied by Spatz (1964) and MacPhee (1977b).

In tupaiids and primates of modern aspect (other than lorises and dwarf lemurs), the branches of the lateral internal carotid are enclosed in bony tubes as they run along the walls of the tympanic cavity. The vessels are homologous in the two groups, but the canals enclosing them are apparently not (MacPhee,

1977b). The canal surrounding the stapedial artery of *Tupaia glis* is formed very early, as a result of cartilage derived from the tegmen tympani migrating along the entire intratympanic part of the vessel. The resulting cartilaginous tube is replaced endochondrally from the petrosal ossification center. Nothing like this mode of stapedial canal development has been reported for any other mammal. The canal surrounding the promontory artery of *Tupaia* is derived from the petrosal, but as a periosteal outgrowth (i.e., without preformation in cartilage). In tupaiines (but not *Ptilocercus*), this canal is fenestrated for at least part of its length in the adult, and the canal surrounding the stapedial artery also retains a hiatus where the artery goes through the stapedial crura (Fig. 4). The canal surrounding the lateral internal carotid trunk is of entotympanic origin in all tupaiids. In non-cheirogaleid lemurs, the stapedial and lateral internal carotid canals are complete and originate periosteally from the petrosal surface underlying the arteries. The proximal portion of the promontory artery is housed in a tube, but the artery is uncovered where it crosses the anterior part of the promontorium. Developmental evidence thus suggests that the last common ancestor of tupaiids and primates lacked bony tubes surrounding the carotid branches (with the possible exception of the proximal part of the promontory artery). Such tubes are absent in *Plesiadapis tricuspidens* (Russell, 1964) and are unconfirmed for *Phenacolemur jepseni* (Szalay, 1972); yet they have been developed in parallel in several insectivoran lineages (Van Valen, 1965; Rich and Rich, 1971).

Similar arguments apply to the arrangement of intrabullar septa in tupaiids and lemurs, which Saban (1956/57, 1963) has cited as a special resemblance between the two groups. Since the bones forming these septa are not homologous in the two groups (MacPhee, 1977b), their last common ancestor is unlikely to have had ossified septa. Some of the differences in septal composition between tupaiids and lemurs reflect more profound differences in the formation of the tympanic cavity's bony roof. In lemurs, the embryonic piriform fenestra in this roof (Fig. 5) is filled by an extension of the tegmen tympani and by an epitympanic petrosal process which is produced from the anterior end of the promontorium and the anterior part of the bulla. The anterior septum ("septum principal" of Saban, 1963) develops from parallel ridges which appear along the line of contact of these two petrosal outgrowths. In *Tupaia*, the epitympanic process is formed by the promontorium alone (since a rostral tympanic process is not present), and it does not expand across the piriform fenestra to meet the tegmen tympani. The piriform fenestra never actually closes in *Tupaia*, although it is occluded in the late fetus (and the adult) by an epitympanic expansion of the entotympanic across the anterior part of the tympanic roof (Fig. 8b). The tupaiid anterior septum develops from this part of the entotympanic (cf. *Dendrogale*, Fig. 4). This comparison suggests that the last common ancestor of tupaiids and lemurs not only lacked an osseous anterior septum, but also retained a persistently open piriform fenestra like that seen in many lipotyphlans (McDowell, 1958).

4. Other Cranial Features

The bony mosaic pattern of the medial orbital wall in tupaiids conforms closely to that seen in *Lemur,* and Saban (1956) listed certain features of this mosaic among the traits linking the tupaiids to the primates. Butler (1956) regarded the orbital mosaic as one of the characters that indicates affinities between Primates and the Menotyphla (including tupaiids). Jones (1929) and Le Gros Clark (1959) have argued that, given other evidence for the inclusion of tupaiids in Primates, the tupaiids' lemur-like orbital mosaic implies special affinities to Lemuriformes.

The traits cited in these assessments are features of the palatine and lacrimal bones. In Lipotyphla (other than *Potamogale*), the orbital lamina of the palatine is diminutive and is ordinarily separated from the frontal by a maxillary-orbitosphenoid contact, whereas in tree shrews and many primates the lamina is large and contacts the frontal. In tree shrews and *Lemur* the palatine extends to the lacrimal, and the lacrimal touches the zygomatic in the anterior orbital margin.

Most of these traits appear to be primitive for Theria or Mammalia. The orbital lamina of the palatine is large and contacts the frontal in *Morganucodon, Ornithorhynchus,* marsupials, carnivores, ungulates, macroscelidids, and dermopterans (Muller, 1934; Haines, 1950; Butler, 1956; McDowell, 1958; Russell, 1964; Kermack and Kielan-Jaworowska, 1971). The lacrimal reaches the zygomatic in most of the same groups and in reptiles and many rodents as well (Hogben, 1919; Romer, 1956). The orbital lamina of the palatine is, however, small in some early Tertiary primates (Simons and Russell, 1960; Russell, 1964; Wilson, 1966), and is reportedly quite diminutive in *Plesiadapis tricuspidens* (Russell, 1964). It is possible that the palatine was small in the ancestral primate; but in either event, the large orbital lamina of the palatine cannot be interpreted as a tupaiid-primate synapomorphy. Similar reasoning applies to the palatine-lacrimal contact, which occurs in some of the therian groups mentioned above but is unknown in early Tertiary primates (*contra* Gregory, 1920, and Szalay, 1976: cf. Stehlin, 1912; Le Gros Clark, 1934; Piveteau, 1957; Russell, 1964). Among extant primates other than the genus *Lemur* it occurs in a minority of cases or not at all (Cartmill, 1978). This contact may be partly dependent on the approximation of the anterior orbital margin to the posterior edge of the palate. Relatively minor changes in craniofacial growth patterns could easily result in the appearance or disappearance of this contact in parallel in different lineages.

Most of the supposed tupaiid-primate synapomorphies in arrangement of cranial foramina seem dubious. The superior orbital fissure and foramen rotundum are separate in other groups of therian mammals (e.g., marsupials, carnivorans, rodents), and this may be a therian symplesiomorphy. If not, it is a tupaiine-primate convergence, since the two foramina are confluent in *Ptilocercus* and many prosimians (Mivart, 1864, 1867/68). The zygomatic foramen is large in some tupaiines and lemuriforms; it is present but diminutive in

Ptilocercus, *Dendrogale*, and *Anathana* (Mivart, 1867/68; Gregory, 1910; Lyon, 1913) and in many lemuriforms, including *Lemur fulvus* and *L. rubriventer* specimens (our observations). Its occurrence clearly depends in part on the presence of a large zygomatic bone with a well-developed postorbital process; it is accordingly not found in lipotyphlans that lack a zygomatic arch or in *Plesiadapis* (Russell, 1964). Its rather haphazard distribution makes its systematic value doubtful.

In marsupials and many lipotyphlans and rodents, the two orbits are connected by a suboptic foramen that lies on or near the orbitosphenoid-alisphenoid suture below the optic canal (Mivart, 1867/68; Gregory, 1910; Butler, 1956). A similar but more anteriorly placed foramen is found in macroscelidids (Butler, 1956). Butler (1948) found this foramen occupied by an interorbital vein in *Erinaceus*. The suboptic foramen is absent in *Tupaia glis*, *Ptilocercus*, Dermoptera, and prosimians, and so its absence might be taken as synapomorphous for a tupaiid-primate group or the Archonta. However, an interorbital foramen in approximately the macroscelidid position occurs in *Dendrogale* (e.g., USNM 320779) and some specimens of *Tupaia tana* (Fig. 13); furthermore, the large and closely-appositioned optic foramina of tupaiids might easily transmit a homologous interorbital vein which was not partitioned off by bone from the optic nerve. (Dried skulls of some Megachiroptera seem to have an unossified meningeal partition here.) It will take further comparative dissection to resolve these problems of homology.

The complete postorbital bar of tupaiids is unlikely to represent a tupaiid-primate synapomorphy, since plesiadapoids lacked such a bar (Russell, 1964; Wilson and Szalay, 1972; Kay and Cartmill, 1977), Van Valen (1965) rejects this as a synapomorphy for similar reasons, and also on the grounds that the bar in the *Ptilocercus* skull described by Mivart (1867/68) is incomplete. Mivart's

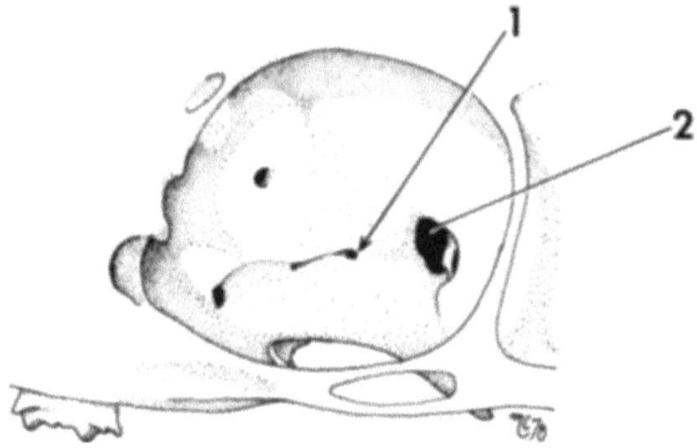

Fig. 13. Left orbit of a specimen of *Tupaia tana* in the senior author's collection, showing a small interorbital foramen (1) which lies anterior to the optic foramen (2) and may be homologous with the supposed suboptic foramen in macroscelidids. A probe passed through this foramen traverses a midline orbitosphenoid (or presphenoid) air sinus and emerges through the contralateral foramen. Specimen provided by Dr. W.P. Luckett.

figures suggest to us that the zygomatico-frontal suture in his specimen sprang open bilaterally during preparation. We regard the complete postorbital bar as a tupaiid specialization convergently arrived at in primates of modern aspect. Cartmill (1972) proposed that the tupaiid bar was acquired in an ancestor which, like *Ptilocercus* (but unlike plesiadapoids), had pronounced orbital convergence resembling that seen in prosimians, and that it was retained in tupaiines despite a secondary laterad reorientation of the optic and orbital axes.

Paulli (1900) and Le Gros Clark (1925) studied serial sections of the nasal region of *Tupaia* spp. and concluded that tree shrews and typical Malagasy lemurs share a derived reduction of the ectoturbinals from three to two. Even if this were true, the supposed resemblance would probably not represent a synapomorphy, since at least three ectoturbinals persist in *Daubentonia*, as Le Gros Clark himself later noted (Le Gros Clark, 1971, p. 269: cf. Kollmann and Papin, 1925). However, a recent study by Woehrmann-Repenning and Meinel (1977) has shown that a third ectoturbinal is in fact retained in *Tupaia glis*, and that the nasal fossa of this tupaiid does not differ from those of *Sorex*, *Crocidura*, *Talpa*, or *Erinaceus* in the number and arrangement of turbinals or any other significant aspect of gross anatomy.

5. *Discussion*

Figure 14 is a provisional character analysis of the 23 features of cranial morphology that seem most relevant to the question of tupaiid affinities. As far as possible, we have grouped these traits in a way which facilitates comparison of tree shrews with plesiadapoids and primates of modern aspect. All the traits are known to occur in *Tupaia glis*. Most of them are found in all other tupaiines, but this is not certain for traits 3, 6, 9, 11, and 21.

A. Tree Shrews and Primates

Traits 1–6 are found in tupaiids but not in primates. Only the first of these is both uniform in tupaiids and clearly derived with respect to the eutherian morphotype; it is thus the only tupaiid autapomorphy we can identify with confidence. Traits 4–9 are primitive for Eutheria, and are accordingly irrelevant to the problem of tupaiid affinities. The primitive eutherian condition is not known for traits 3, 10–14, and 19; these seven traits therefore shed no light on the affinities of tupaiids. If a case is to be made for tupaiid-primate affinities on the grounds of cranial anatomy, it must be based on the remaining eight features. These are almost certainly derived relative to the eutherian morphotype; they occur in at least some tupaiids and primates, and are therefore either synapomorphies or convergences. These derived resemblances between

Table 2. Abbreviations

AC	aperature of the cochlear fenestra
ACa	auricular cartilage
ACF	anterior carotid foramen
AI	Anatomisches Institut, J.-W.-Goethe-Universität, Frankfurt/M
AL	alisphenoid canal
AP?	ascending pharyngeal artery (?)
AS	alisphenoid
AST	anterior septum
AT	auriculotemporal nerve
AV	aperture of the vestibular fenestra
b	buccal artery
BF	basicapsular fenestra
BO	basioccipital
BS	basisphenoid
CAT	cartilage of the auditory tube
CC	common carotid artery
CEn	caudal entotympanic
CH	chorda tympani
Co	cochlea
CT	cavum tympani (tympanic cavity)
CTPP	caudal tympanic process of the petrosal
d	ramus to digastric m.
DT	deep temporal artery
EAM	external acoustic meatus
Ec	ectotympanic
EO	exoccipital
ep	ramus to external pterygoid m.
ER	epitympanic recess
ES	suture between entotympanic and petrosal
EX	external carotid
FA	facial artery
FF	foramen faciale (for nerve VII)
FI	foramen of inferior petrosal sinus
FM	fibrous membrane of the tympanic cavity
FO	foramen ovale
FSM	foramen stylomastoideum
G	glenoid fossa
GF	Glaserian fissure (canal of stapedial ramus inferior)
IA	inferior alveolar artery and nerve
IC	(lateral) internal carotid artery
ICF	(lateral) internal carotid foramen
IO	infraorbital artery
jf	jugular foramen
LA	lingual artery
LI	lateral internal carotid artery
LN	lingual nerve
LP	laryngeopharyngeal arterial trunk
LSC	lateral semicircular canal
m	ramus to masseter m.
Ma	malleus
MC	Meckel's cartilage
ME	medial internal carotid (medial entocarotid)
mg	meningeal ramus of stapedial ramus superior
MM	membranous meatus
MPIH	Max-Planck-Institut für Hirnforschung, Frankfurt/M.
O	occipital artery
OR	orbital ramus of stapedial ramus superior
P	petrosal
PA	posterior auricular artery
PF	piriform fenestra
PGF	postglenoid foramen
PGP	postglenoid process
PLF	posterior lacerate foramen (= jf)
PP	petrosal plate (RTPP + CTPP)
Pr	promontorium of pars cochlearis
Pra	promontory artery
Pt	pterygoid bone
RC	Reichert's cartilage
REn	rostral entotympanic
RI	ramus inferior of stapedial artery
RTPP	rostral tympanic process of the petrosal
s	superior ramus of stapedial artery
SA	stapedial artery
sfor	foramen for stapedial ramus superior
SM	stapedius muscle
sm	ramus to sternocleidomastoid m.
spl	ramus to splenius capitis m.
Sq	squamosal
ST	ramus to submandibular gland
SU	superficial temporal artery
t	ramus to temporalis m.
tf	transverse facial artery
TM	tunica mucosa of the tympanic cavity
TPA	tympanic process of the alisphenoid
TPB	tympanic process of the basisphenoid
TR	temporal rami of stapedial ramus superior
TT	tegmen tympani
Ty	tympanic membrane
USNM	United States National Museum, Washington, D.C.
V	vertebral arteries
XII	hypoglossal nerve (nerve XII)

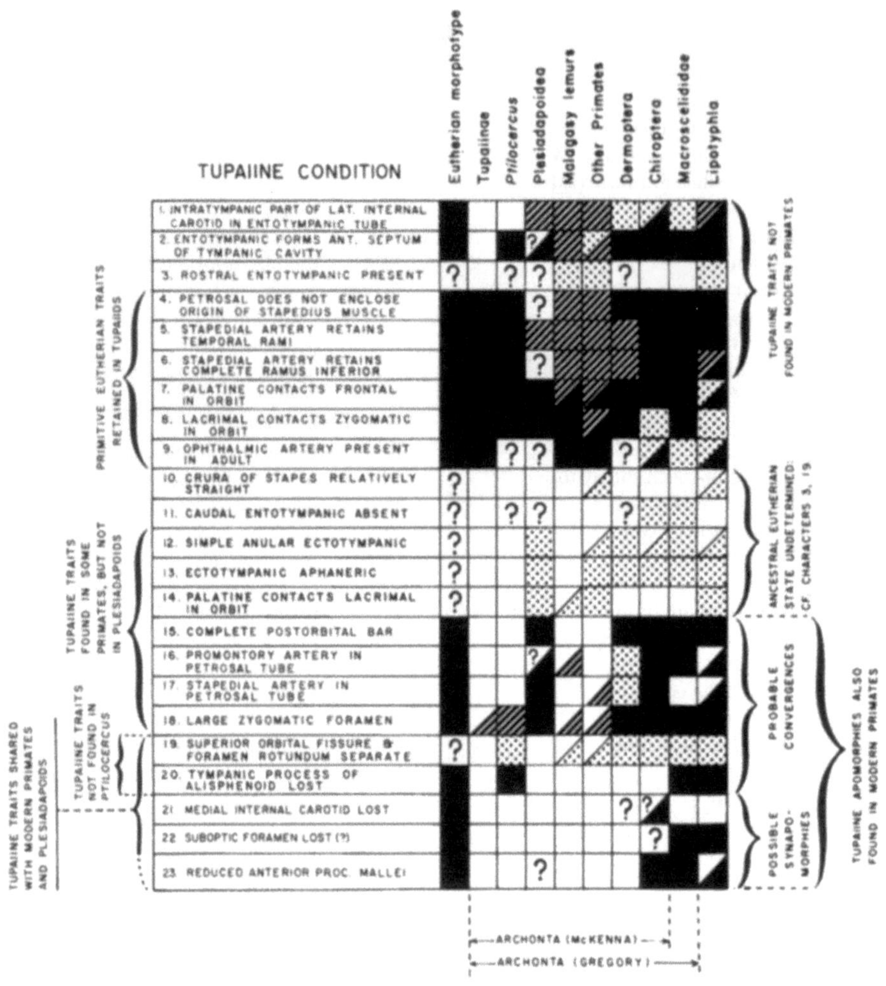

Fig. 14. Provisional character analysis of cranial traits of Tupaiinae and possible sister groups. Shading conventions are as follows:
1. Black = primitive eutherian state
2. White = tupaiine state (where different from 1)
3. Oblique lines = lemurid state (where different from 1 and 2)
4. Dots = other states (differing from 1, 2, and 3)
5. ? = character state unknown or uncertain
Use of two different types of shading in some squares indicates that character states are variable within the relevant group.

tree shrews and primates fall into two categories: (A) those which are not found in plesiadapoids (traits 15–18), and (B) those which are (traits 20–23). Category A may also include traits 12–14, since (whatever their polarity) they are absent in *Plesiadapis* and its allies. Similarly, category B may also include trait 19.

For the four (or more) traits in category A, plesiadapoids either retain the primitive eutherian state or exhibit a derived state different from that shared by tree shrews and lemurs. Therefore, if any of the traits in category A is a tupaiid-primate synapomorphy, then the features of the teeth and auditory region which are thought to link plesiadapoids to modern primates (but are not seen in tupaiids) must be mere convergences. Accepting this conclusion would leave us with little excuse for retaining plesiadapoids in the order Primates. We believe that most students of primate evolution will choose instead to interpret category A traits as convergences between tupaiids and primates.

The five traits which might belong in category B can be interpreted as tupaiid-primate synapomorphies without excluding plesiadapoids. But two of these (traits 19 and 20) are almost certainly convergences, since the derived states (however we interpret morphocline polarity) are not shared by tupaiines and ptilocercines. (Similar reasoning would justify the exclusion of trait 18 from category A.) We are left, then, with three possible tupaiid-primate synapomorphies: (1) loss of the suboptic foramen (which is not certainly absent in all tupaiines nor indeed certainly present in the ancestral Eutheria), (2) absence of the medial internal carotid (a vessel of obscure homologies, absent in all other Insectivora), and (3) reduction of the anterior process of the malleus (one of the commoner trends in eutherian middle-ear evolution). The last trait is unknown in plesiadapoids, and may prove to belong in category A. We conclude that cranial morphology provides no warrant for the thesis that tree shrews and primates share a unique common ancestor.

B. Tree Shrews and Other Mammals

1. Tree Shrews and Lipotyphlans

Cranial morphology also provides no evidence for close ties between tupaiids and lipotyphlans. None of the characters distinctive of at least some lipotyphlans (e.g., tympanic process of the basisphenoid, loss of the ophthalmic artery in the adult, reduction of the zygomatic and palatine) are seen in tupaiids. The lack of even a rudimentary entotympanic in extant Lipotyphla is a conspicuous and important difference from tree shrews. Certain derived states for traits 16, 17, and 23 (Fig. 14) occur in both groups, but some lipotyphlans are persistently primitive in these features, and those which resemble tupaiids therefore probably do so in a convergent manner. Similar reasoning applies to traits 10 and 12, if tupaiids are not persistently primitive in these respects. Trait 11 may eventually prove to be a tupaiid-lipotyphlan synapomorphy, but we doubt it.

2. Tree Shrews and Elephant Shrews

Traits 3, 10, 14, 17, and 21 link tree shrews and elephant shrews, but we cannot tell which state of the first three of these characters is primitive. The stapedial artery's tube is formed endochondrally in tree shrews but periosteally in elephant shrews (MacPhee, 1977b), which suggests convergence. The loss (or relocation) of the medial internal carotid may represent a synapomorphy linking tupaiids, macroscelidids, lipotyphlans, and primates together; or a synapomorphy for two or three of them, convergently developed in the remainder; or a quadruple convergence. We see no reason for choosing one of these alternatives; indeed, we are not quite certain that the tupaiid condition is derived for Eutheria. The construction of the bulla in macroscelidids and tupaiids is strikingly different, suggesting that their last common ancestor lay close to the basal eutherians.

3. Archonta

McKenna (1975) has recently revived Gregory's superorder Archonta, which includes tree shrews, bats, colugos, and primates. Neither McKenna nor Gregory have provided a workable diagnosis of this group, although Szalay (1977) has identified several possible synapomorphies of the archontan ankle joint. Szalay remarks that Gregory's concept of Archonta "was so genuinely based on phylogenetic reasoning that later students, using a more patristic approach to phyletics, often found it impossible to accept." While we share Szalay's esteem for Gregory's work, Gregory's (1910) discussion of the affinities of the groups comprising the Archonta does not seem to us to employ any systematic distinction between primitive and derived characters. For example, among the "considerations" indicating "the derivation of the order (Primates) from . . . Insectivores resembling in many ways *Tupaia* and *Ptilocercus*," Gregory lists the ancestral primate's tritubercular upper molars, tuberculo-sectorial lower molars, internal carotid course which "corresponds with that in the Tupaiidae," and several other traits that Gregory himself regarded as primitive for Eutheria. It remained for Le Gros Clark (1925) to put speculations about tupaiid affinities on a genuinely phylogenetic footing.

At present, the published evidence that the Archonta constitute a monophyletic group is restricted to a few traits shared by possible early dermopterans and plesiadapoids (Rose and Simons, 1977), presumed synapomorphies of the ankle joint in primates, tupaiids, and dermopterans (Szalay, 1977; Szalay and Drawhorn, this volume), and immunological characteristics of serum albumins and transferrins which link primates to tree shrews and colugos (Cronin and Sarich, 1975, 1978, this volume; cf. Dene *et al.*, this volume). No cogent evidence has so far been presented on the affinities of bats to other archontans. Current defenders of the phyletic coherence of Archonta have had little to say about cranial morphology, although Szalay (1977) attempted to deal with the marked differences in archontan ear regions by raising the possibility that the petrosal bulla of primates may be a secondarily fused entotympanic, or the entotympanic of tree shrews a secondarily detached petrosal lamina. As shown above, both possibilities seem incompatible with embryological evidence. McKenna (1975) mentions that tree shrews differ from other archontans in "possessing an ossified entotympanic bulla that only fuses to the petrosal late in ontogeny." In fact, there is firm evidence for the existence of both rostral and caudal entotympanics in bats (Kampen, 1905, 1915; Fawcett, 1919; Klaauw, 1922; Frick, 1954); the same situation is said to obtain in colugos (Klaauw, 1930), though the evidence is less conclusive. Among archontans, it is the primates rather than the tree shrews which are exceptional in bulla construction.

Examination of juvenile *Cynocephalus* skulls reveals a configuration quite unlike anything found in tupaiids or inferred for the ancestral primates. The bulla is formed almost wholly by the ectotympanic, and does not incorporate a rostral tympanic process of the petrosal. The basicranial circulation appears to be essentially dog-like. No canals or foramina for any part of the internal carotid's lateral branch can be distinguished. A canal which runs from the jugular foramen forward between the petrosal and basioccipital and emerges endocranially medial to foramen ovale may transmit a medial internal carotid, or an inferior petrosal sinus—or both, as in the dog (Miller *et al.*, 1964). The postglenoid foramen is tiny or absent. The petrosquamous sinus runs across the floor of the subarcuate fossa and is drained via a sigmoid sinus through the jugular foramen. However, much of the petrosquamous drainage leaves the skull via a peculiar canal beginning in the depths of the subarcuate fossa and taking a sinuous course between the tympanic cavity and the anterior wall of the voluminous mastoid air sinus to emerge on the posteromedial aspect of the bulla. We suspect that this canal is homologous with mastoid foramina in other eutherians (e.g., *Rattus*). The total vascular pattern is not easily derivable from that seen in any extant insectivorans or primates.

Although we have not investigated bat morphology, what we know about primates, tree shrews, and colugos indicates that the last common ancestor of these three groups had a largely unossified bulla and retained complete lateral and medial internal carotid systems. With respect to cranial anatomy, the archontan morphotype is therefore indistinguishable from the eutherian morphotype. The evidence we have personally examined supports Simpson's (1945) assessment of Archonta as "almost surely an unnatural group."

Cranial morphology, then, does not support any of the hypotheses that have been advanced concerning the phyletic affinities of tree shrews. The features of the tupaiid head that are not simply retentions from the ancestral eutherians appear to be unique to tupaiids or convergences with other groups of placental mammals. Of course, this does not prove that tupaiids have no special affinities to other Eutheria. If, for example, Szalay (1977) has correctly identified apomorphies of the tarsus that demonstrate tupaiid affinities to colugos and prosimians, all we can say is that the last common ancestor of these animals retained a cranial morphology that was remarkably primitive, in view of that ancestor's postcranial specializations. But this is perfectly possible. We remain reluctant to accept this conclusion only because we are not yet persuaded that the resemblances identified by Szalay are apomorphous, and because phyletic reconstructions based exclusively on the morphology of a single organ have generally proved unreliable in the past. Our dismally conservative conclusion is that for the present tupaiids are still best regarded as members of a paraphyletic order Insectivora or as a separate eutherian order *incertae sedis*. The two assertions are virtually equivalent.

ACKNOWLEDGMENTS

Grateful thanks are due to Drs. H. W. Setzer and R. W. Thorington, Jr. (U.S. National Museum), Prof. Dr. R. Hassler and Dr. H. Stephan (Max-Planck-Institut für Hirnforschung), Prof. Dr. W. B. Spatz (Universität Freiburg im Breisgau), Prof. Dr. D. Starck (J.-W.-Goethe-Universität), and Dr. D. G. Steele (University of Alberta) for permission to study collections in their care and for other favors, and to Mr. S. Saylor (University of Winnipeg) for photographic assistance. The senior author thanks Dr. Thorington and his colleague Dr. T. L. Strickler for allowing him to dissect the ear regions of skulls under their care. The senior author's research is made possible by an award (5–K04–HD00083–O2) from the U.S. National Institutes of Health; that of the junior author was conducted while he was a Canada Council Doctoral Fellow. We thank K. Brown, Dr. R. F. Kay, Dr. W. L. Hylander, and J. R. Wible for their comments on the manuscript, and Dr. M. Archer for information on the metatherian ear region.

6. *References*

Archer, M. 1976. The basicranial region of marsupicarnivores (Marsupialia), interrelationships of carnivorous marsupials, and affinities of the insectivorous marsupial peramelids. *Zool. J. Linn. Soc. London* 59:217–322.

Beer, G.R. de. 1937. *The Development of the Vertebrate Skull*. Clarendon Press, Oxford.

Boulay, G.H. du, and Verity, P.M. 1973. *The Cranial Arteries of Mammals*. Heinemann Medical Books, London.

Buchanan, G. D., and Arata, A.A. 1969. Cranial vasculature of a neotropical fruit-eating bat, *Artibeus lituratus*. *Anat. Anz.* 124:314–325.

Bugge, J. 1972. The cephalic arterial system in the insectivores and the primates with special reference to the Macroscelidoidea and Tupaioidea and the insectivore-primate boundary. *Z. Anat. Entwickl.* 135:279–300.

Bugge, J. 1974. The cephalic arterial system in insectivores, primates, rodents and lagomorphs, with special reference to the systematic classification. *Acta anat.* 87 (supplement 62):1–160.

Butler, P.M. 1948. On the evolution of the skull and teeth in the Erinaceidae, with special reference to fossil material in the British Museum. *Proc. Zool. Soc. London* 118:446–500.

Butler, P.M. 1956. The skull of *Ictops* and the classification of the Insectivora. *Proc. Zool. Soc. London*, 126:453–481.

Carlsson, A. 1922. Über die Tupaiidae und ihre Beziehungen zu den Insectivora und den Prosimiae. *Acta zool. (Stockholm)* 3:227–270.

Carlsson, A. 1926. Über den Bau des Dasyuroides byrnei und seine Beziehungen zu den übrigen Dasyuridae. *Acta zool. (Stockholm)* 7:249–275.

Cartmill, M. 1972. Arboreal adaptations and the origin of the order Primates, pp. 97–122. *In* R.H. Tuttle (ed.). *The Functional and Evolutionary Biology of Primates*. Aldine-Atherton, Chicago.

Cartmill, M. 1975. Strepsirhine basicranial structures and the affinities of the Cheirogaleidae, pp. 313–354. *In* W.P. Luckett and F.S. Szalay (eds.). *Phylogeny of the Primates*. Plenum Press, New York.

Cartmill, M. 1978. The orbital mosaic in prosimians and the use of variable traits in systematics. *Folia primatol.* 30:89–114.

Cartmill, M., and Kay, R.F. 1978. Cranio-dental morphology, tarsier affinities, and primate suborders, pp. 205–214. *In* D.J. Chivers and K.A. Joysey (eds.). *Recent Advances in Primatology*, Vol. 3. Academic Press, London.

Cronin, J.E., and Sarich, V.M. 1975. Molecular systematics of the New World monkeys. *J. Hum. Evol.* 4:357–375.

Cronin, J.E., and Sarich, V.M. 1978. Primate higher taxa: The molecular view, pp. 287–289. *In* D.J. Chivers and K.A. Joysey (eds.). *Recent Advances in Primatology*, Vol. 3. Academic Press, London.

Davis, D.D., and Story, H.E. 1943. The carotid circulation in the domestic cat. *Zool. Ser. Field Mus. Nat. Hist.* 28:1–47.

Doran, A.H.G. 1878. Morphology of the mammalian *Ossicula auditus*. *Trans. Linn. Soc. London (Zool.)* 1:371–497.

Fawcett, E. 1919. The primordial cranium of *Miniopterus schreibersi* at the 17 millimetre total length stage. *J. Anat.* 53:315–350.

Fleischer, G. 1973. Studien am Skelett des Gehörorgans der Säugetiere, einschliesslich des Menschen. *Säugetierk. Mitt.* 21:131–239.

Frick, H. 1954. Die Entwicklung und Morphologie des Chondrokraniums von *Myotis* Kaup (Beitrag zur Kenntnis der Morphologie des Chiropterkraniums III). George Thieme Verlag, Stuttgart.

Gingerich, P.D. 1976. Cranial anatomy and evolution of early Tertiary Plesiadapidae (Mammalia, Primates). *Univ. Michigan Pap. Paleontol.* 15:1–141.

Gregory, W.K. 1910. The orders of mammals. *Bull. Amer. Mus. Nat. Hist.* 27:3–524.

Gregory, W.K. 1920. On the structure and relationships of *Notharctus*, an American Eocene primate. *Mem. Amer. Mus. Nat. Hist.* 3:49–243.

Grosser, O. 1901. Zur Anatomie und Entwickelungsgeschichte des Gefässsystemes der Chiropteren. *Anat. Hefte* 17:203–424.

Haines, R.W. 1950. The interorbital septum in mammals. *Zool. J. Linn. Soc. London* 41:585–607.

Henson, O.W., Jr. 1961. Some morphological and functional aspects of certain structures of the middle ear in bats and insectivores. *Univ. Kansas Sci. Bull.* 42:151–255.

Henson, O.W., Jr. 1974. Comparative anatomy of the middle ear. *Handbk. Sensory Physiol.* 5 (1):40–110.

Hogben, L.T. 1919. The progressive reduction of the jugal in the Mammalia. *Proc. Zool. Soc. London* 1919:71–78.

Hunt, R.M. 1974. The auditory bulla in Carnivora: An anatomical basis for reappraisal of carnivore evolution. *J. Morph.* 143:21–76.

Jones, F.W. 1929. *Man's Place among the Mammals*. E. Arnold, London.

Jones, F.W. 1949. The study of a generalized marsupial (*Dasycercus cristicauda* Krefft). *Trans. Zool. Soc. London* 26:409–501.

Jones, F.W., and Lambert, V.F. 1939. The occurrence of the lemurine form of the ectotympanic in a primitive marsupial. *J. Anat.* 74:72–75.

Kay, R.F., and Cartmill, M. 1977. Cranial morphology and adaptations of *Palaechthon nacimienti* and other Paromomyidae (Plesiadapoidea, ?Primates), with a description of a new genus and species. *J. Hum. Evol.* 6:19–53.

Kampen, P.N. van. 1905. Die Tympanalgegend des Säugetierschädels. *Geg. morph. Jb.* 34:321–722.

Kampen, P.N. van. 1915. De phylogenie van het entotympanicum. *Tijds. Nederl. Dierk. Ver.*, 2e ser., 14:xxiv.

Kermack, K.A., and Keilan-Jaworowska, Z. 1971. Therian and non-therian mammals. *Zool. J. Linn. Soc. London* 50 (suppl. 1):103–116.

Klaauw, C.J. van der. 1922. Über die Entwickelung des Entotympanicums. *Tijds. Nederl. Dierk. Ver.* 18:135–174.

Klaauw, C.J. van der. 1929. On the development of the tympanic region of the skull in the Macroscelididae. *Proc. Zool. Soc. London* 1929:491–560.

Klaauw, C.J. van der. 1930. On mammalian auditory bullae showing an indistinctly complex structure in the adult. *J. Mamm.* 11:55–60.

Klaauw, C.J. van der. 1931. The auditory bulla in fossil mammals, with a general introduction to this region of the skull. *Bull. Amer. Mus. Nat. Hist.* 62:1–352.

Kollman, M. and Papin, L. 1925. Études sur les Lémuriens. Anatomie comparée des fosses nasales et de leur annexes. *Arch. Morphol.* 22:1-60.

Le Gros Clark, W.E. 1925. On the skull of *Tupaia*. *Proc. Zool. Soc. Lond.* 1925:559–567.

Le Gros Clark, W.E. 1926. On the anatomy of the pen-tailed tree-shrew *(Ptilocercus lowii)*. *Proc. Zool. Soc. London* 1926:1179–1309.

Le Gros Clark, W.E. 1934. On the skull structure of *Pronycticebus gaudryi*. *Proc. Zool. Soc. London* 1934:19–27.

Le Gros Clark, W.E. 1959. *The Antecedents of Man. An Introduction to the Evolution of the Primates* (1st edition). Edinburgh University Press, Edinburgh.

Le Gros Clark, W.E. 1971. *The Antecedents of Man. An Introduction to the Evolution of the Primates* (3rd edition). Edinburgh University Press, Edinburgh.

Lyon, M.W., Jr. 1913. Treeshrews: An account of the mammalian family Tupaiidae. *Proc. U.S. Natl. Mus.* 45:1–188.

MacPhee, R.D.E. 1977a. Ontogeny of the ectotympanic-petrosal plate relationship in strepsirhine prosimians. *Folia primat.* 27:245–283.

MacPhee, R.D.E. 1977b. Auditory regions of strepsirhine primates, tree shrews, elephant shrews, and lipotyphlous insectivores: An ontogenetic perspective on character analysis. Ph.D. dissertation, Univ. of Alberta.

MacPhee, R D.E. 1979. Entotympanics, ontogeny and primates. *Folia primat.* 31:23–47.

Matthew, W.D. 1909. The Carnivora and Insectivora of the Bridger Basin, Middle Eocene. *Mem. Amer. Mus. Nat. Hist.* 9:289–567.

McDowell, S.B. 1958. The Greater Antillean insectivores. *Bull. Amer. Mus. Nat. Hist.* 115:113–214.

McKenna, M.C. 1963. The early Tertiary primates and their ancestors. *Proc. XVI Internatl. Congr. Zool.* 4:69–74.

McKenna, M.C. 1966. Paleontology and the origin of the primates. *Folia primat.* 4:1–25.

McKenna, M.C. 1975. Toward a phylogenetic classification of the Mammalia, pp. 21–46. *In* W.P. Luckett and F.S. Szalay (eds.). *Phylogeny of the Primates.* Plenum Press, New York.

Miller, M.E., Christensen, G.C., and Evans, H.E. 1964. *Anatomy of the Dog*. W.B. Saunders, Philadelphia.

Mivart, St. G. 1864. Notes on the crania and dentition of the Lemuridae. *Proc. Zool. Soc. London* 1864:611–648.

Mivart, St. G. 1867/1868. Notes on the osteology of the Insectivora. *J. Anat. Physiol.* 1:281–312, 2:117–154.

Muller, J. 1934. The orbitotemporal region of the skull of the Mammalia. *Arch. Neerl. Zool.* 1:118–259.

Padget, D.H. 1948. Development of the cranial arteries in the human embryo. *Contrib. Embryol. Carnegie Inst.* 32:205–261.

Paulli, S. 1900. Über die Pneumaticität des Schädels bei den Saugethieren. Eine morphologische Studie. III. Über die Morphologie des Siebbeins und die der Pneumaticität bei ben Insectivoren, Hyracoideen, Chiropteren, Carnivoren, Pinnipedien, Edentaten, Rodentiern, Prosimiern und Primaten, nebst einer zusammenfassenden Übersicht über die Morphologie des Siebbeins und die der Pneumaticität des Schädels bei den Säugethieren. *Geg. morph. Jb.* 28:483–564.

Piveteau, J. 1957. *Traité de Paléontologie*. VII. *Primates: Paléontologie Humaine*. Masson et Cie., Paris.

Reinbach, W. 1952. Zur Entwicklung des Primordialcraniums von *Dasypus novemcinctus* Linné *(Tatusia novemcincta* Lesson) II. *Z. Morphol. Anthropol.* 45:1–72.

Rich, T.H.V., and Rich, P.V. 1971. *Brachyerix*, a Miocene hedgehog from western North America, with a description of the tympanic regions of *Paraechinus* and *Podogymnura*. *Amer. Mus. Novit.* 2477:1–58.

Romer, A.S. 1956. *Osteology of the Reptiles*. Univ. Chicago Press, Chicago.

Rose, K.D., and Simons, E.L. 1977. Dental function in the Plagiomenidae: Origin and relationships of the mammalian order Dermoptera. *Contrib. Mus. Paleont. Univ. Mich.* 24:221–236.

Roux, G.H. 1947. The cranial development of certain Ethiopian "insectivores" and its bearing on the mutual affinities of the group. *Acta Zool. (Stockholm)* 28:165–397.

Russell, D.E. 1964. Les Mammifères paléocènes d'Europe. *Mém. Mus. natl. d'Hist. nat. (Paris)*, sér. C, 13:1–321.

Saban, R. 1956/57. Les affinites du genre *Tupaia* Raffles 1821, d'après les caractères morphologiques de la tête osseuse. *Ann. Paléont.:* 42:169–224, 43:1–44.

Saban, R. 1963. Contribution à l'étude de l'os temporal des Primates. Description chez l'Homme et les Prosimiens. Anatomie comparée et phylogénie. *Mém. Mus. natl. d'Hist. nat. (Paris)*, sér. A, 29:1–378.

Segall, W. 1969. The auditory ossicles (malleus, incus) and their relationships to the tympanic: in marsupials. *Acta anat.* 73:176–191.

Segall, W. 1970. Morphological parallelisms of the bulla and auditory ossicles in some insectivores and marsupials. *Fieldiana (Zool.)* 51:169–205.

Shindo, T. 1914. Zur vergleichenden Anatomie der arteriellen Kopfgefässe der Reptilien. *Anat. Hefte* 51:267–356.

Simons, E.L. 1974. Notes on early Tertiary prosimians, pp. 415–433. *In:* R.D. Martin, G.A. Doyle, and A.C. Walker (eds.). *Prosimian Biology*. Duckworth, London.

Simons, E.L., and Russell, D.E. 1960. Notes on the cranial anatomy of *Necrolemur*. *Breviora Mus. Comp. Zool.* 127:1–14.

Simpson, G.G. 1945. The principles of classification and a classification of mammals. *Bull. Amer. Mus. Nat. Hist.* 85:1–350.

Spatz, W.B. 1964. Beitrag zur Kenntnis der Ontogenese des Cranium von *Tupaia glis* (Diard 1820). *Geg. morph. Jb.* 106:321–416.

Spatz, W.B. 1966. Zur Ontogenese der Bulla tympanica von *Tupaia glis* Diard 1820 (Prosimiae, Tupaiiformes). *Folia primatol.* 4:26–50.

Starck, D. 1967. Le crâne des Mammifères, pp. 405–549, 1095–1102. *In* P. Grassé (ed.). *Traité de Zoologie*, 16 (1). Masson et Cie., Paris.

Starck, D. 1975. The development of the chondrocranium in primates, pp. 127–155. *In* W.P. Luckett and F.S. Szalay (eds.). *Phylogeny of the Primates*. Plenum Press, New York.

Stehlin, H.G. 1912. Die Säugetiere des schweitzerischen Eocaens, VII (1): *Adapis*. *Abhandl. schweiz. paläont. Ges.* 38:1165–1298. (Not successively paginated)

Steuerwald, E.A. 1969. Review of the phylogenetic position of the tree shrew (*Tupaia glis* Diard), with new observations on the arteria carotis interna. Ph.D. dissertation, Michigan State Univ.

Story, H.E. 1951. The carotid arteries in the Procyonidae. *Fieldiana (Zool.)* 32:477–557.

Szalay, F.S. 1972. Cranial morphology of the early Tertiary *Phenacolemur* and its bearing on primate phylogeny. *Amer. J. Phys. Anthrop.* 36:59–76.

Szalay, F.S. 1975. Phylogeny of primate higher taxa: The basicranial evidence, pp. 91–125. *In* W.P. Luckett and F.S. Szalay (eds.). *Phylogeny of the Primates*. Plenum Press, New York.

Szalay, F.S. 1976. Systematics of the Omomyidae (Tarsiiformes, Primates): Taxonomy, phylogeny, and adaptations. *Bull. Amer. Mus. Nat. Hist.* 156:157–450.

Szalay, F.S. 1977. Phylogenetic relationships and a classification of the eutherian Mammalia, pp. 315–374. *In* M.K. Hecht, P.C. Goody, and B.M. Hecht (eds.). *Major Patterns in Vertebrate Evolution*. Plenum Press, New York.

Tandler, J. 1899. Zur vergleichenden Anatomie der Kopfarterien bei den Mammalia. *Denksch. kais. Akad. Wiss. (Wien), math.-nat. Klasse* 67:677–784.

Tattersall, I. 1973. Cranial anatomy of Archaeolemurinae (Lemuroidea, Primates). *Anthrop. Pap. Amer. Mus. Nat. Hist.* 52:1–110.

Van Valen, L. 1965. Tree shrews, primates, and fossils. *Evolution* 19:137–151.

Wassif, K. 1948. Studies on the structure of the auditory ossicles and tympanic bone in Egyptian Insectivora, Chiroptera and Rodentia. *Bull. Fac. Sci. Fouad I Univ.* 27:177–213.

Webster, D.B., and Webster, M. 1975. Auditory systems of Heteromyidae: Functional morphology and evolution of the middle ear. *J. Morph.* 146:343–376.

Werner, C.F. 1960. Das Mittel- und Innenohr, pp. 1–40. *In* H. Hofer, A.H. Schultz, and D. Starck (eds.). *Primatologia* 2(1), Lfg. 5. Karger, Basel.

Wilson, J.A. 1966. A new primate from the earliest Oligocene, West Texas. Preliminary report. *Folia primatol.* 4:227–248.

Wilson, J.A., and Szalay, F.S. 1972. New paromomyid primate from middle Paleocene beds, Kutz Canyon area, San Juan Basin, New Mexico. *Amer. Mus. Novitates* 2499:1–18.

Woehrmann-Repenning, A., and Meinel, W. 1977. A comparative study on the nasal fossae of *Tupaia glis* and four insectivores. *Anat. Anz.* 142:331–345.

7. *Note Added in Proof*

Since this chapter was submitted, our conjecture that the medial and lateral internal carotids are one and the same has been conclusively demonstrated by Presley (1979; *Acta anat.* 103: 238–244).

Evolution and Diversification of the Archonta in an Arboreal Milieu

4

FREDERICK S. SZALAY and GERRELL DRAWHORN

1. Introduction

This paper attempts to analyze three distinct but closely interrelated biological problems pertinent to the animals included within the Archonta. The tree shrews and relatives (Scandentia), the colugos (Dermoptera), and the archaic primates will be examined, primarily their tarsus, to pose and answer questions about their evolutionary history.

First, as an inseparable investigation related to the last issue, the meaning of the characters employed are analyzed in terms of mechanical (functional) concepts often employed in the study of articulating bones.

Second, the meaning of the various form-function complexes in terms of their biological roles (*sensu* Bock and von Wahlert, 1965) is discussed.

Third, the problem of phylogenetic relationships of these animals is discussed, both as part of the larger context of Eutheria, and their hypothesized relative recency of relationships as well as transformational histories among themselves.

It is important to state at the very beginning that this paper will not

---

FREDERICK S. SZALAY • Department of Anthropology, Hunter College, CUNY, 695 Park Avenue, New York, New York 10021. GERRELL DRAWHORN • Department of Anthropology, University of California, Davis, Davis, California 95616

examine in detail the arguments for or against character polarities proposed in the past for cranial, dental, and soft anatomical features which have bearing on tupaiid and archontan affinities. Admirable reviews along these lines have been provided by Butler (this volume), Luckett (this volume), Cartmill and MacPhee (this volume), and by Novacek (this volume). We have, in the past and also more recently, considered the cranial and dental evidence in both great detail and in considerable breadth in terms of taxonomic comparisons, and our skepticism matched that of the other contributors in regards to the usefulness of these features to place the tupaiids in a phylogenetic perspective. We want to state, however, and emphasize below, that as a methodological conviction of ours, neither primitive nor autapomorphous features of taxa, even if there are several of these, have bearing on the significance and usefulness of even a single complex and unique shared derived feature. When such a derived feature displays a character cline and it is understood both in terms of mechanical function and associated biological roles, it is the character complex to be considered, and the primitive and autapomorphous features are to be ignored.

2. *Materials*

The postcranial material for this study was provided by the American Museum of Natural History, The Chicago Natural History Museum, the United States National Museum, the University of California Museum of Paleontology, the University of Minnesota, Princeton University, and the Museum National d'Histoire Naturelle, Paris. The Cretaceous and Paleogene materials are from collections from the following sites: Bug Creek Anthills, Swain Quarry, Saddle, Walbeck, Mason Pocket, Cernay, Cedar Point, Shotgun, Bitter Creek Station (V-70246, V-70214), Four Mile Quarries (East Alheit, Timberlake, Sand), Dormaal, and Phosphorites of Quercy (classical locs.). For a summary of chronological and geographical relationships of these and other Cenozoic localities, see Szalay and Delson (1979).

3. *Assumptions and Methods*

In formulating our hypotheses of function, biological roles, and subsequently morphocline polarities, a number of assumptions were made. We emphasize here (see also Szalay, 1977a) that the phylogenetic hypothesis arrived at is a consequence of our understanding of morphocline polarities. We largely derived our views on polarities from a mechanical appraisal of the characters we weighted to be important based on criteria of complexity, importance to the organism, and uniqueness.

1. Within the confines of species, specific genotype selection favors those form-function complexes which allow the most important movements, at the required speeds, in a most energy efficient manner which the organism needs to survive in a specific environment.

2. Differences in the relative size and character of articulations of joint surfaces, given limitations imposed by heritage, are related to changing emphasis in mobility-stability.

3. Form and function differences of joint surfaces and muscle and tendon insertions, reflect, at least in part, differences in mobility, stability, and habitual (facultative) contact of homologous character complexes.

4. Any mated system of joints (i.e. articulating bones) has inherent limitations as to mobility and stability.

5. Joint surfaces like all features which evolved are compromises, their form and function resulting from both conflicting selection pressures, as well as the further compromise necessarily superimposed by the constraints of heritage.

6. From either the astragalus or calcaneum, the movements of the astragalus can be recognized which occur during either inversion, level orientation, or eversion. Similarly the movements between the astragalus and navicular, and the calcaneum and cuboid, can be recognized from the astragalus and calcaneum, respectively.

7. In the close-packed position of joints, "the two bones are inseparable by traction and the articular surfaces are most fully in contact. Habitual motions are toward or away from the close-packed position and both involve swings combined with rotation. Ligaments already tight from the swing are twisted to tautness by the rotation, pressing the joint surfaces together and making the bones functionally one. Close-packed surfaces entail maximum terminal congruence while full congruence does not necessarily entail close-pack although it usually does. The close-packing part of the male surfaces is its broadest and often flattest portion. In full congruence the broadest male area contacts the largest possible area on the female surface. The close-packed position is assumed only for special efforts because it is dangerous to the bones which are less resistant to torsion than to compressive or tensile forces. When possible, however, weight is carried habitually near enough to the close-packed position to secure a wide distribution of the weight over the joint surfaces" (Ziemer, 1978, p. 128).

8. Rules of systematic practices have no law-like meaning outside the rules governing the biological evolution of features.

9. Hypotheses of homology are rooted in recognition of morphological and/or functional similarity.

10. Hypotheses of morphocline polarity may be based on (a) the appraisal of the fossil record, (b) the functional and ontogenetic evaluation of the character states of the morphocline, and (c) the heritage aspect of a character complex which channelled the changes seen in the character states, or which persists in a "relic" or "vertical" manner in various character states.

11. The use of "out-group" comparison is to be used only with reservation, as the distribution of groups and characters can be biased by incomplete knowledge of taxonomic diversity or by autapomorphies of sister groups.

12. The varying degree to which the astragalo-calcaneal and calcaneo-astragalar facets do not overlap is significant. These discrepancies, characteristic of groups, denote the degree to which translation takes place in addition to rotation. In the joints between the heel and ankle bone the type and extent of the translation found is indicative of the range of movement found in the foot (Szalay and Decker, 1974). In addition, the close-packed position, when the surfaces are closely fitting, reveals the proximal position of the joint in which weight is carried for optimal stress distribution.

These ideas are partly present either implicitly or explicitly in the works of Gregory (1920), Schaeffer (1947), MacConaill (1953a, b, 1958, 1973), Yalden (1970, 1972), Preuschoft (1971), Jenkins (1974), Szalay and Decker (1974), Szalay (1977b), Ziemer (1978), Fleagle (1977), and numerous others.

The expected and readily levelled criticism, from students who consider dental and cranial form the ultimate arbiter, towards a study which employs bone and joint morphology and the concepts of homologous mechanical function, is the challenge that such similarities are due to convergence of function or habitus. Similarities, of course, in any form-function complex can be convergent if not the result of common inheritance. As in any other system, similarity has to be carefully evaluated to determine its cause. By not using a functional approach, however, such an analysis is hamstrung, as the very questions which may help determine a homology-convergence dilemma are not posed. Inquiry into the mechanical meaning of form, even in the simplest terms, gives additional meaning and therefore a greater perspective to the character states compared.

A case in point is the simple mechanical necessity of inverting the foot when placing it on a branch for body support and propulsion. A survey of a whole host of similar sized eutherians (insectivorans, rodents, carnivorans, and primates) shows that rotation of the foot takes place by either sliding, rolling, or spinning the cuboid on the calcaneal articulation. While this occurs, the navicular, rigidly attached to the cuboid, displaces the astragalus by transmitting forces to the astragalar head. As a result, the astragalus undergoes translation and rotation, and is displaced to various degrees. It usually reaches a close-packed position with the navicular and the calcaneum at the end of this displacement, i.e. when the foot is fully inverted. The relative size and shape of the astragalocalcaneal and calcaneoastragalar facets show this condition from which the differences in the above mechanical and behavioral circumstances can be assessed. Now, in spite of the similarity of the mechanical solution in a host of eutherians, there is no mistaking of taxon specific morphologies, in spite of any convergence in habitus and subsequently mechanics. As cross-lophs, mesostyles, hypocones, reduction of trigonids, molarization of premolars, papillation of enamel, and other dental features can be most often easily gleaned to be either homologies or convergences, so can the morphology of tarsal bones be decided, in spite of clearly present similarities, either to be homologies or convergences. Any hypothesis pertaining to similarity, of course, is just that, a hypothesis. But consideration of function gives a biological insight into a character complex which will not result from a mere computer-like recording of form. It is perhaps bold to say, but seems plausible, that since the beginning of the science of form the improvements in data gathering, namely the recognition of features, were brought about by insight and perspectives which focused on some aspects of ontogeny and mechanics (physiology). Systematists who claim to use purely form, professing no interest in development and function, are most assuredly often employing characters from the communal pool of literature and personal communication which have their roots in biologically dominated studies. This, we believe, is (or should be) the difference between a biologist's (i.e. an evolutionist's) view of form and that of the abstractions of spherical geometry alone.

It is doubtful that analysis of form alone, without functional consideration, will result in understanding the path of selection. Even a minimal appraisal of functional systems, however, can provide answers by helping to recognize key biological roles performed, food exploited, or substrates utilized in a particular manner. The classical notion of using only "taxonomic" characters to explore evolutionary history has its major drawback in the unexplored relationship of most taxonomic features to function.

The astragalus and the calcaneum act both in tandem and as links in a chain of elements that provide support and propulsion for the animal (i.e. the tibia-fibula proximally and the navicular-cuboid distally). A biomechanical analysis of an isolated unit in this chain is limited, and reference to more complete tarsal complexes of living forms or to articulated or associated material of extinct animals is a requisite for a clear evaluation of potential ranges of mobility in isolated material. Ranges of potential movement in some late Cretaceous eutherians and in *Plesiadapis* have been approximated by Szalay and Decker (1974).

Although characters (and phylogenetic hypotheses based on them) can be and are usually based mostly on form, the biomechanical/functional information available in structural analysis greatly aids polarity determination. Features which are important to the functioning of the individual are more likely to be monitored by natural selection. The acquisition of such characters carries a greater biological weight than those that may drift into or out of the population, unseen by natural selection.

The functional capabilities of any given feature may be projected to a constellation of biological roles. Biological role (*sensu* Bock and von Wahlert, 1965) is one of the most difficult levels to approach in the hierarchical framework utilized in this study. The complex interactions of any feature with the environment may result in a single or several roles. In contrast to the biological roles that a structure may play, any given biological role may be fulfilled by several different types of structure in different organisms. The appearance of these 'paradaptations' (see Bock, 1977), in which either the form or function in the features of two organisms differ yet the same biological role is adequately fulfilled, suggests that convergent characters (though superficially similar) may never realize an identical morphological relationship.

Terminology used in this paper is based upon that of Szalay and Decker (1974). For explanation and abbreviations of the anatomy of the astragalocalcaneal complex see Fig. 1.

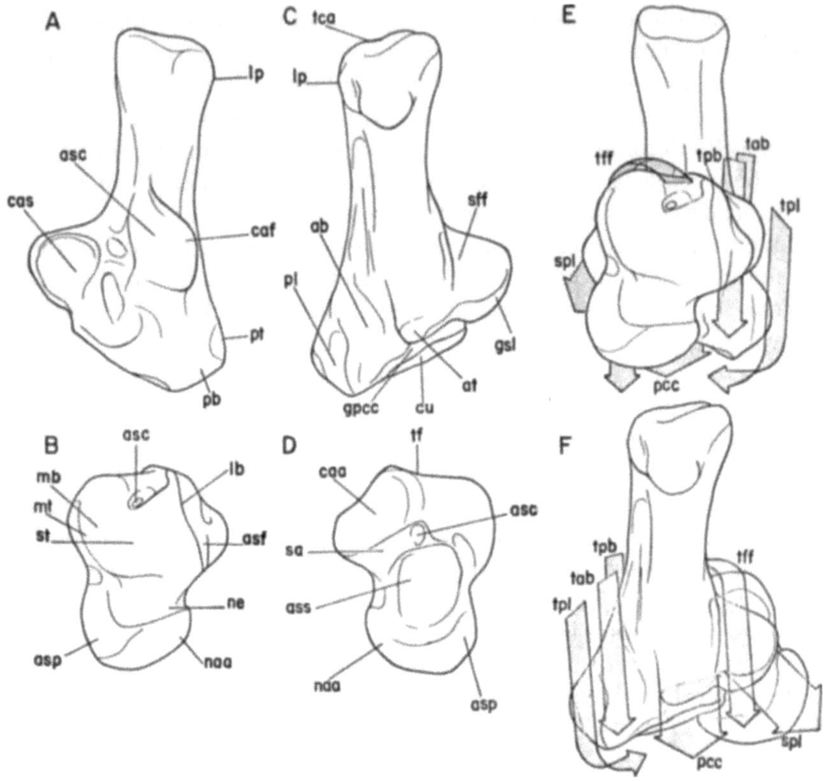

Fig. 1. The calcaneum (A,C) astragalus (B,D), and the astragalocalcaneal complex (E,F) of *Protungulatum* to demonstrate the terminology used in this paper. Dorsal (A,B,E) and plantar (C,D,F) views. Abbreviations on the calcaneum: ab, groove for M. abductor digiti quinti?; asc, astragalocalcaneal facet; at, anterior plantar tubercle; caf, calcaneal fibular facet; cas, calcaneal sustentacular facet; cu, cuboid facet; gpcc, groove for plantar calcaneocuboid ligament; gsl, sustentacular groove for attachment of "spring" ligament; lp, lateral process of tuber calcanei; pb, groove for tendon of M. peroneus brevis; pl, groove for tendon of M. peroneus longus; pt, peroneal tubercle; sff, sustentacular groove for tendon of M. Flexor (digitorum) fibularis; tca, tuber of the calcaneum. Abbreviations on the astragalus: asc, astragalar canal; asf, astragalar fibular facet; asp, astragalar "spring" ligament facet; ass, astragalar sustentacular facet; caa, calcaneoastragalar facet; lb, lateral border (crest) of trochlea; naa, naviculoastragalar facet; ne,neck of astragalus; sa, sulcus astragali; st, superior tibial facet of trochlea; tf, trochlear groove for tendon of M. flexor (digitorum) fibularis. Abbreviations on the astragalocalcaneal complex: pcc, (long) plantar calcaneocuboid ligament; spl, "spring" ligament (calcaneonavicular ligament); tab, tendon or fleshy fibers of abductor digiti quinti or other muscles; tff, tendon M. flexor (digitorum) fibularis; tpb, tendon M. peroneus brevis; tpl, tendon M. peroneus longus.

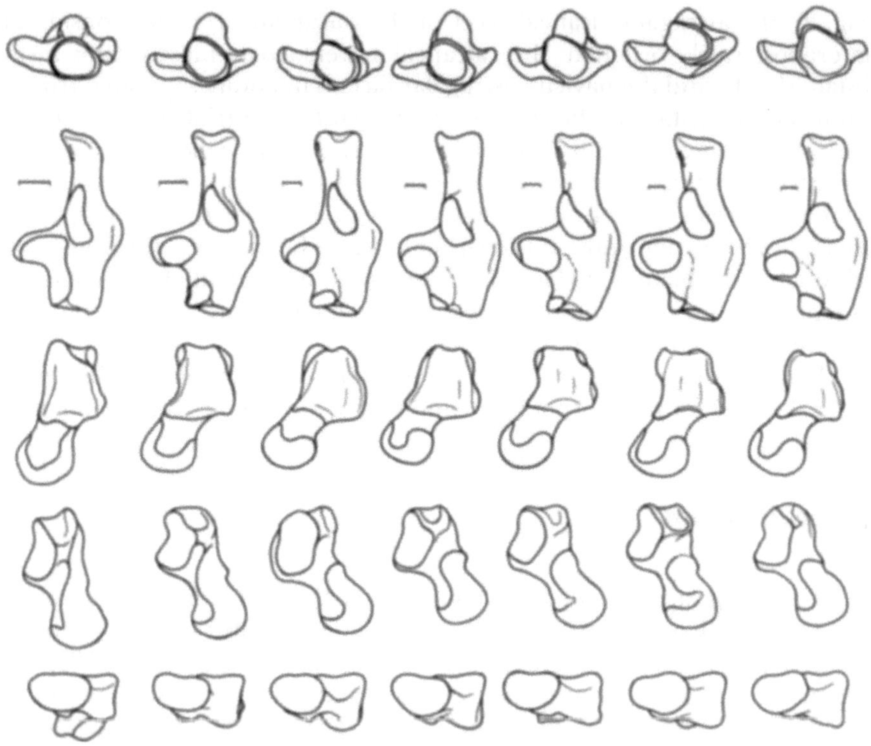

Fig. 2. Comparative morphology of selected features of the calcaneum and astragalus of seven tupaiid species, from left to right: *Ptilocercus lowii* (CNHM 76855), *Tupaia minor* (AMNH 103438), *Tupaia nicobarica* (USNM 111782), *Tupaia glis modesta* (AMNH 26659), *Lyonogale tana* (AMNH 102516), *Urogale everetti* (AMNH 203293), and *Tupaia longipes* (USNM 199162). The sequence, from left to right, is believed to represent the range of habitual substrate preference, from *Ptilocercus* and *Tupaia minor*, which are believed to be almost completely arboreal, to the genera on the right which are increasingly terrestrial. The views, from top to bottom, represent distal and dorsal views of the calcaneum and the dorsal, plantar, and distal views of the astragalus, respectively.

The study of the tarsus in living tupaiids has been crucial in assessing the meaning of subtle differences in morphology of the bone and articular surfaces. The morphological and mechanical changes in the character clines of available living tupaiid astragali and calcanea have shown astonishing correlation with the known (albeit poorly) degree of arboreality or terrestriality of these species (Fig. 2). Unfortunately, as the natural history of the tupaiids is relatively poorly known, species specific posture, locomotor details, or specific habitat preferences are not known with any degree of reliability.

We hypothesize that the common ancestor of known tupaiids was fully arboreal, as *Ptilocercus*. Our reasons are rooted in the morphological analysis of the tarsus. Even in such highly terrestrial forms as *Urogale* or *Lyonogale tana* the foot has joint orientations which are usually associated with and facilitate inversion, rather than correlate with habitual level orientations. For example

(Fig. 2), the astragalocalcaneal facet of the calcaneum remains considerably anteroposteriorly oriented, the calcaneal sustentacular facet retains a small distal "island", and the naviculoastragalar facet is not quite separated from the astragalar sustentacular facet even in the most terrestrial of our series. In contrast, in *Ptilocercus* and *Tupaia minor*, the two forms reported to be almost exclusively arboreal, the astragalocalcaneal facet is long and helical, almost completely anteroposteriorly oriented; the calcaneal sustentacular facet is broadly present not only on the sustentaculum but also on the body of the calcaneum; the plantar surface of the astragalus is dominated by a continuous sustentacular facet beginning almost on the medial and distal extreme and continuous with the naviculoastragalar facet.

It appears to us, then, that the astragalocalcaneal complex of *Ptilocercus* is an ideal model for the morphotype condition of these bones for the Tupaiidae.

4. The Archonta

The living tupaiids, the fossil mixodectids, the extant and extinct colugos, primates, and bats are provisionally accepted as members of the cohort Archonta (see especially Gregory, 1910; McKenna, 1975; Szalay, 1977b). The rationale for aligning the tupaiids, colugos, and primates, discussed in detail by Szalay (1977b), is based upon the shared, derived features of the astragalocalcaneal complex. Many of the Paleogene taxa have been referred to the cohort as a result of their dental relationships (e.g. the Plagiomenidae) with extant taxa, and additional corroborative evidence for such and some new associations is provided by the large sample of fossil tarsals analyzed in this paper. The position of the Chiroptera to this group is yet uncertain. No clear shared, derived states can be discerned between the tarsals of bats and any other taxa examined in this study. The chiropteran ankle has undergone a major functional transformation, probably associated with the extreme reorientation of the femoral-acetabular articulation. The hindlimb is in a permanently super-abducted condition, with the knee joint having been rotated 180° from the primitive eutherian expression. These characteristics are already found in *Icaronycteris index* (Jepsen, 1970), an early Eocene chiropt. The rationale for including the Chiroptera in this group will be discussed later.

The original suggestion of the relationship between colugos, bats, and primates can be traced back to Linnaeus. In 1758, Linnaeus included *Vespertilio* (bats) as one of the four genera of primates. He referred the colugos to another of the orders, designating it *Lemur volans*. The first mention of a tupaiid-primate similarity (but not relationship) is that of Huxley (1872). Gregory (1910) was the first to formally present a hypothesis indicating a phyletic relationship between the tupaiids, primates, dermopterans, and chiropterans. He included the Macroscelididae within this constellation, noting a special relationship between the tupaiids and the elephant shrews by placing them together in the order Menotyphla. More recently, Szalay's (1977b) examination of the tarsal complex of a host of eutherians resulted in a gestalt of characters which suggest a common source for taxa considered archontan (excluding macroscelidids).

The following complex of synapomorphic character states distinguishes the archontan tarsus from the known Cretaceous eutherians. In the astragalus

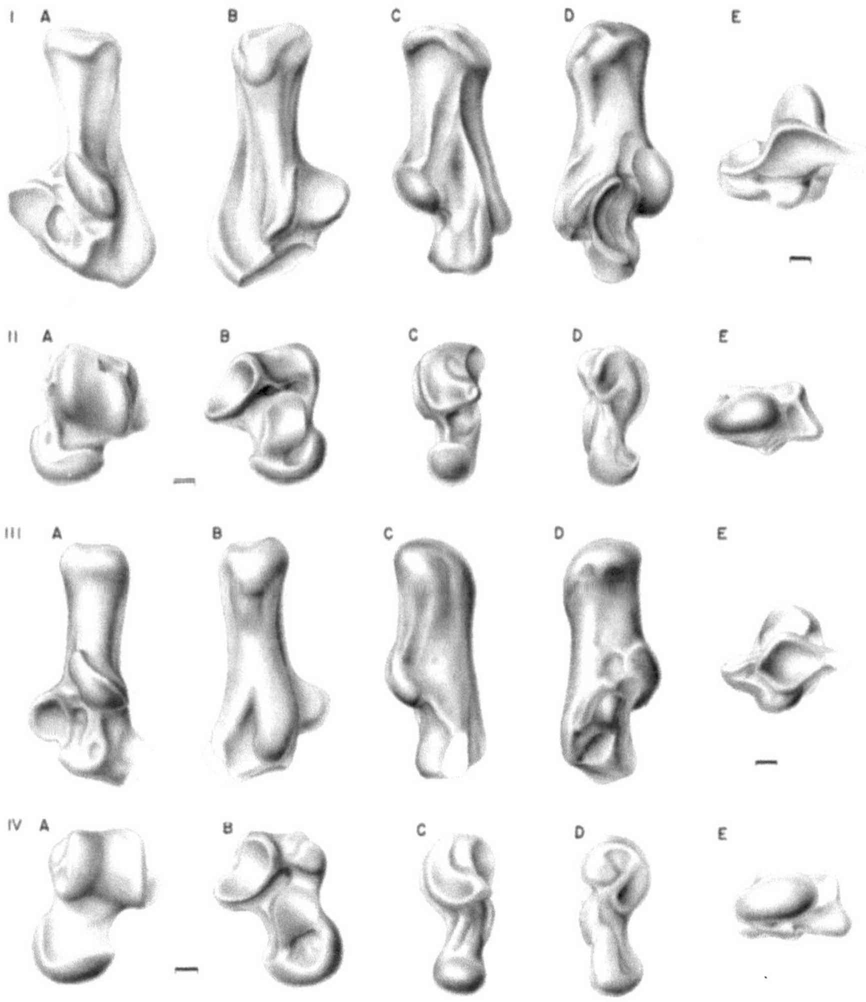

Fig. 3. *Protungulatum donnae*, Maestrichtian late Cretaceous, Montana. I. Left calcaneum, based on UMVP 1823 from Bug Creek Anthills. II. Left astragalus, based on UMVP 1914 and 1915, from Bug Creek Anthills. *Procerberus formicarum*, Maestrichtian late Cretaceous, Montana. III. Left calcaneum, based on UMVP 1825 from Bug Creek Anthills. IV. Left astragalus, based on UMVP 1806, from Bug Creek Anthills. Views from left to right, are dorsal (A), plantar (B), lateral (C), medial (D), and distal (E). Reconstructions with broken lines. Scales represent one mm.

there is a proximo-distally enlongated tibial trochlea, extending onto the neck; presence of a high and sharply crested lateral margin and rounded medial margin of the tibial trochlea; the tibial trochlea is aligned obliquely to the longitudinal axis of the astragalus; trochlear length/astragalar length ratio greater than .60; astragalar sustentacular facet confluent with naviculoastragalar facet, primitively probably both laterally and medially (as a result of the medial expansion of the latter); calcaneoastragalar facet aligned obliquely relative to the longitudinal axis of the astragalus (primitively a transverse orientation). In the calcaneum, the astragalocalcaneal articulation is posteriorly attenuated, forming a relatively small angle to the longitudinal axis of the calcaneum (allowing both rotation and translation); the peroneal process (or tubercle) is large in surface area, reaching a posterolateral position level to the posterior astragalocalcaneal facet; the calcaneal sustentacular facet probably not limited to the sustentacular process but continues onto the body of the calcaneum; the cuboidocalcaneal facet is rounded and concave and aligned transversely to the longitudinal axis of the calcaneum (primitively, it faces medially); the sustentacular process is prominent and distinct from the distal margin of the calcaneum; the calcaneal fibular facet is very reduced or lost; the groove for the plantar calcaneo-cuboid ligament is all but lost from the distal surface of the calcaneum (perhaps shifting to a position on the lateral

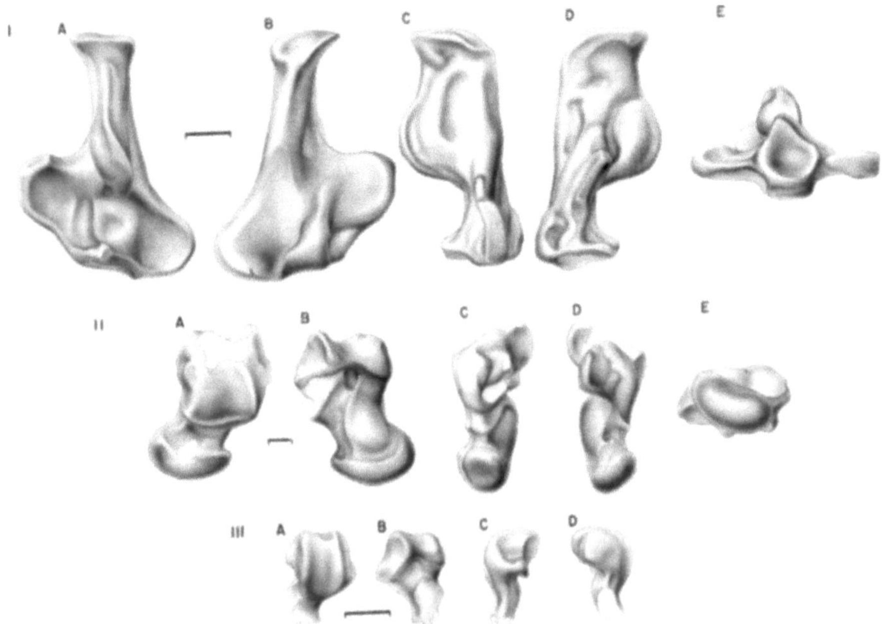

Fig. 4. I. Archontan, Tiffanian Paleocene, Colorado. Left calcaneum, based on AMNH 89530, from Mason Pocket. II. Primate, Tiffanian Paleocene, Wyoming. Left astragalus, based on uncatalogued Princeton University specimen from Cedar Pt. locality. III. Primate, Torrejonian Paleocene, Wyoming. Left astragalus, based on AMNH 92003, from Bison Basin Saddle locality. Views from left to right, are dorsal (A), plantar (B), lateral (C), medial (D), and distal (E). Reconstruction with broken lines. Scales represent one mm.

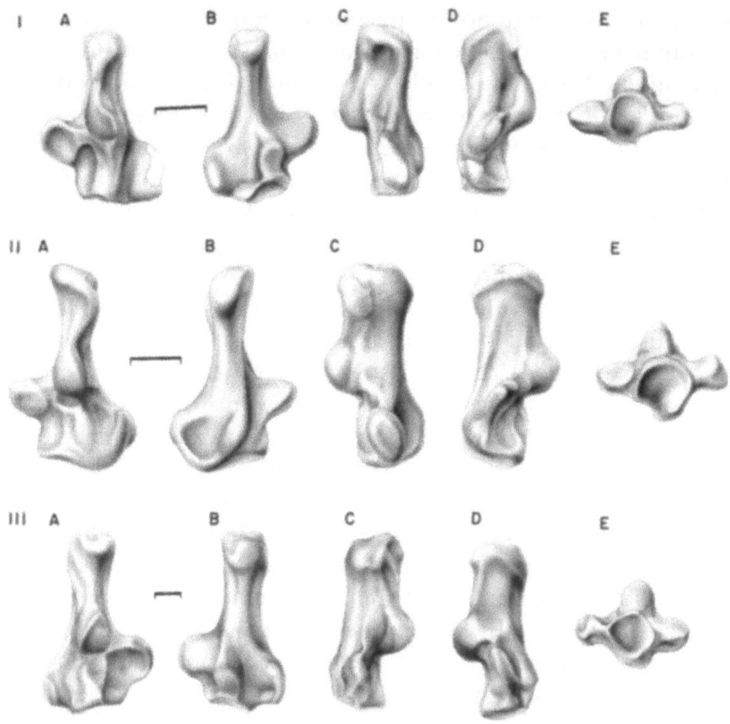

Fig. 5. I. Primate, Torrejonian Paleocene, Wyoming. Left calcaneum, based on AMNH 92002 and 92007, from Bison Basin Saddle locality. This form may represent the same species as the astragalus in Fig. 4, III, possibly a small paromomyid. II. Archontan or primate, Torrejonian-Tiffanian Paleocene, Wyoming. Left calcaneum, based on an unnumbered specimen of the Princeton collection, from the Shotgun Anthill locality. III. Primate, Tiffanian Paleocene, Colorado. Right calcaneum, based on AMNH 89527, from Mason Pocket. Views from left to right, are dorsal (A), plantar (B), lateral (C), medial (D), and distal (E). Reconstruction with broken lines. Scales represent one mm.

surface of the anterior plantar tubercle). This complex and long list of features does not appear on any one known archontan.

The archontan morphotype calcaneum is difficult to reconstruct as the differences from a form like *Protungulatum* (Fig. 3) were probably not great. Specimens like AMNH 89530 (Fig. 4), a Shotgun locality calcaneum (Fig. 5), or AMNH 88815 (Fig. 8) all represent structurally an archontan morphological complex on which either the primate, tupaiid, or dermopteran features are not clearly discernable. All of these specimens have a nearly rounded cuboid facet with a slight depression, an astragalocalcaneal facet more aligned with the calcaneum long axis than in *Protungulatum*, and a well developed sustentacular facet on the body of the calcaneum.

The astragalus in the extant *Ptilocercus* and a number of Paleocene probable plesiadapiforms would appear to be representative of the condition expected in the archontan morphotype. In these the expected archontan elements are present but a number of primitive eutherian or perhaps condylarth elements are also retained. Some specimens retain a pronounced convex astragalar sustentacular facet, an astragalar canal that appears on the superior trochlea, a fibular shelf, and a relatively short (vis-a-vis the body and head) astragalar neck. Of course, other plesiadapiform primates have lost these primitive conditions or have developed unique character states that make them poor candidates to be representative of the basal archontan. For example, *P. gidleyi* AMNH 17379 (Fig. 9) from Mason Pocket has exaggerated development in the lateral margin of the tibial trochlea and the lateral edge of the calcaneoastragalar articulation.

We can infer several functional transformations that must have occurred in the transitional period between archontans and non-archontans. While Szalay and Decker (1974) referred to this modification in reference to the primates, most of the same generalities are now known to be applicable to the ancestry of the Archonta (but see some important exceptions below).

It has been generally suspected, on grounds of the dental and basicranial evidence, that the primates derived from an ancestor that could be classed as either a condylarth or erinaceotan insectivoran. That viewpoint is not contradicted by the known postcranial evidence. Aside from a purely descriptive discussion of the transformation of the archontan lineage from a *Protungulatum*-like ancestor, it is worthwhile, indeed our goal, to determine the reasons for this change. What selection pressures acted upon some populations of Cretaceous eutherians which "directed" morphological change along a path that would result in the forms we know existed in the Paleocene and Eocene? It is towards these selective forces which were acting on the function of the pes that we will now direct our discussion.

The primitive eutherians had a limb morphology adapted to habitual plantigrade locomotion upon a flat or slightly inclined substrate. Haines (1958) arrived at this conclusion after his studies of Paleocene condylarth mani described by Matthew (1937). A biomechanical analysis of the Cretaceous eutherian tarsals by Szalay and Decker (1974) not only corroborated Haines' conclusion, but recognized some diagnostic features in the astragalocalcaneal complex indicative of either habitual inversion or eversion. In that same study, it was implied that while the overall phenetic similarity between the plesiadapiform primates and the *Protungulatum*-like astragalocalcaneal complex is considerable, the functional differences between the two groups are separated by a strongly defined morphological boundary. The addition of the material in this investigation corroborates the hypothesis that the Archonta were derived from a form with *Protungulatum*-like pes, although the functional (and supposedly adaptational) lacuna remains unbridged.

As described by Szalay and Decker (1974), the *Protungulatum* pes was primarily adapted for maintaining support and propulsion upon a stable and horizontal surface. Moderate eversion and a limited degree of inversion are quite adequate for a foot adjusting to terrestrial conditions. While terrestriality was probably the predominant habitat for Cretaceous placentals, it is not to be misconstrued that there was no diversity in locomotory patterns present. Some differences are known. While the plantar surface of the foot was somewhat more laterally oriented in *Protungulatum,* the foot of *Procerberus* and *Cimolestes* was relatively more medially aligned. The calcaneofibular articulation of the primitive eutherian pes restricted mobility of the upper ankle joint. This restrictive element was reduced when fibular contact was increasingly limited

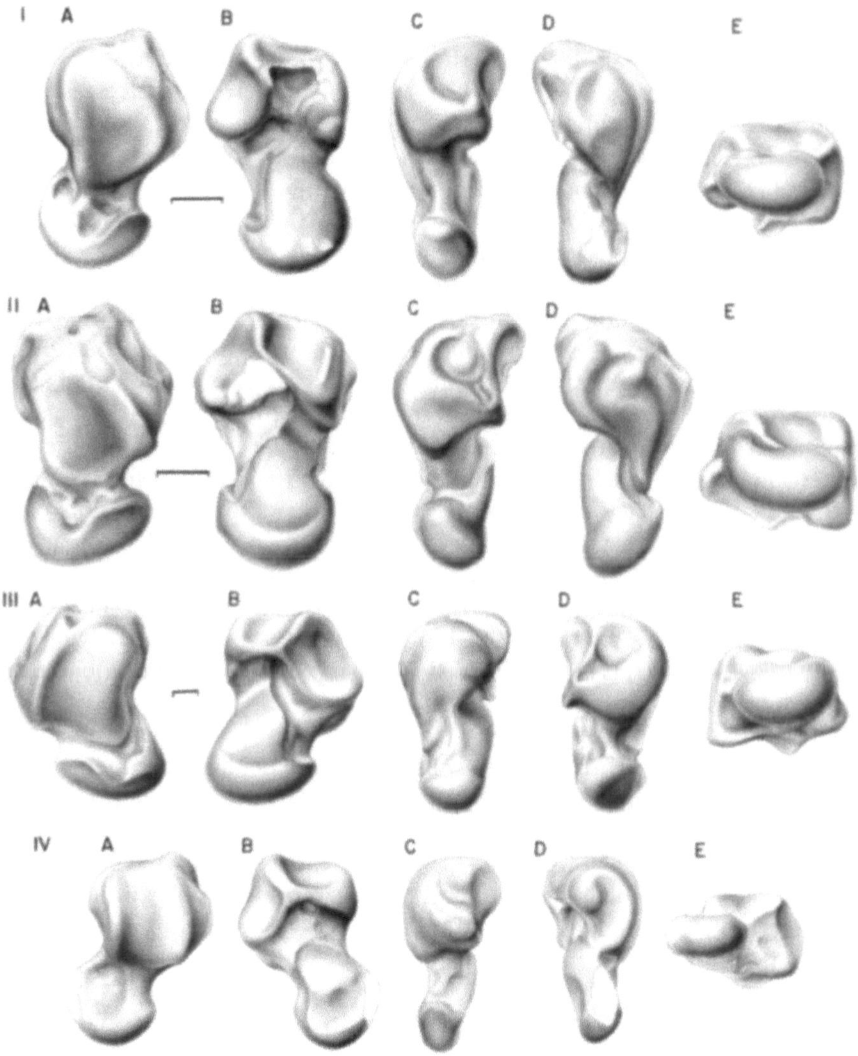

Fig. 6. I. Primate, Torrejonian Paleocene, Wyoming. Left astragalus, based on AMNH 92004, from Swain Quarry. II. Primate, Torrejonian-Tiffanian Paleocene, Wyoming. Left astragalus, based on an uncatalogued Princeton specimen, "Shotgun 1," from the Shotgun Anthill locality. III. *Plesiadapis*, cf. *P. rex*, Tiffanian Paleocene, Wyoming. Right astragalus, based on uncatalogued Princeton University specimen, from Cedar Point locality. IV. *Saxonella crepaturae*, Thanetian Paleocene, Germany. Left astragalus, based on an uncatalogued specimen of Halle University, from the Walbeck locality. Views from left to right, are dorsal (A), plantar (B), lateral (C), medial (D), and distal (E). Reconstruction with broken lines. Scales represent one mm.

to the astragalus, as in the Archonta, or, independently, in Leptictimorpha and derivatives (Szalay, 1977b). In the primitive condition, the fibular facet of the calcaneum acted to oppose the resultant forces transmitted from the lower ankle joint.

The transversely broad tibial trochlea of *Protungulatum* reflects the orientation of the lower ankle joint. The latter allows only simple rotational movements along a path lateral to the long axis of the calcaneum. In order to maintain the close association of the tarsals to receive the stresses during contact with the substrate, mobility of the lower ankle is restricted.

In contrast to the inferred primitive eutherian morphotype, quasi-represented by *Protungulatum,* the archontan calcaneum bears a helical or posteriorly expanded astragalocalcaneal articular surface. The translational component is increased posteriorly and a helical movement is added when the astragalus is pushed dorsal and posteriorad while the foot is inverted and the navicular is moved along with the cuboid, as it is usually locked to the latter by closely fitting facets. Under ordinary circumstances the ability to move the astragalus along the described path increases the range of inversion of a given tarsus. An increase in the stability, accompanying the increased fore and aft mobility of the lower ankle joint in the Archonta, is realized by the increased postero-medial surface in the astragalocalcaneal articulation. A similar increase in the homologous region and its effect upon the potential degree of rotation during inversion and eversion have been found in a number of arboreal eutherians (e.g. , *Potos flavus,* the kinkajou, the viverrid *Arctictis binturong,* and many sciurid rodents). The shift of the calcaneoastragalar facet in the Archonta to a position more oblique to the longitudinal axis rather than lying more transverse to the axis is clearly related to the more longitudinal orientation of its "mate" upon the calcaneum.

Also associated with the increased capacity to invert the pes is a change in the location of the groove for the plantar calcaneocuboid ligament and the anterior plantar tubercle. The migration of the groove from a broad, distal, and transverse situation to a condition where it lies medially indicates a shift in the calcaneocuboid joint axis. Propulsive forces would have been more efficient along a medial axis rather than one more laterally directed. In an inverted position a medially aligned calcaneocuboid ligament might also act to more efficiently brace the pes when it encountered irregular substrate supports.

The calcaneocuboid articulation of primitive archontans is quasi-circular, when observed distally, and slightly concave. It would appear to be laid down upon a short, broad-based cone with an axis oriented slightly anteriorly, laterally, and dorsally with respect to the upper joint axis. *Protungulatum* has a calcaneocuboid facet that would have been projected upon a long and broad-based cone with an axis anterior and lateral to the articulation. In comparison to the primitive eutherians, the calcaneocuboid joint of the Archonta allows a considerably greater degree of medio-lateral rotation along the long axis of the foot.

The head of the astragalus in the Archonta also suggests a greater potential for inversion than that observed in the late Cretaceous eutherians. In *Protungulatum, Procerberus,* and *Cimolestes* the lateral portion of the astragalar head is more robust than the medial half. In archontans this state is transformed, the medial head being equally or more robust than the lateral portion of the naviculoastragalar facet. As a result of this medial expansion in articular surface in the Archonta, the astragalar sustentacular articulation is absorbed by the margin of the articular facet for the "spring" (calcaneonavicular) ligament. The archontan foot, measured by this increased medial expansion of the head of the astragalus, was more capable of coping with stresses in inversion than was the pes of the other Cretaceous eutherians.

In summary, the pedal morphology of *Protungulatum,* which appears to represent a structural stage ancestral to that of the Archonta, was adapted to substrates with horizontal to slightly inclined slopes and which were stable underfoot. A diversity of substrates could accommodate this type of functional system, from open stubble grassland to forest-floor litter environments.

The conditions of a forest floor frequently present obstacles to small terrestrial animals. These barriers (fallen logs, tree trunks, roots, dead tree branches fallen to the ground, etc.) must be often circumvented for more preferred substrate. Adaptation to habitual level orientation by a pes cannot optimally accommodate pathways that are uneven, discontinuous, and sloping, conditions which would present no problem for a tarsus capable of a greater degree of inversion. It is likely that it was such a forest litter environment in which the terrestrial forebears of the Archonta lived. For a discussion of the floral-animal communities of the Bug Creek Anthills of Montana, Van Valen and Sloan (1977) should be consulted.

It was probably an adaptation to the irregularities present on the forest floor that resulted in the foundation for the increased development of hindfoot inversion, progressively more necessary in scansorial and eventually arboreal forms. Once the potential for locomotion in a modestly para-arboreal stratum (low bushes, brush, tree trunks, logs, etc.) was attained, it would have been exploited, at least on an irregular basis. While the intitial changes may have been locomotory (and this is highly speculative), there may have been shifts in dietary adaptations that followed almost immediately. It is likely that the dental specializations of the ancestral archontan arose as a response to the dietary opportunities made available by initial and subsequent locomotory adaptations related to arboreality (see Szalay, 1968).

A particularly interesting and very likely archontan calcaneum from Mason Pocket is AMNH 89530 (Fig. 4). This form has an exceptionally large peroneal process, and an anteroposteriorly shallow distal end, both features found in the Eocene dermopts. The modest size of the anterior plantar tubercle is also characteristic of the tupaiids, in sharp contrast of this feature to both plesiadapiforms and primitive euprimates. AMNH 89530, unlike dermopts, has no facet for a prominent cuboid pivot.

A left calcaneum from the late Paleocene Shotgun locality (Fig. 5II) is enigmatic as to allocation within the Archonta. Its relatively very large peroneal process, unemarginated distally, in contrast to plesiadapiforms, suggests a form allied to the Wasatchian calcanea allocated to the Dermoptera. Yet is may represent an early archontan offshoot, not dissimilar to the archontan morphotype.

AMNH 88815 from the Wasatchian (Fig. 7II), although broken, is a discernably enigmatic form, again possessing a number of features more primitive than usually found in the subdivision of the Archonta. Although the peroneal process is primate-like, and the form may represent *Phenacolemur,* the plantar margin of the calcaneocuboid facet preserves a remnant of the groove for the plantar calcaneocuboid ligament on the plantar surface of the sustentaculum. There is a groove, albeit a very faint one, for the tendon of M. flexor (digitorum) fibularis. Unlike in other archontans, the body of the calcaneum seems to lack the continuation of the sustentacular facet. The latter condition is so unique that it may be a derived trait. Several astragali from the same locality (Fig. 8) probably belong to the same form as the calcaneum. Lack of a well developed connection between the calcaneoastragalar and naviculoastragalar

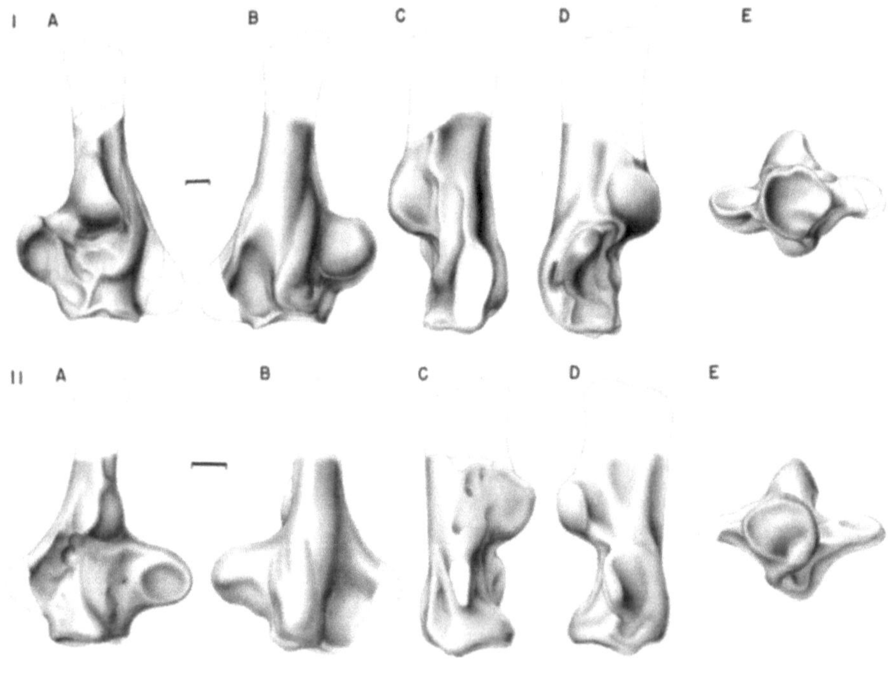

Fig. 7. I. Primate, Wasatchian, Early Eocene, Colorado. Left calcaneum, based on AMNH 88808 and UCMP 80253, from East Alheit Pocket. This form may represent one of the larger *Phenacolemur*. II. Archontan, Wasatchian, Early Eocene, Colorado. Right calcaneum, based on AMNH 88815, from East Alheit Pocket. This form either represents a skeletally modified, smaller *Phenacolemur*, or possibly a tupaiid. Views, from left to right, are dorsal (A), plantar (B), lateral (C), medial (D), and distal (E). Reconstructions with broken lines. Scales represent one mm.

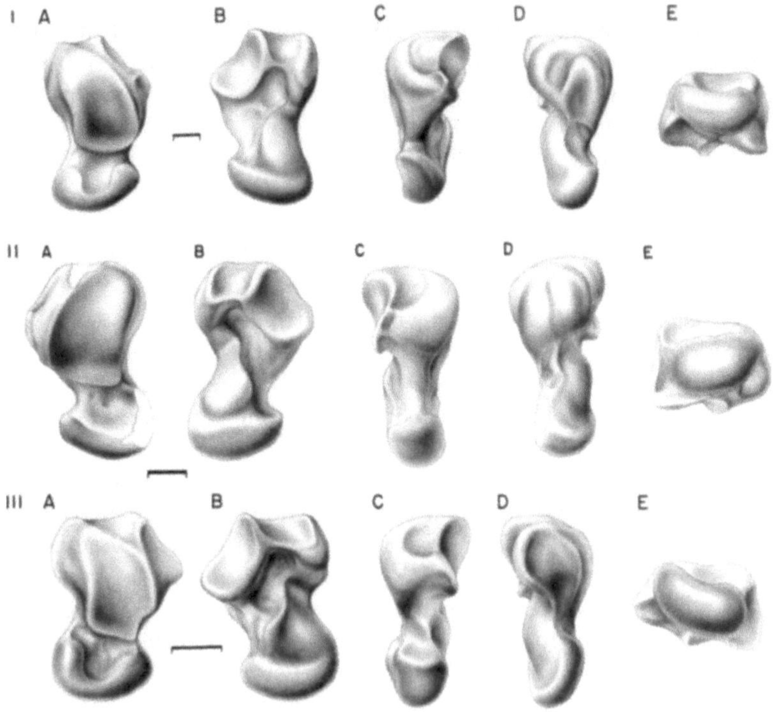

Fig. 8. I. Primate, Wasatchian, Early Eocene, Colorado. Left astragalus, based on AMNH 88809, from East Alheit Pocket. This form may represent the same species as the calcaneum in Fig. 7 I, possibly a larger *Phenacolemur*. II. Archontan or primate, Wasatchian, Early Eocene, Colorado. Right astragalus, based on AMNH 29130, 20131, and 88812, from East Alheit Pocket. This form may represent the same species as the calcaneum of Fig. 7, a smaller *Phenocolemur*, or a tupaiid. III. Primate, Torrejonian Paleocene, Wyoming. Left astragalus based on AMNH 92015, from Bison Basin Saddle locality. Views, from left to right, are dorsal (A), plantar (B), lateral (C), medial (D), and distal (E). Reconstructions with broken lines. Scales represent one mm.

facets, and lack of an extension on the sustentacular facet, support the hypothesis that the two tarsals belong to the same species, considering, in addition, their shared archontan features and appropriate size relationships. Although these bones are noted under the Archonta, they may have belonged to one of the small phenocolemurines that were prevalent during the Wasatchian (see McKenna, 1960; Bown and Rose, 1976; Szalay and Delson, 1979).

A. Scandentia

The branching relationships of the undoubted Archonta remain unclear. The colugos, despite their unusual locomotor pattern and distinctive herbivorous diet, are relatively unstudied. Because of the role the tree shrews have played as a "model" of early primate behavior (and we predict that they will justifiably continue to do so) and the controversy over their systematic relationships, it is appropriate to discuss the tupaiids first. Like the several early dissenters (Evans, 1942; Roux, 1947; Haines, 1955, 1958) from the views of Gregory (1910), Le Gros Clark (1926, 1959), Simpson (1945), and many others, the majority of opinions which denied any special relationship between the Tupaiidae and the primates have been voiced during the past decade (Van Valen, 1965; McKenna, 1966; Campbell, 1966a, b, 1974; Szalay, 1968; Martin, 1968a,b). Campbell (1974) in his review points out that he believes much of the evidence associating the tree shrews with the primates is based upon (1) erroneous analysis, (2) the result of convergent acquisition of characters emanating from the scansorial/arboreal livelihood shared between the two groups, (3) characters that are more broadly distributed in other taxa, and (4) retention of primitive character states.

Like all others who have criticized the earlier assessments of primate-tupaiid relationships, however, Campbell did not offer any alternative hypotheses of tupaiid relationships. His suggestion was to place the tree shrews back into the "Insectivora," a taxon with about as vacuous phylogenetic validity as possible. Further serious general criticisms can be leveled at Campbell's (1974) generally thoughtful paper, as well as all of the other papers in this volume discussing tupaiid relationships. The development of autapomorphies by tupaiids is not evidence for lack of recency of relationships, as it is one of the assumptions of evolutionary theory that non-repetitive changes will occur in characters. In general, the longer two related taxa are separated in time the more genetic differences will accumulate, particularly in characters under different selective forces. It would be completely unexpected that the genetic base of two features undergoing mutational and/or selectionally guided change would remain identical in form after more than 60 million years of independent evolution in their respective lineages. Stability in form, however, is likely only in those features whose biological roles have not been greatly changed from a common ancestry; i.e., they have been and are heavily canalized. We hasten to note that this argument cannot be turned around and used to account for the proposed homologies as mere convergence of the features in question. Differentiation between homology and convergence is still to be decided on the basis of detailed, intricate comparisons, utilizing taxa with divergent habits and hypothesized relationships based on features other than the ones under investigation. Convergent similarities, because they are based upon different genetic frameworks and independent mutational events (stochastically acquired), cannot be identical in structural composition. The features may be functionally similar but phenetic, and, subsequent to analysis, phylogenetic relationships *can* be discernable with relatives with different adaptations (see Szalay, 1977a). Thus, rodents adapted to specific arboreal locomotion show functional elements that suggest the acquisition of arboreality in their limb structure, yet these forms (all rodents) still bear the unmistakable stamp of rodent phylogenesis.

Taking Campbell's (1974) valid criticisms of the approach to primate-tupaiid relationships to heart, extensive comparisons to arboreal marsupials, edentates, rodents, and carnivorans were made (see Szalay, 1977b). Consequently we feel confident that the similarities suggested as synapomorphies between the Tupaiidae, Dermoptera, and Primates are homologous. It may be that many of the so-called "primitive" characters delineated by Campbell will be found to be limited to the Archonta, or, also to the Ungulata. As techniques, phrasing of questions, and perceiving of problems will improve in the study of neuroanatomical and myological structures, special relationships will be found between structures present in the arboreal squirrels, marsupials, and archontans that will help reflect their phylogeny by pointing out true convergences.

Interestingly, immunological and protein sequence data suggest a similar general hypothesis of archontan monophyly to the one we advocate. Sarich and Cronin (1976; also Cronin and Sarich, this volume) noted that detailed immunological assays of albumin and transferrin suggest a tupaiid-primate-dermopteran relationship forming a clade distinct from other placental groups. They also suggest, based upon their results and globin sequence data, that the Tupaiidae branched initially from the remaining archontan stock about 75–80 million years ago. *Cynocephalus* (=*Galeopithecus*), according to them, remains indistinct from the initial primate radiation about 65 million years ago. Dene et al. (1976, this volume), however, basing their observations on a general immunological assay of serum proteins, arrive at slightly different conclusions. They confirm a special cladistic relationship among the Archonta, but suggest that the colugos branched off first, with tupaiids branching from the primates at a later point. The primate radiation is designated as distinct from these prior divergences.

As earlier established by Le Gros Clark (1926), *Ptilocercus* probably represents a generally more primitive taxon of its family than any of the tupaiines. This general statement, of course, must be carefully evaluated character by character. Based upon the sum of such comparisons, however, we believe the generalization holds true. The dentition is perhaps most primitive among living tree shrews (but see Butler, this volume). The astragalocalcaneal complex of *Ptilocercus* also fits into this pattern (see Fig. 9). It retains a large number of primitive archontan characteristics that are reduced, transformed, or lost in the tupaiines. The following archontan character complex is retained, however, in the astragalus: a long, narrow tibial trochlea; sharp lateral crest and rounded medial margin of the tibial trochlea; trochlea extends onto the neck of the astragalus; a relatively large trochlear length/astragalar length ratio; an astragalosustentacular facet confluent with the naviculoastragalar facet; and an obliquely aligned calcaneoastragalar articulation. The following archontan character complex is retained in the calcaneum: a longitudinally oriented posterior astragalocalcaneal articulation; a reduced, although still prominent peroneal tubercle, the posterior margin of which lies even with the astragalocalcaneal facet; a calcaneal sustentacular facet that extends onto the body of the calcaneum; a sustentacular process that is distinct from the distal margin of the calcaneum; and a cuboid facet that is concave and rounded.

Despite these primitive archontan characteristics the following derived conditions of the astragalus are shared among the Tupaiidae: the concave calcaneoastragalar facet is aligned almost exactly along the long axis of the

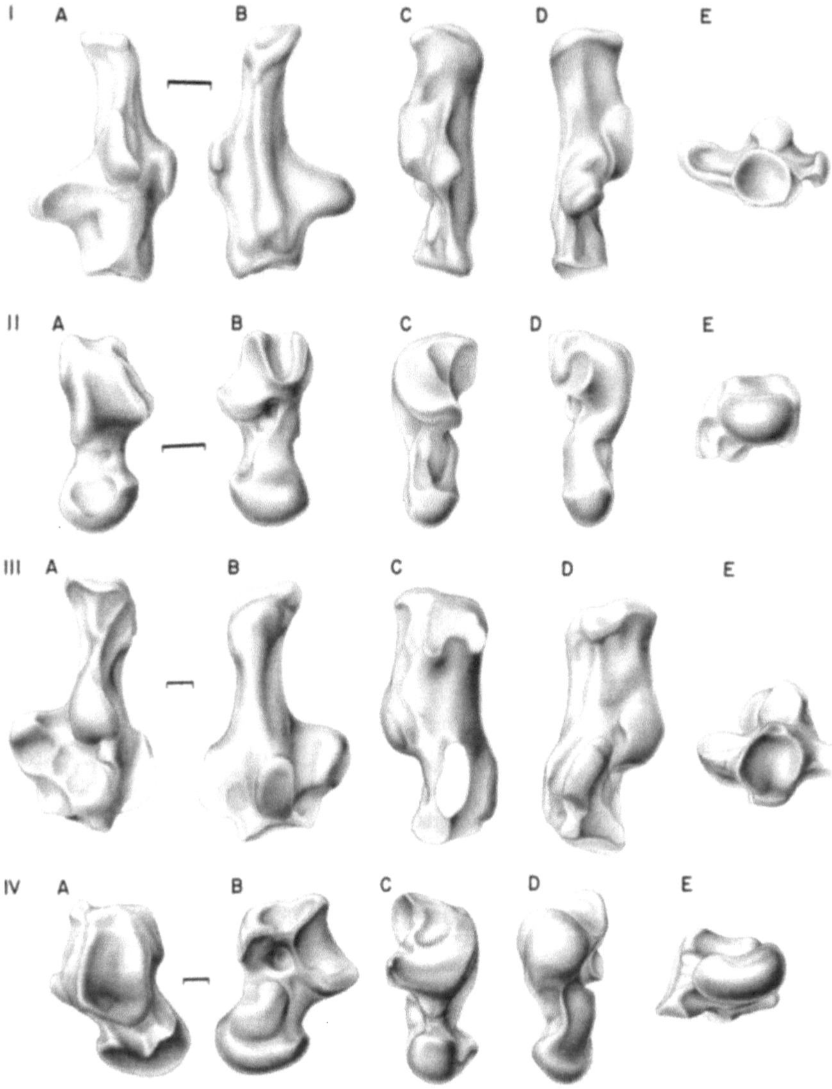

Fig. 9. *Ptilocercus lowii*. Recent, North Borneo. I, left calcaneum; II, left astragalus, both based on CNHM 76855. *Plesiadapis gidleyi*, Torrejonian Paleocene, Wyoming. III, left calcaneum; IV, right astragalus, both based on AMNH 17379, from Gidley Quarry, associated dentition and skeleton. Views from left to right, are dorsal (A), plantar (B), lateral (C), medial (D), and distal (E). Scales represent one mm.

astragalus; sulcus astragalus does not approach the trochlear groove for the tendon of M. flexor (digitorum) fibularis; groove for the tendon of M. flexor fibularis is aligned parallel to the long axis of the astragalus and located upon a ventrally projecting medial body; lateral and medial margins of the trochlea are posteriorly equidistant; the length of the astragalus is relatively greater in contrast to the squat bones in the Paleogene archontans. The calcaneum has lost the primitively large peroneal process found in early dermopts and archaic primates. The anterior plantar tubercle, greatly reduced, has receded more distally on the calcaneum than it probably was on the ancestral archontan.

As noted above, some structural variation of the astragalocalcaneal complex within the Tupaiidae tends to reflect their reported habitual locomotory repertoire. The calcaneum of *Ptilocercus,* generally accepted as the most arboreal member of the order, has an extremely elongated astragalocalcaneal articulation which forms a narrow angle with the long axis of the calcaneum (Fig. 9). In the tupaiines, in general, this articular facet is more transversely aligned. The arc of rotation of the tupaiine astragalus appears to have increased and translation is somewhat less extreme than in *Ptilocercus*. A well-defined morphocline is apparent in this feature, clearly shown in Fig. 2. It is significant that this morphocline represents varying degrees of arboreality-terrestriality and that, as noted above, the most terrestrial forms, which we judge to be the most derived, are also thoroughly adapted for the reduced demands of inversion and greater transverse stability.

A similar transformation can be discerned in the character of the distal articular surfaces of the calcaneum. The sustentacular process in *Ptilocercus* bears a large, continuous, crescent-shaped facet. On the other extreme, however, in *Tupaia longipes,* one of the most terrestrial of tupaiines, this facet is reduced and separated into two entities. In tupaiines, there is a distinct protuberance of the body just dorsal to the cuboid facet. When the astragalus is rotated upon the calcaneum in the tupaiines, the effect of these features upon the mobility of the subtalar joint becomes apparent. The two features act to "embrace" the astragalus, forcing the head to abbreviate lateral and medial movements and rotate within the confines of the concave sustentacular articulation. Stabilization of the subtalar joint probably results from the restriction of the extremes of medio-lateral mobility in the more terrestrial tupaiines. In *Ptilocercus,* the topographical relief of the body (or interosseous sinus) is low and presents little resistance to the distal movements of the astralagar head of *Ptilocercus.* Correlated with the sustentacular facet of the calcaneum, of course, is the conformation of the astragalar sustentacular facet. As shown on Fig. 9, in *Ptilocercus* this facet is continuous with the navicular facet distally. These facets are secondarily progressively more isolated from each other in the more terrestrial tupaiines.

While generalities about the locomotory capabilities of tupaiids can only be made with strong reservations, due to the incomplete and difficult nature of the field observations, it is likely that broad categories of habitat can be established. It is generally accepted that *Urogale everetti* and *Lyongale tana* are probably at least as, or more, terrestrial than other tupaiids, yet they are known to be capable climbers (Steinbacher, 1940; Wharton, 1950b; Polyak, 1957; Davis, 1962; Sorenson and Conaway, 1964). Evidence for the climbing abilities of *Dendrogale* and *Anathana* are almost totally lacking. Sorenson and Conaway (1964) and Sorenson (1970) attempted to synthesize the known evidence and their own experimental work and recognized the following groupings of several species of *Tupaia:* (1) *T. minor* is the most arboreal, and (2) *T. gracilis,* (3) *T. glis longipes,* (4) *T. glis/T. montana*, (5) *T. chinensis/T. palawanensis* mostly terrestrial, with (6) *Lyonogale tana* being the most terrestrial. Kloss (1903, 1911) reported that both *T. glis longicauda* and *T. nicobarica* are strongly arboreal, while Bartels (1937) suggested that *T. javanica* spends much of its time in the canopy.

Sorenson (1970) noted that both *T. minor* and *T. gracilis* are capable of hypersupinating the hind limb in order to descend inclined substrates, and Jenkins (1974) found that *T. glis* is also capable of this movement. *T. chinensis* and *Lyonogale tana* apparently do not utilize this ability frequently, which Sorenson suggested to be indicative of a more terrestrial overall adaptation. Of the tupaiids, only *Ptilocercus lowii* is believed to be almost wholly arboreal (Le Gros Clark, 1926; Davis, 1962). It seems certain that the greater overall mobility of the astragalus upon the calcaneum than that of other tupaiids reflects this.

In the tupaiids the anterior plantar tubercle seems to be displaced proximally from the primitive archontan position adjoining the cuboid facet. The plantar tubercle has retreated to a position along the long axis of the calcaneum but lying opposite the sustentacular process in *Ptilocercus.* In the tupaiines the posterior (proximal) placement is even more extreme, with the tubercle lying in a position ventral to the posterior astragalocalcaneal articulation. As a result of the distal and superior tubercle that has developed in the tupaiines, the cuboid facet has taken on a semitrapezoidal shape in distal profile. In *Ptilocercus* the cuboid facet is reminiscent of the rounded primitive condition.

The tupaiids do not have a grasping hallux. Jenkins (1974) noted, however, that the hallux of *Tupaia glis* acted independently of the other toes on uneven and arboreal substrates. The hallux is so positioned as to lie across the top of the branch while the remaining digits fall on the lateral side. The hallux acts to counter slippage off the substrate and possibly maintains contact between the foot and substrates with narrow diameters on sharp, apical crests and steeply inclined sides. There is no evidence that the tupaiid hallux actually forms an anatomically adapted grasping unit, although the tendon of M. abductor hallucis brevis is accentuated.

One major distinction observed within the osseous structure of the archontan tarsals that may indicate differences in the action of the hallux may be structures associated with the tendon of M. flexor (digitorum) fibularis. The flexor fibularis tendon acts to initiate digital and plantar flexion in the foot. The capability to perform these actions becomes increasingly important when the long axis of the pes is placed transversely to the long axis of a branch, or along a slender branch. In tupaiids, the pes is generally aligned to the long axis of the branch. It does not have to cope with maintaining maximal support and contact through plantar and digital flexion around a substrate of limited diameter.

B. Dermoptera

While Linnaeus referred to *Cynocephalus volans* as *Lemur volans,* suggesting a close "essentialistic" relationship between primates and colugos, Gregory (1910) was the first modern systematist to formalize a special relationship between the latter, tupaiids, and primates. Simpson (1937) discussed the problem of special affinity between these, and recently, Van Valen (1966), Russell *et al.* (1973), Rose (1973), Rose and Simons (1977), Szalay (1975, 1977b), and McKenna (1975) have either passingly noted or discussed the systematic position of the dermopterans. Only one family of dermopterans survives today. The Cynocephalidae is represented by two genera and two species (see Szalay, 1969, p. 241), *Cynocephalus volans* and *Galeopterus variegatus.*

A number of characters typically found in the Archonta characterize the dermopteran astragalocalcaneal complex in addition to the derived dermopteran features. On the astragalus, the calcaneoastragalar facet is aligned obliquely to the longitudinal axis of the bone; the tibial trochlea is proximodistally elongated and aligned obliquely to the long axis; the astragalar sustentacular facet is laterally confluent with the naviculoastragalar facet; and the lateral margin of the tibial trochlea is more posteriorly extended than the medial margin. On the calcaneum, the posterior astragalocalcaneal articulation forms a relatively small angle to the long axis; the calcaneal sustentacular process continues onto the body of the calcaneum (i.e., not restricted to the sustentacular process); the sustentacular process is distinct from the distal margin of the calcaneum; the cuboid facet is round and concave; the cuboid facet is aligned transversely to the long axis of the calcaneum (i.e., not obliquely as in known Cretaceous eutherians); calcaneal fibular facet is lost; the groove for the plantar calcaneocuboid ligament is not retained on the distal margin of the calcanum. The aforementioned characteristics clearly associate the living colugos with the Archonta.

Colugos have many unique dermopteran structures distinct from other extant members of the Archonta. The following derived complex of characters appears to distinguish the gliding colugos from their arboreal relatives. On the astralagus, there is an extension of the sulcus astragali into the trochlear groove for the tendon of M. flexor (digitorum) fibularis; ossification (buttressing) medial to the sulci astragali on the ventral body (occasionally extending onto the neck); and a naviculoastragalar articulation symmetrical mediolaterally to the axis of the astragalus. On the calcaneum the most prominent derived feature is the deep, slightly medially offset cuboid pivot. The tuber of the calcaneum is transversely narrowed; the posterior astragalocalcaneal facet is long, only slightly convex, and lies parallel to the long axis of the calcaneum;

and the astragalocalcaneal articulation lies almost wholly posterior to the sustentacular process.

This combination of unique features is present in a number of Early Eocene forms, and these specimens have been assigned to a dermopteran-complex of two undetermined species. While most of these Eocene tarsals

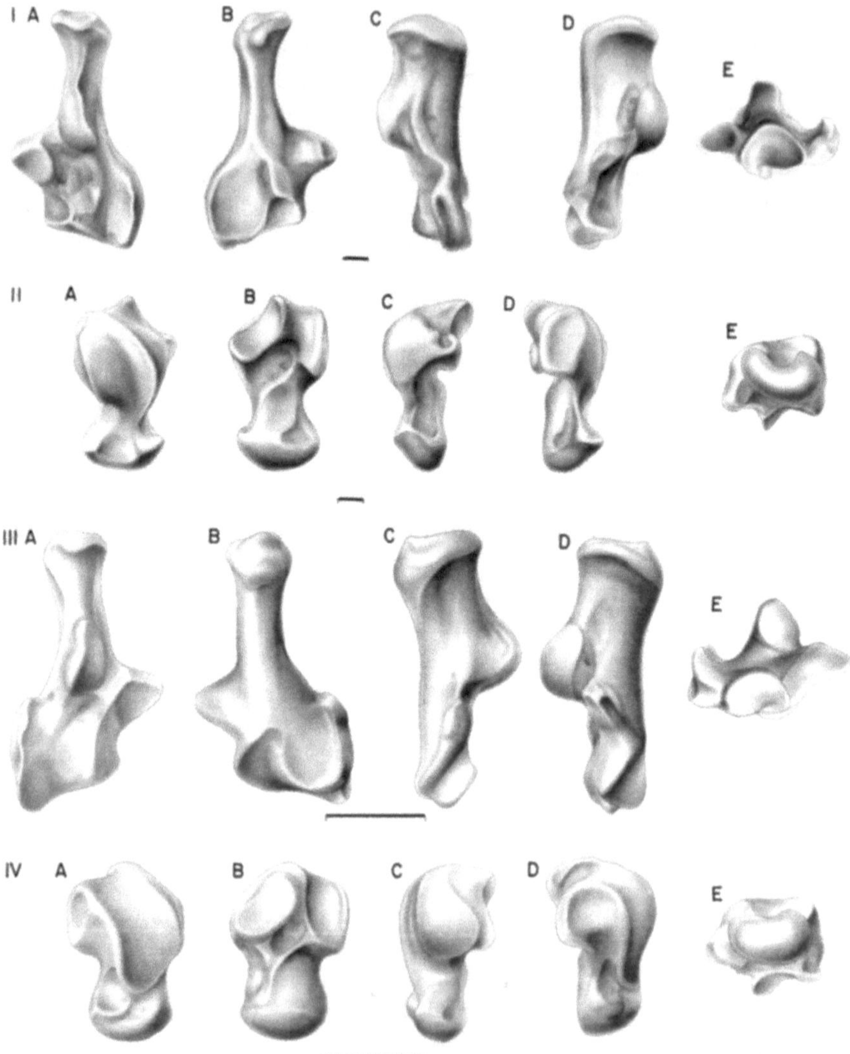

Fig. 10. Dermopteran. Wasatchian, Early Eocene, Colorado. I. Left calcaneum, based primarily on AMNH 89518, from Sand Quarry. II. Left astragalus, based on UCMP 60252, from Sand Quarry. III. Right calcaneum, based on UCMP 113287 and others, from Bitter Creek locality V70246. IV. Left astragalus, based on UCMP 113242 and others from Bitter Creek locality V70246. Views from left to right, are dorsal (A), plantar (B), lateral (C), medial (D), and distal (E). Scales represent one mm.

belonged to species which may have had specialized characteristics that eliminate them from consideration of direct ancestry to the living dermopterans, these forms are clearly assignable to that order. The peroneal process is exceptionally well developed in what we recognize as archaic dermopterans but is nearly completely lost in the living species. This may indicate that the living colugo lineage branched off quite early. The colugos from the Eocene display two characteristic forms of the peroneal process, different from the reduced condition (Fig. 10).

Recent colugos and the two types of dermopts, assigned to the order based on their tarsal remains (Fig. 10), share a number of significant derived features which we believe to be dermopteran specializations. The combination consists of an anteroposteriorly long groove for the tendon of the M. flexor (digitorum) fibularis on the plantar side of the astragalus, a concave astragalar sustentacular facet, a laterally conspicuously grooved peroneal process, and a cuboid facet on the distal part of the calcaneum which is deep and cone-shaped.

While the astragali assigned to dermopterans from the Eocene are generally similar to extant colugos, there are some important distinctions. The living species have reduced considerably the sharpness of the lateral trochlear margin. The crests of extant flying lemurs are distinguishable only on the midportion of the body of the astragalus. The archaic dermopts carry a lateral crest that extends so far distally that it often merges with the supero-lateral margin of the naviculoastagalar facet (Fig. 10). Ventral to the distal position of the medial astragalar margin a shallow depression often appears in these Eocene astragalar specimens. The archaic dermopterans also have an astragalar head bearing a "half-doughnut" shaped naviculoastragalar facet when viewed from a distal perspective. In the living colugos the naviculoastragalar articulation is more rounded, close to being circular in outline when observed in distal aspect.

The calcaneocuboid pivot of the Eocene dermopts is either a stage in the evolution of the condition found in extant colugos, or perhaps slightly differently evolved. Apparently, the primitive colugo pattern was an anteromedially spiralling concave facet towards an apex aligned medial to the anterior plantar tubercle.

The early dermopterans have the distal margin of the calcaneum lying somewhat oblique to the long axis of the bone, this being the primitive archontan condition, though it is clearly progressive compared to the highly oblique pattern of *Protungulatum.* It should also be pointed out that the peroneal process contributes to much of this angulation found in the Eocene colugos. Since the peroneal process is reduced in the extant members of the order, it is likely that at least a part of the convergent appearance of transversely aligned distal margins of the calcaneum in primates and colugos can be attributed to this transformation. It should be noted that the tupaiids also retain a slight oblique angle of the cuboid facet.

The dermopterans, both living and extinct, have generally low topographic relief on both the neck of the astragalus ventrally and the interosseous sinus and sustentacular facet of the calcaneum superiorly. This shallow topography allows a gliding motion of the astragalus above the calcaneum. The gliding movements of the distal articulations, together with the long, planar, and proximo-distally aligned astragalocalcaneal and oblique calcaneoastragalar facets allow extremes of inversion as a result of the translational component emphasized in the lower ankle joint. Restriction of mobility lies chiefly in the need to maintain naviculocuboid contact. The presence of the cuboid pivot acts to stabilize this latter association without unduly restricting mobility.

What selectional forces shaped this highly mobile ankle joint which derives its basic pattern from an archontan ancestry? The observations of Wharton (1950a) remain our only detailed natural history information on colugo posturing and locomotion. He found that colugos were slow moving, nocturnal herbivores that were apparently incapable of moving along the superior surface of branches. Despite this they were highly capable arborealists, moving within the forest canopy with sloth-like deliberation. A colugo hangs suspended by the limbs below a superstrate and proceeds below it in a slow, quadrupedal, "foot-over-foot" progression. In normal locomotion within the foliage, three extremities will remain in contact with the superstrate at all times.

The colugos usually hang by all four feet, but they are capable of hanging solely by the forelimbs. In order to prevent fouling the uropatagium with feces, a colugo will hang suspended beneath a branch with the mani and deftly flip the gliding membrane out of the way of the falling excrement. It is likely that hyper-supination is necessary for throwing the uropatagium out of the way when the colugo must rid its bowels. Hyper-supination may be necessary when a colugo sleeps during the day. Several are often found together in the hollows of trees, clinging to the walls in a heads-down posture. The pattern in all of these activities requires the mechanical necessity of extreme inversion. In particular, when hanging by all fours, the feet habitually face medially.

It is likely that, much like *Tupaia glis* (Jenkins, 1974), colugos can hyper-abduct the femora and align the tibiae nearly parallel to the body. This flexibility of the hind limbs would aid in the spreading of the gliding membrane. Colugos are known to be capable of gliding distances of over 100 meters between trees (Burton, 1950; Tate, 1947). They emit a high pitched shriek upon releasing the superstrate. The possibility that colugos echolocate cannot be ruled out, especially since judging distance to landing sites and avoiding obstructions must somehow occur in this nocturnal browser.

Inversion serves the important function of turning the plantar surface of the pes to face the sides of the overlying branch. Most weight-bearing is by the large, laterally compressed claws. Tensional stress on these structures from the resultant gravitational force upon the body is reduced by the maintenance of three or four limbs in contact with the superstrate. The weight of the animal is thereby borne equally by the extremities. Although the plantar surface of the foot must face the surface of the branch, it is not known to us the extent to which the volar surface of the pes makes contact with the bark.

The general similarity in the locomotor pattern of the colugos (in the trees) to the phalangers, lorisines, and tree sloths is striking. It should be pointed out that some similar biomechanical solutions have developed in these groups to enable them to move under a superstrate. Perhaps the most similar forms adaptationally in some respects to the colugos are the species of the edentate family Bradypodidae. Like the flying lemurs, the two- and three-toed tree sloths move slowly underneath a superstrate. Both are nocturnal herbivores and spend the day in a state of torpor. They appear restricted to movements below a superstrate and are nearly helpless on the ground. Like the colugos, tree sloths have strong, laterally compressed claws which act as hooks that are placed onto appropriate stable surfaces. The tarsals of *Bradypus* and *Choloepus* have an extreme degree of mobility, accomplished convergently by strikingly different means. The upper ankle joint permits up to 180° of flexion and supination. The astralagus has a great amount of mediolateral mobility, allowing inversion and eversion upon a relatively immobile calcaneo-navicular structure. The astragalonavicular articulation is best considered a nearly perfect "ball and socket" joint. While much of the capability to dorsiflex is not

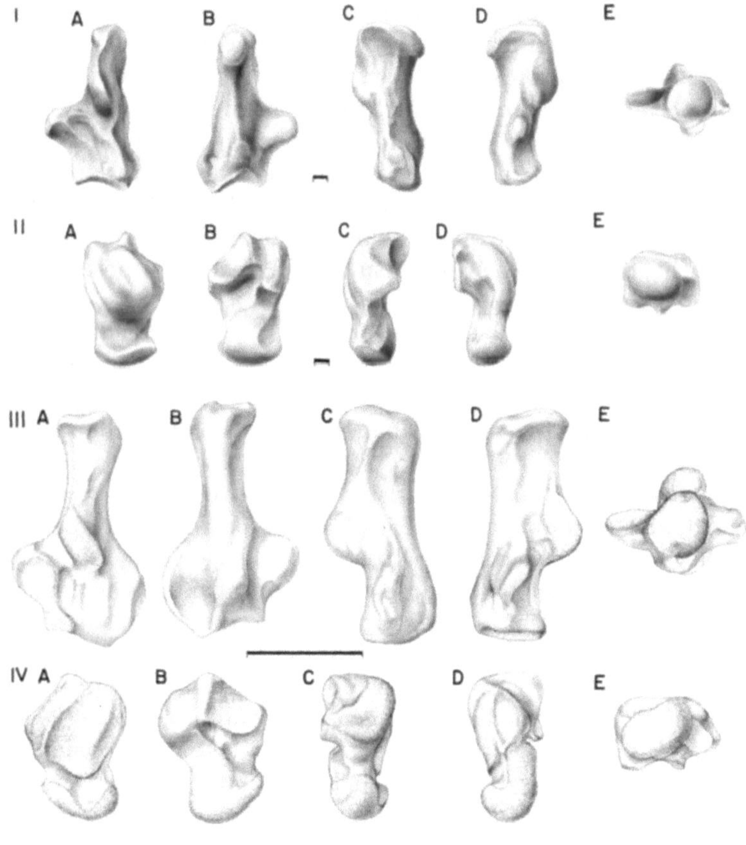

Fig. 11. *Cynocephalus volans*. Recent, Philippines. I. Left calcaneum. II. Left astragalus. *Plesiadapis tricuspidens*. Thanetian Paleocene, France. III. Left calcaneum, based on MNHN R-611. IV. Right astragalus, based on MNHN R-610. Scales represent one mm.

present in the ankle of colugos, the same basic biomechanical functions are solved. While the colugo calcaneum and cuboid-navicular retain their independent activities, a similar measure of inversion is generated through the development of a cuboid pivot, an elongated astragalocalcaneal facet parallel to the long axis of the calcaneum, a spherical naviculoastragalar articulation, and relatively unrestricted mobility of the sustentacular articular surfaces.

Despite the behavioral similarities that can be described there is relatively little phenetic similarity between the astragalocalcaneal units of tree sloths and colugos. Selection has had to act on different genetic frameworks to construct functionally analogous solutions for similar behavioral requirements.

It is of interest to note here that while the naviculoastragalar facets of colugos and *Bradypus* are circular in distal aspect and probably convergently resemble the last common ancestor of extant primates in this respect, the early Eocene dermopterans have an astragalar head that is transversely broad and concave on its superior surface. This is possibly the primitive dermopteran condition because the astragalar head of some plesiadapiforms (Fig. 6) has a concave depression on the superior surface. This condition is also found in the phalangerid marsupials and in the lorisine primates. The similarities among these three groups are related to the increased range of inversion-eversion permitted by the expanded navicular-astragalar range of contact. In the Lorisinae, expansion appears to have been primarily in a superomedial direction if the galagine condition is primitive. The expansion may have also allowed greater potential for an inverted pes to dorsiflex/adduct. Why a slow-locomotor would benefit from this type of articulation rather than a spherical astragalar head is not clearly understood at present and poses a problem for future research.

At the present time speculations of the biological role of the pedal morphology of the Eocene colugos can only be based on what we know of their living counterparts. It is possible that the primitive dermopterans were slow and moved below a superstrate by hanging from small and hook-like, laterally compressed claws.

They were most likely at least partially folivorous. *Plagiomene,* an early Eocene 'dental' dermopteran, already had the beginnings of peculiar incisor morphology of the living colugos (Rose, 1973). The lower incisors of *Plagiomene* are procumbent and bilobate. The lower incisors of extant colugos are similarly splayed outwardly but have added and narrowed the cusps in order to form a comb-like structure on each of the first two lower incisors. The third lower incisor of *Cynocephalus* has four cusps present and would appear to represent a morphological intermediate between the condition of the lower incisors found in *Plagiomene* and the I_1–I_2 of the extant Cynocephalidae. Gregory (1951, quoting H.C. Raven) reported that colugos use their incisors "in scraping the green coloring out of leaves." Winge (1941) suggested that they were used in the procurement of leaves. Wharton (1950a) felt that they were used in grooming the fur. Many of the plagiomenid teeth show apical wear on the cusps characteristic of a folivorous diet.

Even if the early Eocene colugos were incapable of slow quadrupedal locomotory behavior it is valuable to suggest patterns in which this could be acquired in a normal evolutionary transition. Squirrels often move below a branch in apparent disregard of the gravitational pull upon their bodies. Squirrels rarely progress in this manner for any great distance and apparently prefer to locomote on the superior side of limbs and branches. This movement usually occurs when they are moving from a primarily vertically oriented

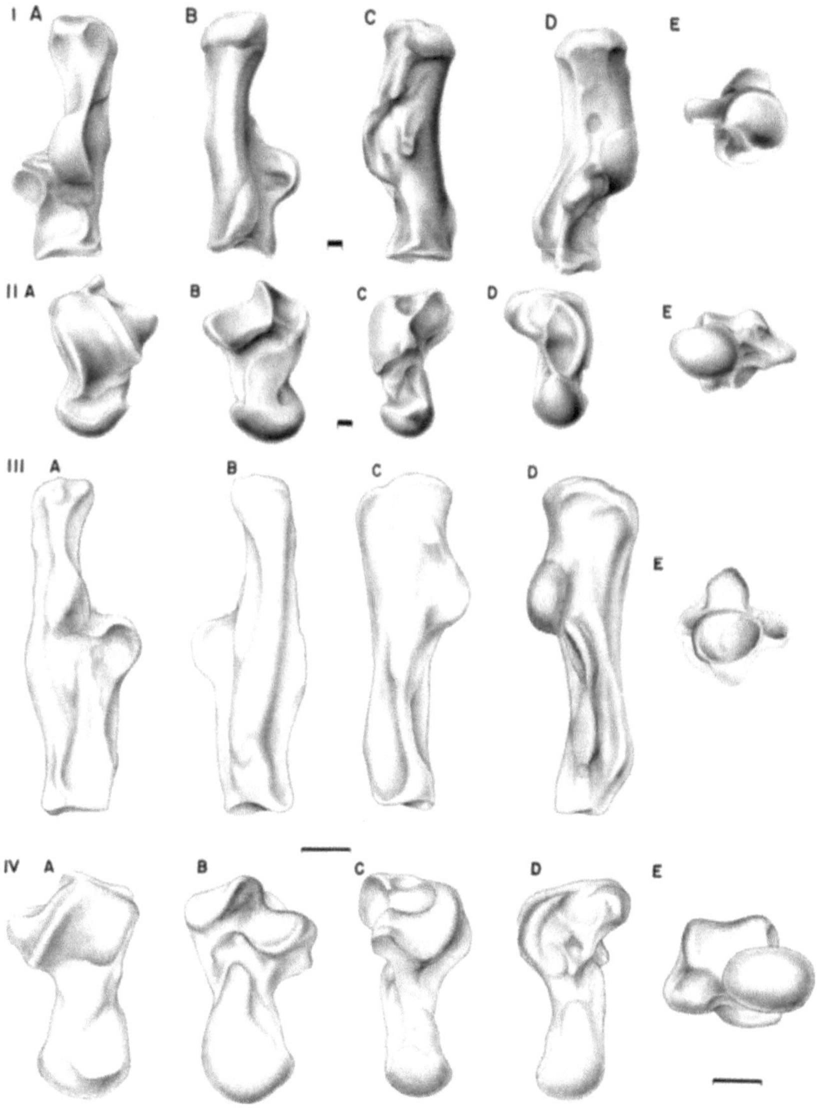

Fig. 12. *Adapis parisiensis*. ?Late Eocene, France. I, left calcaneum; II, left astragalus, both from the Phosphorites of Quercy, Montauban Collections. *Teilhardina belgica*. Sparnacian, Early Eocene, Belgium. III, right calcaneum; IV, right astragalus, both from Dormaal. Views from left to right, are dorsal (A), plantar (B), lateral (C), medial (D), and distal (E). Scales represent one mm.

substrate (such as a tree trunk) to a surface more diagonally inclined. Such surfaces as a limb jutting off from the main trunk of a tree would provide very little difference in forces that must be countered. Climbing the inferior surface of a steeply inclined limb compared to a vertically aligned tree trunk is merely a change in the degree of gravitational force that must be resisted.

If the competition in the arboreal habitat was extreme with every potential dietary/locomotory outlet being filled by increasingly more specialized groups, then expanding the resource base would be very beneficial. One such avenue is the exploitation of the resources that are available from the inferior surface of branches and limbs. Organisms that habitually remained on the superior side of branches would have limited accessibility to these resources, either due to limitations in discovering them or because of restrictions in their ability to gain access to a dietary item on the inferior side of a substrate.

It can be easily illustrated that many features characteristic of progression below a superstrate can be acquired through gradual anagenetic transformation. Natural selection would favor organisms with cautious movements and those that retained a number of extremities on stable and tested footholds. Strong, laterally compressed claws or grasping extremities would also prove favorable in such superstrate locomotion.

The occurrence of the astragalocalcaneal complexes (Fig. 10) we identify as dermopteran and dental remains of microsyopids from the same localities opens up intriguing possibilities concerning the previous suspected relationships of the dentitions. Neither the Four Mile localities (McKenna, 1960) nor the Bitter Creek sites are known to have dental remains of plagiomenids, the only Paleogene group with generally acknowledged special dental similarity to living colugos (Rose, 1973). These localities on the other hand, are known to have yielded microsyopids of two size ranges, those of *Microsyops* (*Cynodontomys*) and *Niptomomys*. Did the pair of astragalocalcaneal complexes described here as dermopteran belong to the same animals whose dental remains have been identified as *Microsyops* and *Niptomomys*, respectively, or to some dentally hitherto unsuspected dermopteran taxa, or, as a third alternative, have they not been dentally sampled in these collections?

If the microsyopid connection proves to be correct when associated specimens are found, then the enigmatic nature of microsyopids will become clarified. In that case, the microsyopids would firmly prove to be dermopteran and nonprimate, although archontan, on dental, basicranial, and pedal evidence.

C. Primates

Many of the tarsal characters employed by Szalay and Decker (1974) as shared derived characters for the primates are in fact archontan features. Two significant pedal features (and several other cranial and postcranial ones) however, appear to unite the Plesiadapiformes and the euprimates (=Strepsirhini and Haplorhini); these are related to the position of the tendon of M.

flexor (digitorum) fibularis. While the groove for the tendon, lying upon the posterior trochlea, is reduced in breadth in the astragali of the tupaiids and colugos, it remains primitively wide in the primates. On the plantar surface of the sustentacular process of the calcaneum the groove for the tendon of M. flexor (digitorum) fibularis is excavated. The importance of this characteristic should not be underestimated. The capability to digital flex and plantar flex is extremely significant in grasping with the hallux. It is impossible to oppose the hallux and the remaining digits without digital/plantar flexion. Apparently the M. flexor (digitorum) fibularis does not play an active role in the colugos and tupaiids.

In actively grasping forms such as the lorisines and some phalangerid and didelphid marsupials (e.g. *Caluromys, Cercartetus,* and *Marmosa*), the largest muscle associated with the upper ankle joint is the M. flexor (digitorum) tibialis and the underlying M. flexor (digitorum) fibularis. They generally send slips to all five toes. Flexion must occur in all digits as a unit. Since the four postaxial digits apply an equal force to that of the hallux in a stable grip, the first digit must be supplied with a relatively larger slip of the common flexor of the digits. Lewis (1964) suggested that a M. flexor accessorius (or quadratus plantae) appears in the primitive mammalian pes. It originated on the lateral surface of the calcaneum and inserts into the tendon of M. flexor (digitorum) fibularis. When the hallux becomes highly divergent the M. flexor accessorius correctly aligns the M. flexor (digitorum) fibularis to the hallux.

The increase in size of the plantar and digital flexors and the presence of the M. flexor accessorius clearly are involved in the development of a divergent and opposable hallux. It was noted by Szalay and Decker (1974) that the plantar groove on the calcaneum might also serve to maintain the alignment of the M. flexor (digitorum) fibularis, thus serving an analogous function to the M. flexor accessorius. It thus appears possible that the presence of the deeply excavated sustentacular groove for the tendon of M. flexor (digitorum) fibularis is related to the existence of a grasping hallux in all the primates (including the plesiadapiforms). The presence of claws in *Plesiadapis* is irrelevant to this adaptation. There are several marsupials which have grasping feet and retain well developed claws.

Many of the identifications in this paper of tarsals as primate (see Figs. 4–8), especially plesiadapiform, are based on what we consider homologous similarity with the tarsals of *Plesiadapis gidleyi* (Fig.9) which are associated with the dentition. As discussed above and under the various archontan groups, the major diagnostic features of the astragalocalcaneal complex of plesiadapiforms lie in the groove for the tendon M. flexor (digitorum) fibularis, and perhaps the shape of the peroneal process.

D. *Chiroptera*

The interrelationships of the bats and the other members of the Archonta are as yet ill-defined. Whether the relationship is a real one has been examined on the basis of three sources of information. Gregory (1910) noted Leche's (1886) observation that the patagium of the colugos and fruit bats is innervated by the same nerves and supported by the same muscles. This pattern is unlike that of the gliding Sciuridae, Anomaluridae, and Phalangeridae. The colugo patagium, like that of bats, surrounds the body, is attached to the tail via a uropatagium, and extends between the fingers (the thumb is free in the chiropterans). Is this detailed similarity homologous, or is it a homoplastic resemblance due to the limited materials available to form a wing-like membrane? We believe this detailed, unique, and functionally elaborate pattern of similarity to be shared and derived between bats and colugos.

While this line of evidence points to special relationship between the colugos and bats, another source of information does not corroborate. The immunological data of both Sarich and Cronin (1976) and Dene *et al.* (1976, this volume) would place the bats outside the clade which includes the primates, colugos, and tupaiids.

The possibility that the bats are an independently arboreal group from the Archonta cannot be ruled out. Certainly the ability to habitually climb precedes volancy, and the Archonta are the earliest known placental climbers adapted to an arboreal existence. The condition of most chiropteran features precludes comparisons with other taxa and where comparisons can be made the features are usually symplesiomorphies. The common ancestor of the Microchiroptera and Megachiroptera did not necessarily have a microchiropteran-like dentition. Consequently, searching for an ancestor with microchiropteran-like dentition may result only in identifying various insectivorous chiropterans.

5. Overview

The adaptation to an arboreal milieu has long been considered one of the significant factors that led to the early Cenozoic radiation of the primates. Interpretations of the transition from terrestriality to arboreality have been based primarily upon neontological evidence rather than reference to the fossil record. A few direct investigations of the paleontological material that bears upon this problem have been made (Simpson, 1935; Decker and Szalay, 1974; Szalay and Decker, 1974; Szalay *et al.*, 1975). These studies have indicated that the morphological attributes associated with mechanics necessary for arboreal locomotion were fully evolved in the plesiadapiform primates. Furthermore, the astragalo-calcaneal complex of the primates is biomechanically distinct from various late Cretaceous eutherians which possessed a tarsus adapted primarily for a horizontal substate.

We argue further in this paper that tupaiids, dermopterans, and plesiadapiforms as well as euprimates share unique, homologous, complexly derived tarsal features indicating both monophyly and arboreality for the ancestry of the Archonta—derived from terrestrial non-archontans. Although the shared and derived characters of an archontan ancestor left clear marks on all subsequent modifications of its derivatives, important autapomorphies distinguish both the archaic and derivative primate and archontan stocks. The primate specialization, not attained by the other archontan groups can be summed up as the hypertrophy of the tendon of the M. flexor (digitorum) fibularis, as suggested by the clearly derived condition of the groove for this tendon on primate calcanea. The emphasis of this groove for the increased size and improved stabilization of the tendon for the M. flexor (digitorum) fibularis points to modification for strong grasping, and perhaps some opposability of digits beyond the condition of an archontan ancestor.

The dermopterans, on the other hand, have placed strong emphasis on a deep cuboid pivot which both facilitates rotation of the foot and strongly stabilizes the calcaneocuboid joint. All the joint modifications described under the Dermoptera point to an adaptive complex which allows relatively free movements between the calcaneum and astragalus, the latter and the navicular and calcaneum and cuboid, without emphasizing the usual relief on the tarsal complex which is necessary for stabilization associated with weight bearing, even in the most arboreally adapted quadrupeds. It appears that these modifications evolved to facilitate freedom of movement of the foot while primarily subjected to tensile forces.

It appears that neither cranial nor soft anatomical features evaluated to date corroborate a special affinity of the tupaiids to primates or the concept of the Archonta (see other papers in this volume and many references therein). It is to be emphatically stated, however, that not one of the features examined by these and past authors, to our knowledge, contradicts our hypothesis. None of these studies are able to show that any of the synapomorphies between tupaiids and other eutherians are more recent than the tarsal synapomorphies proposed for the Archonta (this paper, and Szalay, 1977b).

We must specially address ourselves, however, to Novacek's (this volume) thorough and important contribution, as he has found various postcranial features of tupaiids, as he has seen them, not particularly indicative of the hypothesis to which we so firmly committed ourselves in this paper. Our disagreement with Novacek's contribution is deep both on some seemingly small, but nevertheless important, methodological points and in our view of assessing a form-function complex of selected aspects of the tarsus across virtually all groups of eutherians and metatherians.

In terms of methodological differences in our outlook, perhaps the most serious ones lie in the manner of use of autapomorphs, the way we perceive the use of the morphotype concept, and in the difference between his cladistic and in our essentially transformational view of character states. Contrary to what Novacek states, cladistic methodology, unlike a phylogenetic one, does not consider assessment of character transformation as essential, but rather concentrates on a system of usually unweighted counting of characters designated as either synapomorphs, plesiomorphs, or autapomorphs.

Novacek objects to several features of the character complex cited by Szalay (1977b) for the Archonta in that they are really not genuine synapomorphies for that superorder because they also occur in other taxa. As far as *verbal* characterization of the character states is concerned, this is probably true. The visually ascertained gestalt of the astragalocalcaneal character complex, however, carries a different meaning. A complex feature is the result of a unique developmental system, and convergence of single aspects of such a character complex does not detract from the value of shared similarities of entire homologous complexes.

In assessing proposed hypotheses, Novacek uses the complexes of derived features of advanced members of supraspecific taxa to "falsify" hypotheses based on the shared derived similarity of the *ancestor* of that group to another taxon. For example, the *transformation* of certain features of the astragalar tibial trochlea of *Notharctus* from a well established morphotype condition for the entire order (see our review, Szalay, 1977b; Szalay and Decker, 1974) does not invalidate the presence of archontan features in Paleogene primate astragali. In other words, the very usefulness of the concept of the morphotype complex rests on the assumption that the ancestral condition was the one which transformed to the non-primitive conditions of a monophyletic group. Now as plesiadapiforms, strepsirhines, and haplorhines are recognized as

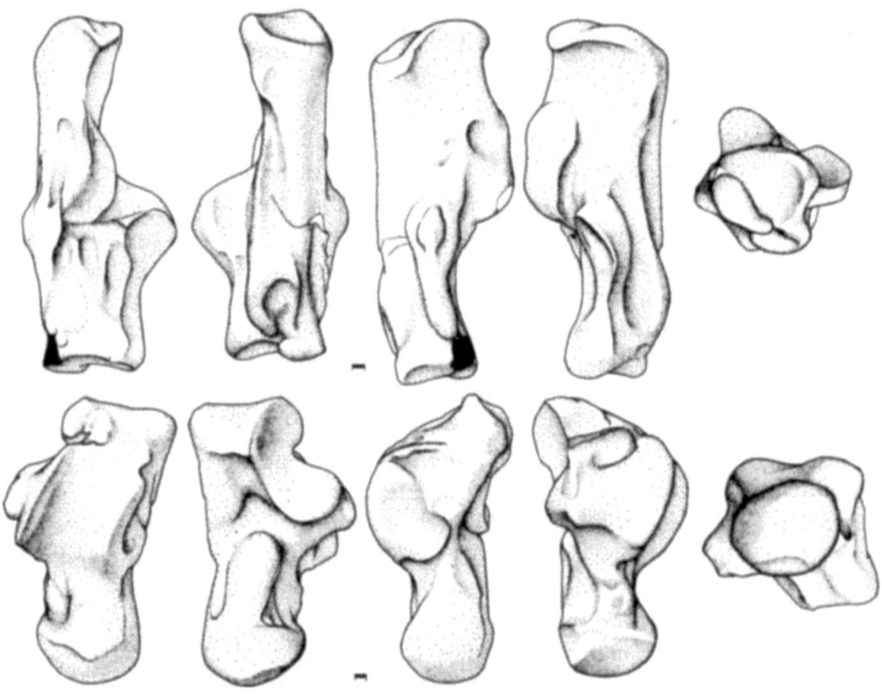

Fig. 13. *Notharctus* sp. Bridgerian, Eocene, Wyoming. Right calcaneum (above) and astragalus (below), based on AMNH 91663. Views from left to right are dorsal, plantar, medial, lateral, and distal.

primates on grounds other than the tarsus, the astragalus of *Notharctus* (Fig. 13) can be firmly assessed, based on tarsal characters alone, to be derived from a clearly more primitive state, similar to the plesiadapiform condition. Its strepsirhine, adapid, or *sui generis* features therefore do not "falsify" the primate morphotype similarity to the tupaiids and dermopterans.

In closing, it should be noted that a morphotype hypothesis, as in the case of primates vs. *Notharctus*, allows a *transformational* assessment of a particular character complex in a full evolutionary sense. This approach is decidedly both *phylogenetic* and *transformational*, and unfortunately not employed by stringently cladistic systematists.

ACKNOWLEDGMENTS

We are especially grateful to Drs. D. Baird, F.A. Jenkins, Jr., M.C. McKenna, W. Matthes, D.E. Russell, R.E. Sloan, and D.E. Savage for graciously allowing the study of miscellaneous Cretaceous and Paleogene postcranials in their care or in collections made by them.This research was supported by a CUNY PSC-BHE Research Award to the senior author.

6. *References*

Bartels, M. 1937. Zur Kenntis der Berbreitung und der Lebensweise javenischer Säugetiere. *Treubia* 16: 149–164.

Bock, W.J. 1977. Adaptation and the comparative method, pp. 57–82. *In* M.K. Hecht, P.C. Goody, and B.M. Hecht (eds.). *Major Patterns in Vertebrate Evolution*. Plenum Press, New York.

Bock, W.J., and von Wahlert, G. 1965. Adaptation and the form-function complex. *Evolution* 19: 269–299.

Bown, T.M., and Rose, K.D. 1976. New Early Tertiary primates and a reappraisal of some Plesiadapiformes. *Folia Primat.* 26: 109–138.

Burton, M. 1950. *Wildlife of the World*. London.

Campbell, C.B.G. 1966a. The relationships of the tree shrews: Evidence of the nervous system. *Evolution* 20: 276–281.

Campbell, C.B.G. 1966b. Taxonomic status of tree shrews. *Science* 153:436.

Campbell, C.B.G. 1974. Phyletic relationships of the tree shrews. *Mamm. Rev.* 4: 125–143.

Davis, D. D. 1962. Mammals of the lowland rain-forest of North Borneo. *Bull. Raffles Mus.* 31: 1–129.

Decker, R.L., and Szalay, F.S. 1974. Origins and function of the pes in the Eocene Adapidae (Lemuriformes, Primates), pp. 261–291. *In* F.A. Jenkins, Jr. (ed.). *Primate Locomotion*. Academic Press, New York.

Dene, H.T., Goodman, M., and Prychodko, W. 1976. Immunodiffusion evidence in the phylogeny of the primates, pp. 171–196. *In* M. Goodman, R.E. Tashian, and J.H. Tashian (eds.). *Molecular Anthropology*. Plenum Press, New York.

Evans, F.G. 1942. The osteology and relationships of the elephant shrews (Macroscelididae). *Bull. Am. Mus. Nat. Hist.* 80: 85–125.

Fleagle, J. G. 1977. Locomotor behavior and skeletal anatomy of sympatric Malaysian leaf monkeys *(Presbytis obscura and Presbytis melalophos). Yrbk Phys. Anthrop. 1976*. 20: 440–453.

Gregory, W.K. 1910. The orders of mammals. *Bull. Am. Mus. Nat. Hist.* 27: 1–524.
Gregory, W.K. 1920. On the structure and relations of *Notharctus*, an American Eocene primate. *Mem. Amer. Mus. Nat. Hist.* n.s. 3: 51–243.
Gregory, W.K. 1951. *Evolution Emerging*. MacMillan, New York.
Haines, R.W. 1955. The anatomy of the hand of certain insectivores. *Proc. Zool. Soc. Lond.* 125: 761–777.
Haines, R.W. 1958. Arboreal or terrestrial ancestry of placental mammals. *Quart. Rev. Biol.* 33: 1–23.
Huxley, T.H. 1872. *A Manual of Anatomy of Vertebrated Animals*. D. Appleton, New York.
Jenkins, F.A., Jr. 1974. Tree shrew locomotion and the origins of primate arborealism, pp. 85–115. *In* F.A. Jenkins, Jr. (ed.). *Primate Locomotion*. Academic Press, New York.
Jepsen, G.L. 1970. Bat origins and evolution, pp. 1–64. *In* W. A. Wimsatt (ed.). *Biology of Bats*, Vol. 1. Academic Press, New York.
Kloss, C.B. 1903. *In the Andamans and Nicobars*. John Murray, London.
Kloss, C.B. 1911. On a collection of mammals and other vertebrates from the Trengganu archipelago. *J. Fed. Malay States* 4: 175–212.
Leche, W. 1886. Uber die säugethiergattung *Galeopithecus*. Eine morphologische Untersuchung. *Kongl. svenska vet. Akad. Handl.* 21: 1–92.
Le Gros Clark, W.E. 1926. On the anatomy of the pen-tailed tree-shrew *(Ptilocercus lowii)*. *Proc. Zool. Soc. Lond.* 1926: 1179–1309.
Le Gros Clark, W.E. 1959. *The Antecedents of Man*. Edinburgh Univ. Press, Edinburgh.
Lewis, O.J. 1964. The evolution of the long flexor muscles of the leg and foot. *Intern. Rev. Gen. Exp. Zool.* 1: 165–185.
Linnaeus, C. 1758. Systema naturae per regna tria naturae, secundum classes, ordines genera, species cum characteribus, differentris, synonymis, locis. Editis decima, reformata. *Stockholar, Laurentii Salvii* 1: 1–824.
MacConaill, M.A. 1953a. The movement of bones and joints. V. The significance of shape. *J. Bone Jt. Surg.* 35B: 290–297.
MacConaill, M.A. 1953b. Close-packed position of joints and its practical bearing. *J. Bone Jt. Surg.* 35B: 486.
MacConaill, M.A. 1958. Mechanical anatomy of motion and posture, pp.47–89. *In* S. Licht (ed.). *Therapeutic Exercise*. E. Licht, New Haven.
MacConaill, M.A. 1973. A structuro-functional classification of synovial articular units. *Irish J. med. Sci.* 142: 19–26.
Martin, R.D. 1968a. Towards a new definition of primates. *Man* 3: 377–401.
Martin, R.D. 1968b. Reproduction and ontogeny in tree-shrews *(Tupaia belangeri)* with reference to their general behaviour and taxonomic relationships. *Z. Tierpsychol.* 25: 409–532.
Matthew, W.D. 1937. Paleocene faunas of the San Juan Basin, New Mexico. *Trans. Amer. Philo. Soc.* 30: 1–510.
McKenna, M.C. 1960. Fossil Mammalia from the early Wasatchian Four Mile Fauna, Eocene of Northwest Colorado. *Univ. Calif. Publ. Geol. Sci.* 37:1–130.
McKenna, M.C. 1966. Paleontology and the origins of the primates. *Folia primat.* 4: 1–25.
McKenna, M.C. 1975. Toward a phylogenetic classification of the Mammalia, pp. 21–46. *In* W.P. Luckett and F.S. Szalay (eds.) *Phylogeny of the Primates*. Plenum Press, New York.
Polyak, S. 1957. *The Vertebrate Visual System*. Univ. Chicago Press, Chicago.
Preuschoft, H. 1971. Body posture and mode of locomotion in early Pleistocene hominids. *Folia primatol.* 14: 209–240.
Rose, K.D. 1973. The mandibular dentition of *Plagiomene* (Dermoptera, Plagiomenidae). *Brev. Mus. Comp. Zool. Harvard* 411: 1–17.
Rose, K.D., and Simons, E.L. 1977. Dental function in the Plagiomenidae: Origin and relationships of the mammalian order Dermoptera. *Contrib. Mus. Paleont. Univ. Michigan* 24: 221–236.
Roux, G.H. 1947. The cranial development of certain Ethiopian insectivores, and its bearing on the mutual affinities of the group. *Acta Zool.* 28:165–307.

Russell, D.E., Louis, P., and Savage, D.E. 1973. Chiroptera and Dermoptera of the French Early Eocene. *U. Cal. Publ. Geol. Sci.* 95: 1–54.

Sarich, V.M., and Cronin, J.E. 1976. Molecular systematics of the primates, pp. 141–170. *In* M. Goodman, R.E. Tashian, and J.H. Tashian (eds.). *Molecular Anthropology*. Plenum Press, New York.

Schaeffer, B. 1947. Notes on the origin and function of the artiodactyl tarsus. *Amer. Mus. Nov.* 1356: 1–24.

Simpson, G.G. 1935. The Tiffany Fauna, upper Paleocene: II. Structure and relationships of *Plesiadapis*. *Amer. Mus. Nov.* 816: 1–30.

Simpson, G.G. 1937. The Fort Union of the Crazy Mountain Field, Montana and its mammalian fauna. *Bull. U.S. Nat. Mus.* 169: 1–287.

Simpson, G.G. 1945. The principles of classification and a classification of mammals. *Bull. Am. Mus. Nat. Hist.* 85: 1–350.

Sorenson, M.W. 1970. Behavior of tree shrews, pp. 141–193. *In* L.A. Rosenblum (ed.). *Primate Behavior*. Academic Press, New York.

Sorenson, M.W., and Conaway, C.H. 1964. Observations of tree shrews in captivity. *J. Sabah Soc.* 2: 77–91.

Steinbacher, G. 1940. Beobachtungen am Spitzhörnchen und Panda. *Zool. Gart.* 12: 48–53.

Szalay, F.S. 1968. The beginnings of primates. *Evolution* 22: 19–36.

Szalay, F.S. 1969. Mixodectidae, Microsyopidae and the insectivore-primate transition. *Bull. Amer. Mus. Nat. Hist.* 140: 193–330.

Szalay, F.S. 1975. Early primates as a source for the taxon Dermoptera. *Amer. J. Phys. Anthrop.* 42: 332–333.

Szalay, F.S. 1977a. Ancestors, descendants, sister groups and testing of phylogenetic hypotheses. *Syst. Zool.* 26: 12–18.

Szalay, F.S. 1977b. Phylogenetic relationships and a classification of the eutherian Mammalia, pp. 315–374. *In* M.K. Hecht, P.C. Goody, and B.M. Hecht (eds.) *Major Patterns in Vertebrate Evolution*. Plenum Press, New York.

Szalay, F.S., and Decker, R.L. 1974. Origin, evolution and function of the tarsus in Late Cretaceous Eutheria and Paleocene primates, pp. 223–259. *In* F.A. Jenkins, Jr. (ed.). *Primate Locomotion*. Academic Press, New York.

Szalay, F.S., and Delson, E. 1979. *Evolutionary History of the Primates*. Academic Press, New York.

Szalay, F.S., Tattersall, I., and Decker, R.L. 1975. Phylogenetic relationships of *Plesiadapis*—postcranial evidence, pp. 136–166. *In* F.S. Szalay (ed.). *Approaches to Primate Paleobiology*. Contrib. Primat., Vol. 5. S. Karger, Basel.

Tate, G.H.H. 1947. *Mammals of Eastern Asia*. Macmillan, New York.

Van Valen, L. 1965. Tree shrews, primates and fossils. *Evolution* 19:137–151.

Van Valen, L. 1966. Deltatheridia, a new order of mammals. *Bull. Amer. Mus. Nat. Hist.* 132: 1–126.

Van Valen, L., and Sloan, R.E. 1977. Ecology and extinction of the dinosaurs. *Evol. Theory* 2: 37–64.

Wharton, C.H. 1950a. Note on the life history of the flying lemur. *J. Mammal.* 31: 264–273.

Wharton, C.H. 1950b. Notes on the Philippine tree shrew *Urogale everetti*. *J. Mammal.* 31: 352–354.

Winge, H. 1941. *The Interrelationships of the Mammalian Genera* (trans. from Danish by E. Diechmann and G. M. Allen). C.A. Reitzels Forlag, Kobenhavn.

Yalden, D.W. 1970. The functional morphology of the carpal bones in carnivores. *Acta anat.* 77: 481–500.

Yalden, D.W. 1972. The form and function of the carpal bones in some arboreally adapted mammals. *Acta anat.* 82: 383–406.

Ziemer, L.K. 1978. Functional morphology of forelimb joints in the woolly monkey *Lagothrix lagothricha*. *Contrib. Primatol.* 14: 1–130.

The Tupaiid Dentition 5

PERCY M. BUTLER

1. Introduction

In discussions of the relationships of the Tupaiidae to other mammals, the dentition has received relatively little attention. Its evolutionary plasticity makes the dentition very useful in differentiating species and genera, but at higher taxonomic levels its value is diminished because it is particularly subject to parallel evolution. Nevertheless, comparison with fossil groups must inevitably rest very largely on dental studies, as teeth provide such a large part of the evidence on which mammalian paleontological history is based. The purpose of this chapter is to describe the characteristic features of the dentition of living Tupaiidae and the principal variations to be found within the family. Comparisons will then be made with primates and with various families of fossil insectivores that have been proposed as possible tupaiid relatives.

Mivart (1867), in his study of the osteology of the Insectivora, described the dentitions of *Tupaia* and *Ptilocercus*, which he compared with other insectivores such as *Hylomys*, *Erinaceus*, and *Talpa*. Gregory (1910) also described *Tupaia* and *Ptilocercus*, noting parallelisms to a variety of mammals. He regarded the dentition of *Ptilocercus* as "on the whole a rather primitive placental dentition," but mentioned the "omnivorous modification" of its molars as a lemuroid character; *Tupaia*, with its sharply pointed cusps and enlarged styles, emphasized insectivorous features. Later in the same work (p. 321) Gregory stated that the molars of primates "primitively perhaps resembled those of the modern *Ptilocercus* in many characters." Le Gros Clark (1926), who gave a detailed description of the dentition of *Ptilocercus*, followed Gregory in concluding that it is more primitive and more lemurine than *Tupaia*.

PERCY M. BUTLER • Department of Zoology, Royal Holloway College, Englefield Green, Surrey, England

Lyon (1913), concerned with the intra-family taxonomy of the Tupaiidae rather than with their relationships to other mammals, made many observations on the teeth of the various genera and species. He gave a key to genera based on teeth, and used dental characters in distinguishing between species of *Tupaia*. More recently, Steele (1973) made an extensive study of dental variability in the family, listing 43 character traits on which he based a cluster analysis (he was unable to obtain specimens of *Anathana*). These two papers form the most important reference sources on the tupaiid dentition to date. Swindler (1976) described the dentitions of four species, for which he gives statistics of tooth measurements. Photographs and short descriptions of the dentitions of *Tupaia glis ferruginea, Urogale cylindrura*, and *Ptilocercus lowii* were provided by James (1960).

The account in this chapter is based on a study of the large collection of tupaiid skulls in the British Museum (Natural History). I am indebted to Mr. I.R. Bishop and staff of the Mammal Section of the Museum for much-appreciated help.

2. Dental Formula

Compared with the basic placental formula, Tupaiidae have lost one upper incisor and the first premolars of both jaws, giving a formula:

$$\mathrm{I}\,\frac{1.2}{1.2.3}\,\mathrm{C}\,\frac{1}{1}\,\mathrm{P}\,\frac{2.3.4}{2.3.4}\,\mathrm{M}\,\frac{1.2.3}{1.2.3}$$

In *Urogale everetti* I_3 is sometimes absent (Lyon, 1913). That the missing upper incisor is I^3 is indicated by the fact that whereas I^2 bites between I_2 and I_3, there is no tooth biting between I_3 and the lower canine. The upper canine stands exceptionally far back from the premaxillary-maxillary suture, and it resembles a premolar in size and shape. It has a milk predecessor, which is almost never the case with P^1 (Ziegler, 1971). I^1, I^2, and the canine are spaced, even in *Ptilocercus* where the remaining teeth are close together, and the posterior position of the canine in relation to the suture would aid in producing this presumably adaptive effect. In the lower jaw the canine is always at least slightly higher than the adjacent teeth, and in *Urogale* it has a typically caniniform appearance, biting in front of the upper canine where the maxilla is excavated to receive it. Kindahl (1957) found no trace in embryos of *Tupaia javanica* of tooth germs of dI^3, I^3, P^1, or P_1.[1]

[1] After this manuscript was written, a different interpretation of the dental formula was proposed by Maier (1979). The teeth regarded by me as canines are identified by Maier as first premolars; the upper canine is absent and the lower canine is the tooth here called I_3. Maier's interpretation implies that the first premolars of tupaiids have milk predecessors, a situation very uncommon in mammals (Ziegler, 1971). It is moreover difficult to see why P_1 should tend to be larger than P_2. The upper canine stands back from the premaxillary suture in the Cretaceous mammals *Asioryctes* and *Zalambdalestes* (Kielan-Jaworowska, 1975).

There is evidence in the Cretaceous placentals *Kennalestes* and *Gypsonictops* of the former presence of a fifth premolar (McKenna, 1975), and attempts have been made to identify five premolars in some Tertiary placentals (Schwartz and Krishtalka, 1976; Krishtalka, 1976a). All the cases that have been cited have small premolariform teeth in the canine position, like *Ptilocercus*. For example, the erinaceoid *Litolestes ignotus* (Schwartz and Krishtalka, 1976) has 11 teeth in the jaw like other primitive placentals, but the fourth tooth is low-crowned and procumbent and is interpreted as the first of a series of five premolars, the canine being considered absent. In other species of *Litolestes*, however, there is an enlarged canine, followed by four premolars. In view of the fact that canines and adjacent teeth vary greatly in size, even in members of the same family (e.g., Talpidae, Tenrecidae), it seems more likely that in *L. ignotus* the fourth tooth is a canine that has become secondarily reduced and premolariform. No Tertiary placental has yet been recognized to have I 3, C 1, P 5, and until one is found it seems advisable to keep to the traditional interpretation. Identification of canines, incisors, and premolars from their shapes has long been held to be unworkable (Moseley and Lankester, 1868).

Supernumerary teeth occasionally occur in Tupaiidae, as in other mammals. Thus in *Dendrogale frenata* (BM 71.2608) there is a tooth on the left side anterior to P_2, and in another specimen of the same species (BM 6.11.6.5) a small tooth stands in the left maxilla immediately behind the premaxillary suture. It is doubtful whether these can be interpreted as atavisms. Two cases of M^4 were noticed; *Tupaia glis siccata* BM 15.5.5.37 (right side only) and *Lyonogale tana utana* BM 71.2593 (right side; left side not preserved).

3. Dentition

A. Incisors

The approximately vertical upper incisors contrast with the procumbent lower incisors (Fig. 1). An analogous arrangement occurs in *Erinaceus*. It appears to be an adaptation for picking up food objects. Proportionately the largest incisors are those of *Ptilocercus*, which was described by Gregory (1910) as incipiently diprotodont. Diprotodonty is the condition in which a pair of lower incisors, and usually a pair of upper incisors, are much enlarged; it has evolved independently in many groups of mammals. In *Ptilocercus* both the upper incisors are enlarged; I^2 sometimes has two roots like I^3 of some species of *Erinaceus*. In most Tupaiinae the upper incisors are rather small and subequal, but I^1 is larger in some species of *Tupaia (T. nicobarica, T. javanica, T. minor, T. gracilis)*, and in *Urogale* I^2 has developed into a stoutly constructed stabbing tooth.

In the lower jaw of all Tupaiidae I_2 is larger than I_1. I_1 and I_2 lie closely parallel, and a number of species of *Tupaia* have been observed to use them as a fur-comb (Sorenson and Conaway, 1966). It is not known whether *Ptilocercus*

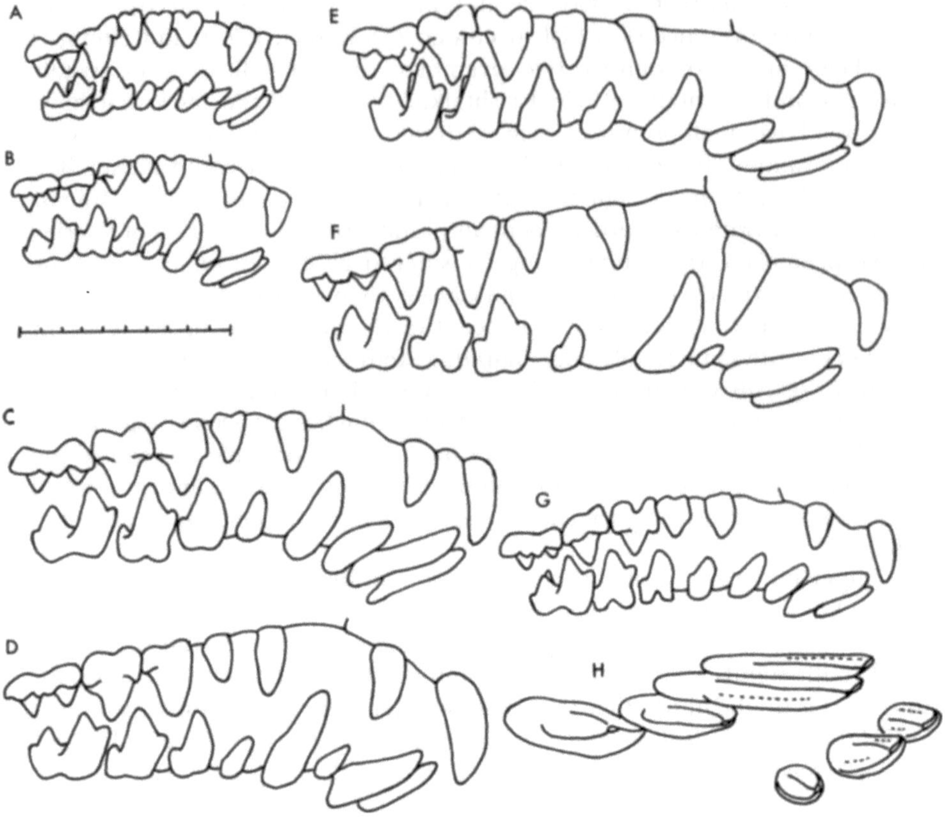

Fig. 1. A–G, lateral view of dentition as far back as M^1 and M_1. A, *Ptilocercus lowii;* B, *Dendrogale frenata;* C, *Tupaia glis ferruginea;* D, *Tupaia nicobarica;* E, *Lyonogale tana;* F, *Urogale cylindrura;* G, *Anathana ellioti.* The scale represents 10 mm. H, *Tupaia glis ferruginea,* lower incisors and canine in dorsal view, and lower incisors in end view, at a higher magnification.

also uses its incisors in this way. I_3, which is smaller and less procumbent, presumably does not form part of the comb. The fur-comb of lemurs, consisting of I_1, I_2, and the canine, is analogous but not homologous with that of tupaiids.

On the upper incisors of *Ptilocercus* the main cusp carries a posterior crest which leads to a posterior basal cusp, better developed on I^2 than on I^1. On I^1 there is also a mesial crest. Among Tupaiinae, the crests of the upper incisors are well developed only in *Dendrogale*, which however lacks the posterior basal cusp. The pattern of the lower incisors is uniform throughout the family. The lingual surface is bounded by a posterior (distal) crest and an anterolingual (mesial) crest. These crests are continuous respectively with the distal and mesial ends of the short incisal edge of the tooth. The lingual surface is raised to form a lingual ridge. Similar arrangements of crests and ridges can be recognized in Tenrecidae and other insectivores, even in Soricidae, and also in many primates.

When the jaws are closed the tips of I_1 and I_2 are in contact with the lingual surface of I^1; I^2 stands laterally to the base of I_2 and immediately anterior to the tip of I_3. When the jaw is lowered and protruded the tip of I^1 wears on its posterior side against the tips of I_1 and I_2, and the tip of I^2 wears in a similar way against the tip of I_3. A few specimens show a wear facet on the lateral side of I_2, near the base, due to contact with the tip of I^2 during lateral excursions of the jaw.

Except for their smaller size, the deciduous incisors resemble the corresponding permanent teeth. In *Urogale* dI^2, though enlarged, is only a little larger than dI^1. dI^2 of *Ptilocercus* has two roots like the canine.

B. Canines

In *Ptilocercus* the upper canine is a two-rooted tooth resembling P^2 and smaller than the incisors. Its main cusp has a strong posterior crest leading to a posterior basal cusp. In *Dendrogale* it is slightly larger than I^2 and P^2; it is again two-rooted, with a posterior crest, but the posterior cusp is not developed. In the other Tupaiinae the upper canine is a simple peg-like or somewhat recurved tooth without a distinct posterior crest and nearly always with a single root. It is somewhat higher than P^2 in some species of *Tupaia*, especially in *T. nicobarica* and *T. javanica*, but in *T. gracilis, T. dorsalis*, and *T. picta*, as well as in *Lyonogale, Anathana*, and *Urogale*, the canine and P^2 are of nearly equal height.

The lower canine always shows some enlargement in comparison with I_3 and P_2, but in *Anathana* the difference is slight. The canine is taller than the molars only in *Urogale* and, to a lesser degree, in *Tupaia nicobarica* and *T. javanica*. It is somewhat procumbent, though less so than the incisors. Except in *Ptilocercus*, it is separated from I_3 by a short gap. The pattern of the lower canine can easily be compared with that of the incisors. There is a posterior crest, especially well developed in *Ptilocercus* and *Dendrogale*, where it ends in a posterior basal cusp. An anterior crest turns lingually at the anterior end to form a rudimentary anterolingual cingulum. The lingual ridge is rounded and not always differentiated.

In *Ptilocercus* the lower canine is close to P_2, and when the jaws close the tip of the upper canine passes laterally to the contact point, occluding with both lower teeth. In Tupaiinae there is a diastema between the lower canine and P_2, very short in *Dendrogale* and comparatively long in *Lyonogale, Anathana*, and *Urogale*. Contact between upper and lower canines is retained in *Dendrogale* and some species of *Tupaia*, including *T. glis*, but it is lost in *T. javanica, T. nicobarica, Lyonogale, Anathana*, and *Urogale*.

The deciduous canines resemble the permanent teeth except for their smaller size. In *Urogale* the lower deciduous canine is less enlarged in comparison with other teeth than the permanent canine.

C. Premolars

P^2 is a simple tooth which resembles the canine in pattern. It is best developed in *Ptilocercus*, where there are two well separated roots and a strong posterior crest leading to a distinct posterior basal cusp. In the Tupaiinae P^2 shows various degrees of simplification: the roots are frequently united, the posterior cusp is absent, and the posterior crest is weak except in *Dendrogale*.

P^3 is a triangular tooth in early stages of molarization (Fig. 2). In Tupaiinae it is much larger than P^2 and similar in length and paracone height to P^4; in *Ptilocercus* it is equal in height to P^2 and much smaller than P^4. The paracone of P^3 is connected to the posterobuccal corner of the crown by a strong crest,

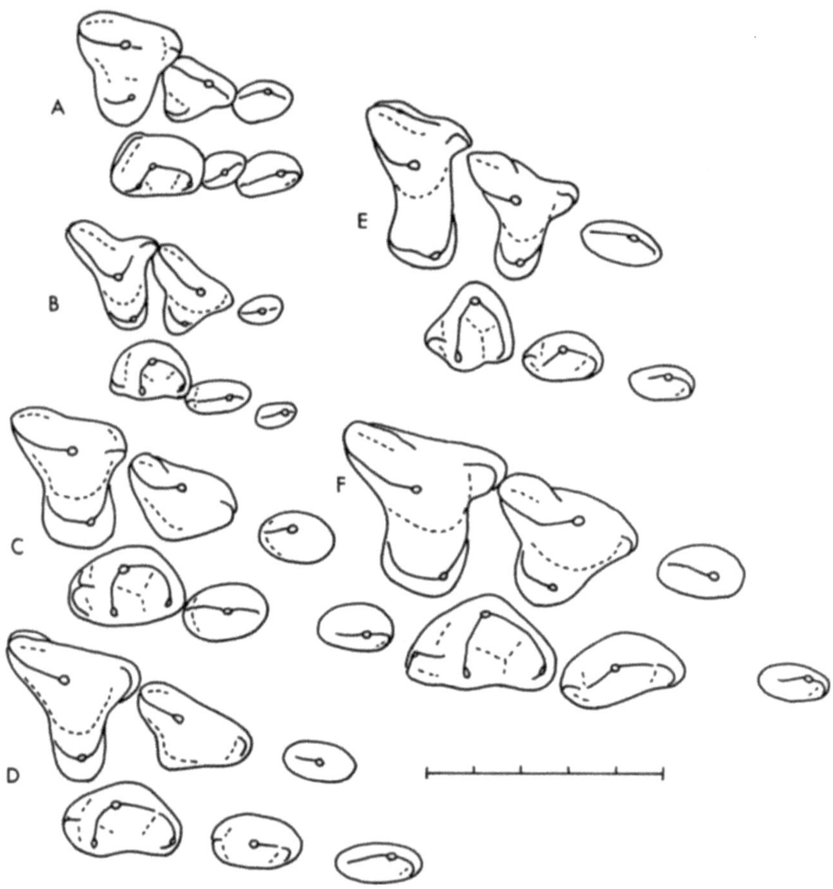

Fig. 2. Upper and lower premolars, crown view. A, *Ptilocercus lowii;* B, *Dendrogale frenata;* C, *Tupaia montana;* D, *Lyonogale tana;* E, *Anathana ellioti;* F, *Urogale cylindrura.* The scale represents 5 mm.

but there is no metacone. The posterior part of the buccal margin is elevated to form a cingulum which however does not extend past the paracone. There are usually three roots, but in some poorly molarized specimens the posterobuccal and lingual roots are united. The protocone varies considerably in its development, even within species. Typically it is a low cusp standing on a lingual shelf, but it may be absent or represented only by a cingulum. In *Ptilocercus lowii* the protocone is present in four specimens from Malaysia but absent in two out of three specimens from Borneo. The protocone is normally present in *Dendrogale*, and particularly well developed in *Anathana*, where P^3 is a transversely widened tooth resembling P^4. Another variable cusp is the parastyle, situated at the anterior corner of the crown. It is absent in *Ptilocercus* and in some specimens of *Tupaia*; when present it may be joined to the tip of the paracone by a crest. A cingulum may link the parastyle with the protocone, as on P^4.

P^4 is less variable than P^3. It is broader than P^3 and always has a separate lingual root supporting a protocone. The crown is dominated by the paracone, the tip of which projects below the level of the molar cusps. It has a strong posterior crest which reaches the posterobuccal corner of the crown, and frequently there is also a less distinct anterior crest to the parastyle. There is a well developed buccal cingulum, usually complete but sometimes broken or indistinct buccally to the tip of the paracone. The lingual border of the tooth is occupied by a cingulum which usually connects with the parastyle and broadens lingually to form the protocone shelf. The protocone is much lower than the paracone and stands somewhat more anteriorly. Metacone and hypocone are absent.

P_2 is a simple, somewhat procumbent tooth, smaller than the lower canine which it resembles in pattern. It normally has a single root, but a second root was observed in a specimen of *Tupaia glis ferruginea* (BM 9.4.1.109). There is a posterior crest, which in *Ptilocercus* and *Dendrogale* leads to a posterior basal cusp, and an anterior crest which at the anterior end turns lingually to form an anterolingual cingulum.

P_3 of *Ptilocercus* is smaller than P_2 and resembles it in pattern; it has a single root. In Tupaiinae P_3 is larger, two-rooted, and better differentiated. It has a small posterior cusp or heel, continuous lingually with a short length of cingulum, representing a rudimentary talonid. A paraconid is frequently developed at the anterior end of the anterior crest, but there is no metaconid; at best this cusp is represented by a posterolingual crest present in *Anathana* and *Tupaia nicobarica*.

P_4 is much larger than P_3 and always has two roots. Its trigonid is recognizably molariform: the metaconid is nearly always differentiated from a posterolingual crest of the protoconid; the paraconid is always present at the anterior end, and becomes more lingually situated when it is larger and more molariform. The trigonid angle is thus more widely open than on the molars but becomes more reduced as the level of molarization rises. The protoconid

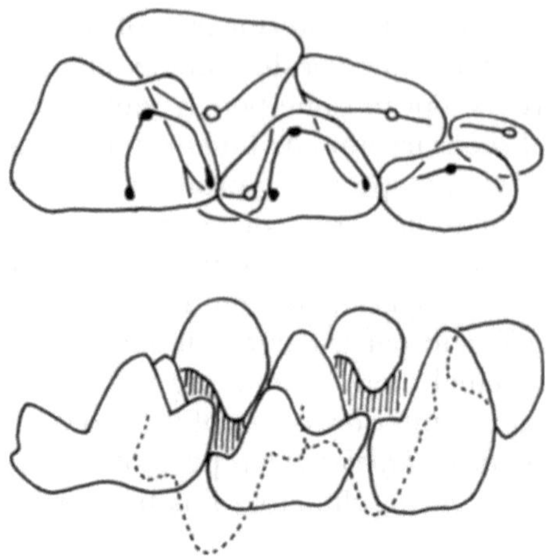

Fig. 3. *Tupaia glis ferruginea.* Premolars in centric occlusion, crown and lingual views.

is similar in height to that of the molars in Tupaiinae, somewhat higher in *Ptilocercus*. In contrast with the trigonid, the talonid retains the simple pattern found on P_3, though it is somewhat larger. *Urogale* is exceptional in that an angulate but very small hypoconid has differentiated from a more lingual cusp, probably the hypoconulid.

In *Ptilocercus* the tooth-rows of both jaws are closed from the canines backwards, and there is an indication of crowding in that P_3 tends to be rotated diagonally to the tooth-row. In Tupaiinae on the other hand P^2 and P_2 are always separated from the adjacent teeth by spaces (diastemata). The spaces are small in *Dendrogale* and short-faced species of *Tupaia (T. minor, T. gracilis)*, and in *Lyonogale* and *Urogale*, where the face is long, the spaces are much greater, especially those anterior to P^2 and P_2.

The posterior premolars, P^4 and P_4, supplemented in Tupaiinae by P^3, are probably the teeth involved in "ingestion by mastication" described by Hiiemae and Kay (1973). They apparently function mainly by holding and puncturing with the tips of the paracone and protoconid, supplemented by shearing between the posterior crest of the upper tooth and the anterior (protoconid-paraconid) crest of the more posterior lower tooth (P^3 against P_4 and P^4 against M_1) (Fig. 3). P_3 of *Ptilocercus* and P_2 of Tupaiinae have little or no occlusal contact because of their reduced size, though in some Tupaiinae P_2 touches the upper canine, as in *Ptilocercus*; in *Ptilocercus* P_2 also touches P^2. The diminutive talonid of P_4, and in Tupaiinae that of P_3, occlude with the tips of the paracones of the corresponding upper premolars, producing steeply inclined, buccally facing wear facets. There is no contact of talonid with protocone, but the protocone shelf of P^4 meets the paraconid of M_1 when the jaws close, and in a similar manner the protocone of P^3, when sufficiently developed, meets the paraconid of P_4. The protocone of P^4 also shears against the posterior surface of the metaconid of P_4.

D. Deciduous Molars

dP^2 is a simple tooth resembling P^2 except that its two roots tend to be more divergent and they are very seldom united.

dP^3 is a triangular tooth like P^3 (Fig. 4). Its protocone and parastyle vary in parallel with the corresponding cusps of P^3. In Tupaiinae the paracone of dP^3 is the most elevated of the upper milk molar cusps and probably has a puncturing function, but in *Ptilocercus* dP^3 is much smaller in comparison with dP^4.

dP^4 is molariform, having well developed metacone and protocone. It differs from M^1 in being narrower, especially anteriorly, where the paracone is less removed from the buccal margin and the parastyle projects more for-

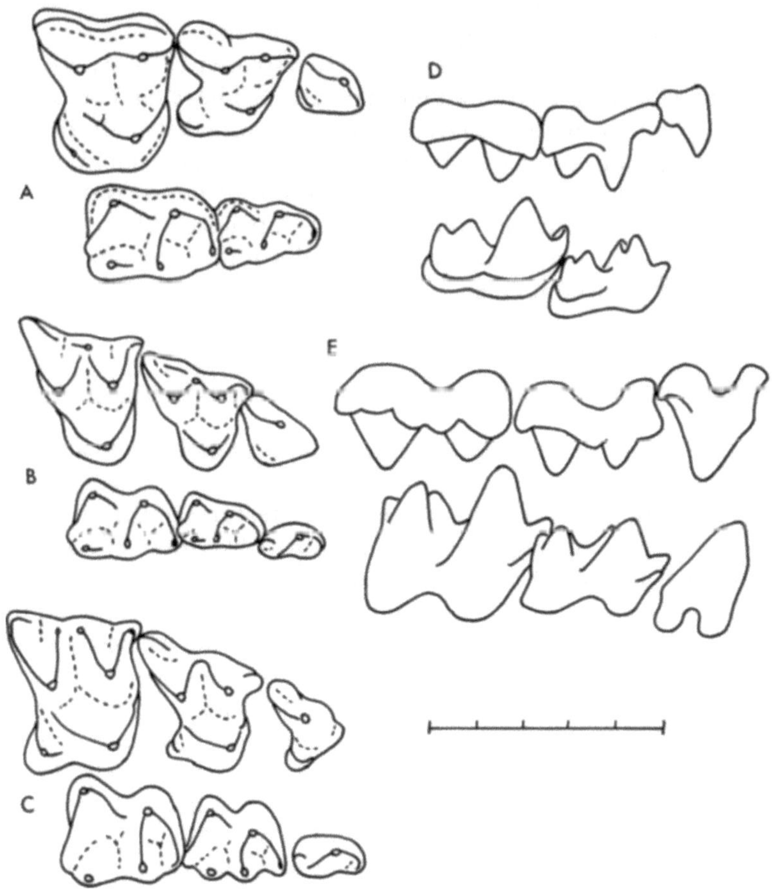

Fig. 4. First permanent molars and deciduous molars. A, *Ptilocercus lowii,* M^1–dP^3 and M_1–dP_4, crown view. B, *Tupaia minor malaccana,* M^1–dP^3 and M_1–dP_3, crown view. C, *Anathana ellioti,* the same. D, *Ptilocercus lowii,* M^1–dP^3 and M_1 –dP_4, buccal view. E, *Tupaia glis,* M^1–dP^3 and M_1–dP_3, buccal view. The scale represents 5 mm.

ward. The metacone is equal to or greater in height than the paracone in Tupaiinae, but in *Ptilocercus* the paracone is more elevated, presumably having a puncturing function like that of P^4. Generic differences in dP^4 reflect differences of molar pattern. Thus in *Ptilocercus* the metacone is connected directly to the paracone by a crest, whereas in Tupaiinae both cusps are joined to the mesostyle. The hypocone is always weaker on dP^4 than on M^1, but it is present in *Ptilocercus, Anathana*, and *Urogale*, the genera in which it is best developed on M^1. dP^4 resembles P^4 in that the buccal cingulum is weak or absent above the paracone; in *Ptilocercus* the cingulum is confined to the metacone region, and in Tupaiinae it is frequently lacking between the parastyle and the mesostyle.

dP_2 is a simple, somewhat procumbent, one-rooted tooth like P_2. dP_3 like P_3 lacks a metaconid, but in Tupaiinae there is usually a posterolingual crest connecting the protoconid with the lingual side of the rudimentary talonid basin. I have not seen an example of dP_3 of *Ptilocercus*.

dP_4 is molariform. It is proportionately narrower than M_1, especially in the trigonid, which has a more widely open angle. The talonid occupies about half the length of the tooth and agrees in structure with that of M_1. In *Ptilocercus* the buccal cingulum, complete on M_1, is confined to the talonid on dP_4.

The milk molars are spaced in the same way as the corresponding premolars. The main puncturing tooth is dP^3 in Tupaiinae, dP^4 in *Ptilocercus*. In both subfamilies the protocone of dP^4 occludes with the talonid on dP_4 in a completely molariform manner.

E. Molars

The terms used in this chapter for description of molar teeth are explained in figures 5 and 6.

The two subfamilies differ in a number of features of the molar pattern. The Tupaiinae have dilambdodont upper molars like those of Soricidae and insectivorous bats. On M^1 and M^2 the paracone and metacone are V-shaped cusps, connected to the styles on the buccal edge by well developed transverse crests. The deep groove between the paracone and the metacone extends to near the buccal margin where it is bounded by the mesostyle. This is a single cusp in *Tupaia minor, T. gracilis, T. javanica*, and *T. nicobarica*, and a ridge-like structure in *Urogale*, while in the remaining Tupaiinae it is represented by two cusps, connected with the paracone and metacone respectively (Figs. 4, 7). Differences in the mesostyle appear to reflect the width and buccal extent of the groove between the paracone and the metacone. This groove receives the hypoconid during occlusion. The metacone is higher than the paracone on M^1, and it is connected to the posterobuccal corner of the crown by a strong shearing crest (metacrista), which occupies about half the posterior edge of the tooth. It shears against the anterior edge of the lower molar trigonid (protoconid-paraconid crest).

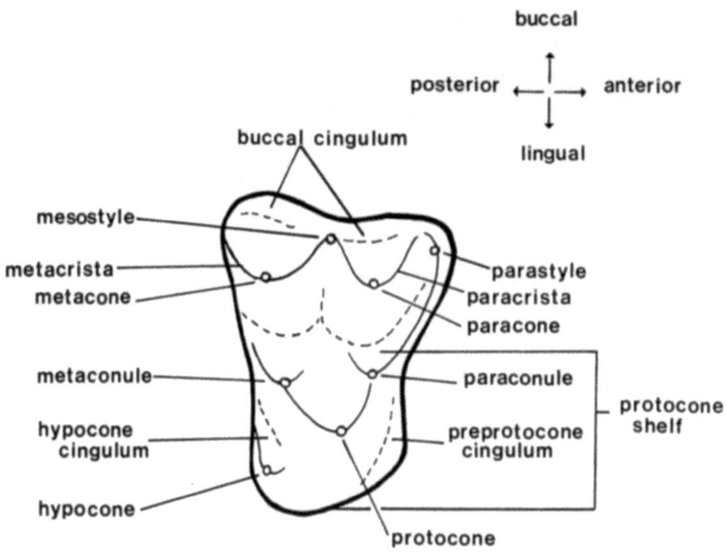

Fig. 5. Diagram of eutherian upper molar to illustrate the terminology used.

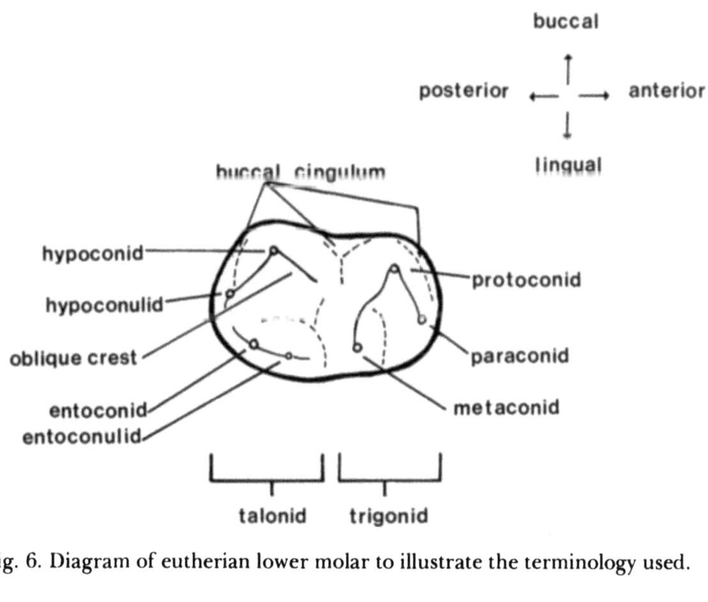

Fig. 6. Diagram of eutherian lower molar to illustrate the terminology used.

The protocone shelf is correspondingly narrow, occupying only about half the width of the tooth or less. The protocone stands on the anterior half of the crown. Its anterior crest follows the anterior border of the tooth and usually extends as a very narrow cingulum past the paracone to the lingual surface of the parastyle. Its posterior crest runs at first almost directly backwards and then curves to meet the posterior edge of the tooth at the base of the metacone. Conules are absent or rudimentary. There is no cingulum anterior to the protocone, except occasionally for a minute rudiment. One specimen of *Anathana ellioti* (BM 942b) has a supernumerary cusp on M^1 and M^2 anterior to the protocone, best developed on the right side.

The hypocone varies in its development. It is best differentiated in *Anathana* (Fig 4) and *Urogale* (Fig. 7), where it stands on a cingulum that occupies a semicircular prominence of the posterolingual outline of the crown. In *Lyonogale* and most species of *Tupaia* the prominence is smaller and the hypocone is hardly if at all differentiated from the cingulum. In *T. javanica, T. nicobarica*, and some specimens of *T. glis* the prominence is only a posterolingual angulation of the outline, and no cingulum or cusp is developed. Finally, in *T. minor* and *Dendrogale* even the angulation is absent.

The upper molars have three roots, of which the lingual root has the

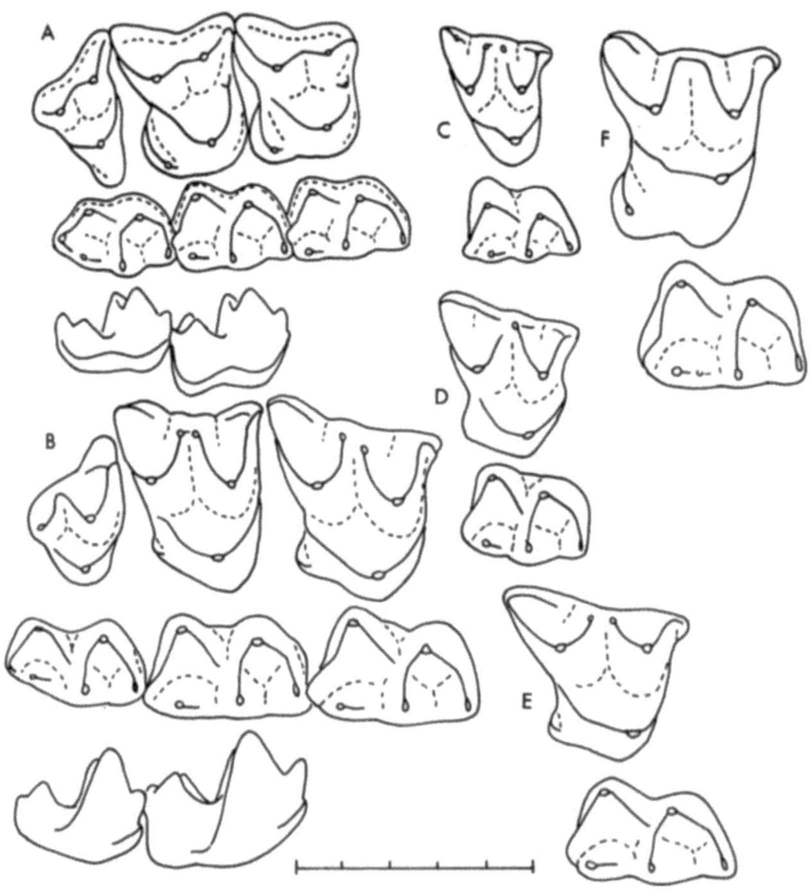

Fig. 7. Permanent molars. A, *Ptilocercus lowii*, upper and lower molars in crown view and M_3–M_2 in buccal view; B, *Tupaia glis*, the same. C–F, M^1 and M_1 in crown view: C, *Dendrogale frenata;* D, *Tupaia javanica;* E, *Lyonogale tana;* F, *Urogale cylindrura*. The scale represents 5 mm.

greatest diameter, especially in *Urogale* and *Anathana* which have the largest hypocones. There are no interradicular crests such as occur in Erinaceidae (Butler, 1948), but skulls from which the teeth have fallen out usually show a foramen, presumably vascular, in the centre of the triangle formed by the alveoli. M^1 and M^2 are proportionately shortest mesiodistally in *Anathana* and longest in *Lyonogale*.

In *Ptilocercus* there is no mesostyle, the buccal edge of the tooth being occupied by a continuous cingulum (Fig. 7). The paracone is a conical cusp, connected directly to the metacone by a crest. The metacone is equal in height to the paracone on M^1 and M^2. It is farther removed from the buccal edge than the paracone, but as in Tupaiinae it is connected to the posterobuccal corner of the tooth by a metacrista; however, this crest occupies only about one-third of the posterior edge of the tooth. The protocone forms about half the tooth width; its anterior crest sometimes develops a paraconule, and continues as a rather broad cingulum up to the parastyle; in some specimens the posterior crest of the protocone is extended past the base of the metacone. There is a cingulum anterior to the protocone. A small but distinct hypocone is developed from the cingulum on a posterolingual lobe of the outline of the tooth.

In all Tupaiidae M^2 differs from M^1 in being proportionately broader buccolingually. The parastyle points more buccally, the mesostyle cuspules of Tupaiinae are less distinctly separated, the metacone is somewhat reduced in height (it is higher than the paracone on M^2 only in *Anathana* and *Urogale*), and the hypocone region is more weakly developed. On M^3, as generally in primitive placentals, the posterobuccal corner of the crown is reduced, so that the buccal edge is rotated posteriorly, the parastyle points directly buccally, the posterior edge of the tooth is short, and the posterior and lingual roots are closer together (Butler, 1939). The metacone is reduced in size and its posterior crest is frequently lost; the mesostyle of Tupaiinae is undivided; the hypocone is usually absent, though a trace of the cingulum remains in *Ptilocercus* and *Urogale*.

Tupaiid lower molars possess a classical trituberculo-sectorial pattern. Trigonid and talonid are of nearly equal length. In most genera the three trigonid cusps are arranged in an approximately equilateral triangle, but in *Anathana* and to a lesser degree in *Urogale* the trigonid is shortened mesiodistally, while in *Lyonogale* it is somewhat narrowed buccolingually on M_1. The paraconid, the lowest of the trigonid cusps, is a trenchant cusp, situated on the lingual side of the tooth in Tupaiinae but slightly less lingual than the metaconid in *Ptilocercus*. The metaconid is placed a little more posteriorly than the protoconid in *Ptilocercus, Dendrogale, Lyonogale*, and most species of *Tupaia*; in *T. montana, T. picta*, and *T. dorsalis*, as well as in *Anathana* and *Urogale*, it is more directly lingual. It is lower than the protoconid except in *Anathana*, where the two cusps are equal in height. The trigonid cusps are lower and blunter in *Ptilocercus* than in the Tupaiinae.

The talonid is wider than the trigonid on M_1. Except in *Anathana*, trigonid and talonid are of equal width on M_2 and the talonid is narrower on M_3; in *Anathana* the talonid is wider on M_2 and equal in width to the trigonid on M_3. The hypoconid and entoconid are lower than the protoconid and metaconid. The anterior crest of the hypoconid (oblique crest) is inclined lingually in Tupaiinae, meeting the trigonid a little buccally to the midline of the tooth on M_1 and in the midline on M_2 and M_3. In *Ptilocercus* the oblique crest is more longitudinal, meeting the trigonid at the base of the protoconid; as a result, the groove on the buccal side between the trigonid and the talonid is shallower in *Ptilocercus* than in Tupaiinae, in correlation with the more buccal position of the paracone on the upper molars. The entoconid is usually somewhat lower than the hypoconid on unworn teeth; it is unusually small in *Tupaia gracilis*, and in *Anathana* it is higher than the hypoconid. On its anterior side an additional cuspule (entoconulid) occasionally develops on M_1 and M_2 (Steele, 1973). The hypoconulid is a low cusp which overhangs the posterior margin of the tooth. It is connected to the hypoconid by a crest, but divided from the entoconid by a groove. In *Ptilocercus* the hypoconulid is situated at one-third of the width of the talonid from the lingual side; in Tupaiinae it is quite near the entoconid, though a little less lingual than that cusp. In all Tupaiidae the talonid of M_3 is narrowed and its entoconid reduced, but only in *Ptilocercus* is the hypoconulid larger than on M_2. *Ptilocercus* also differs from Tupaiinae in that M_3 as a whole is less reduced in size in comparison with M_2.

In *Ptilocercus* a complete buccal cingulum runs from below the paraconid to the hypoconulid, but in Tupaiinae the cingulum is absent except for fragments which occur in some species, below the paraconid and in the reentry between the protoconid and the hypoconid.

4. Molar Function

Mills (1955), from a study of the wear of the teeth in museum specimens of *Tupaia*, showed that molar occlusion involved a transverse movement of the lower molars across the upper molars. He was able to compare the wear facets with those of the gorilla, and in later papers (Mills, 1963, 1966) with other primates, as well as with *Ptilocercus*, Macroscelididae and lipotyphlous insectivores. Subsequently, the wear facets of *Tupaia* molars were described in detail by Kay and Hiiemae (1974) and Maier (1977). Mastication in the living animal was investigated by Hiiemae and Kay (1973), using a cineradiographic technique that had previously been applied to the opossum *(Didelphis)* by Crompton and Hiiemae (1970).

Mastication in *Tupaia* is essentially like that of *Didelphis*, and therefore presumably like that of other mammals whose molar patterns have not departed far from the primitive tribosphenic type. Food is initially subjected to a pulping action by a series of puncture-crushing masticatory cycles, using the tips of the cusps and edges of the crests. Only after the food has been sufficiently softened do the opposing teeth come into close contact. Kay and Hiiemae (1974) confine the term *chewing* to this second stage of mastication. Puncture-crushing results in abrasion, the wearing down of cusps and crests; chewing results in the formation of the attrition facets described by Mills and others. Kay and Hiiemae (1974) distinguish three elements of chewing: shearing, crushing, and grinding. Shearing is performed by the edges of crests which pass each other like the blades of scizzors as the jaws close together. Crushing is produced by pressing together more or less horizontal opposing surfaces. In grinding, opposing surfaces while pressed together slide past each other.

The masticatory cycle can be divided into three parts: a preparatory stroke, a power stroke, and a recovery stroke. In *Tupaia* the preparatory stroke begins with the mouth widely open and with the lower jaw displaced towards the side on which chewing takes place (the active side). The jaw is then moved upwards and medially till at the beginning of the power stroke the buccal surfaces of upper and lower cheek teeth are in vertical alignment. During the power stroke the lower teeth are moved medially, upward and slightly forward in contact with the upper teeth until maximum interpenetration of the cusps (the centric position) is reached. Thereafter the upward movement ceases, but forward movement is increased so that the lower teeth move anterolingually in relation to the upper teeth until they separate for the recovery stroke.

Two phases of the power stroke may thus be distinguished. During the buccal phase of Mills (=Phase I of Hiiemae and Kay, 1973) the lower teeth are moving lingually and upward towards the centric position; during the lingual phase (=Phase II) they are lingual to the centric position and moving more anteriorly, and horizontally or slightly downward. Mills believed that the centric position coincided on both sides of the mouth, the buccal phase on one side being accompanied by a balancing reversed lingual phase on the other. However, Hiiemae and Kay (1973) showed that this is not so; owing to mobility at the symphysis the halves of the mandible can be rotated around their long axes, and the molar regions of the two sides brought closer together, so that the teeth on the balancing side make only transitory contact near the end of the power stroke.

At the beginning of the buccal phase the protoconid of the lower molar meets the parastyle (Fig. 8). The trigonid passes up into the embrasure between two upper molars, its anterior (protoconid-paraconid) crest shearing against the posterior metacone crest (metacrista), and its posterior (protoconid-metaconid) crest shearing first with the paracone-parastyle crest (paracrista) and then with the anterior protocone crest. Soon after the protoconid touches the parastyle the hypoconid meets the mesostyle. It passes up the groove between the paracone and the metacone, its anterior crest (oblique crest) shearing against the paracone-mesostyle crest and its posterior (hypoconid-hypoconulid) crest shearing against the mesostyle-metacone crest. As the lower molar moves lingually the entoconid shears against the posterior crest of the protocone. At the centric position the tip of the hypoconid lies in the centre of the trigon basin, crushing food imprisoned between the lingual surface of the hypoconid and the buccal surface of the protocone. The paraconid meets the small hypocone of the more anterior upper molar. Thus chewing in the buccal phase consists of shearing, supplemented by crushing at the end of the phase.

In the lingual phase the hypoconid moves anterolingually, in the direction of the oblique crest. The crests of the hypoconid are in contact with the crests of the protocone, crossing them at right angles, and the edges of the crests are worn off during the movement. As the buccal surface of the protocone and the lingual surface of the hypoconid are both concave they cannot function by grinding; the grinding surfaces are in fact confined to the crescent-shaped areas produced by wear of the tips of these cusps and the edges of their crests. It would seem that the lingual-phase grinding function is more poorly developed in Tupaiidae than in primates (Butler, 1973).

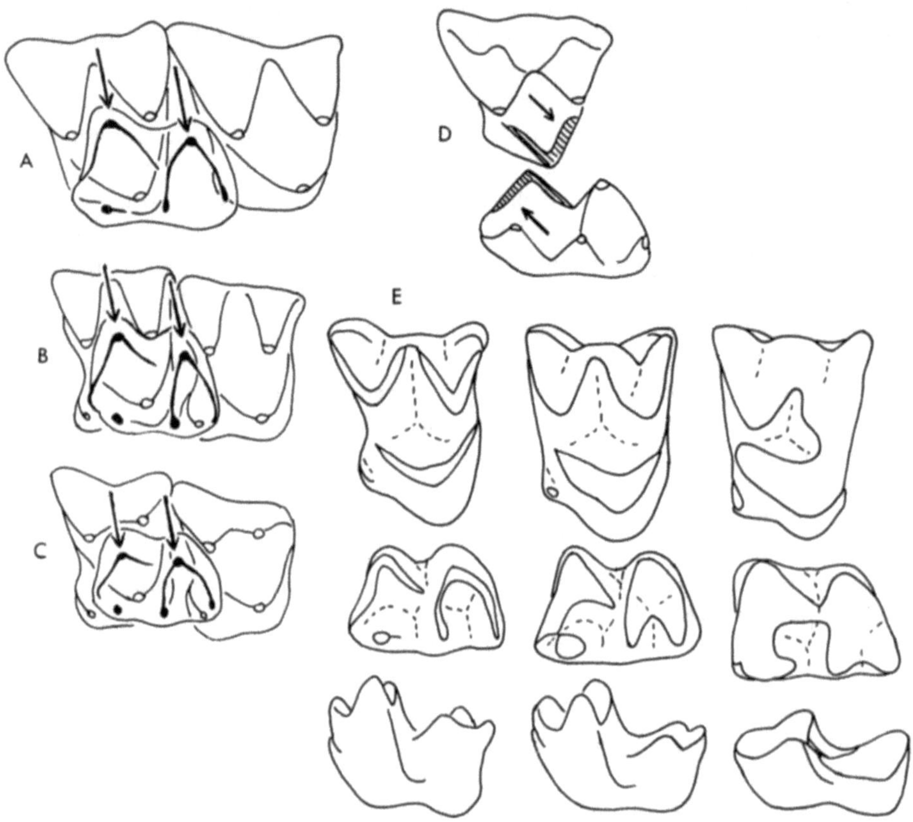

Fig. 8. Molar occlusion. A, *Tupaia glis*, M_2 drawn in position of centric occlusion with M^1 and M^2. B, *Anathana ellioti*, the same. C, *Ptilocercus lowii*, the same. The arrows show the degree of transverse movement of the protoconid and hypoconid in the buccal phase. D, *Tupaia glis*, lightly worn M^1 and M_1 in oblique view to show attrition facets produced in the lingual phase. The arrows show the direction of relative movement. E, *Tupaia glis*, three stages of wear to illustrate abrasion. M^2 and M_2 in crown view and M_2 in buccal view.

Although most attention has been paid to attrition facets, abrasive wear is conspicuous in *Tupaia*. It results in the removal of enamel from the upstanding features of the teeth. On the upper molars a W-shaped strip of dentine is exposed, involving the parastyle, paracone, mesostyle, metacone, and metacrista. Lingually, the protocone and its crests wear into a crescentic area. On the lower molars the trigonid cusps and crests become worn from the top, and a crescentic area of dentine develops on the talonid, incorporating the hypoconid and its crests. It seems probable that abrasion is more important than lingual-phase grinding in producing wear of the protocone and the hypoconid. Abrasive wear begins in *Tupaia* early in the functional life of the tooth. In old individuals the paracone becomes more heavily worn than the metacone, and on the lower molars the buccal cusps wear more than the lingual cusps. The central basin of the upper molar and the deepest part of the talonid basin show little or no wear, even in specimens where the cusps have been largely removed.

Ptilocercus lacks the transversely arranged shearing crests which in Tupaiinae join the paracone and metacone with the parastyle and mesostyle; only the metacrista forms a transverse shearing edge. The total length of shearing crest is thus less than in Tupaiinae. In addition, the cusps are blunter and the edges of their crests more obtuse. Perhaps because of this, abrasion seems less important (to judge from the small sample available). The jaw muscles of *Ptilocercus* are proportionately larger than in Tupaiinae: in the skull the zygomatic arches are more widely expanded, the temporal space is larger, and the coronoid process of the mandible is broader and more vertical. The condyle is also less elevated in relation to the teeth. *Ptilocercus* can probably exert more pressure between its molars, and the bluntness of its cusps is in conformity with this. It is possible also that the temporal muscle is important in biting with the enlarged incisors; in this connection it is interesting that *Urogale*, with its enlarged I^2 and lower canine, has a larger temporal muscle than other Tupaiinae.

5. *Diet*

Information on the diet of Tupaiidae has been reviewed by Martin (1968). All are primarily insectivorous, to judge by food preferences in captivity and stomach contents of specimens killed in the wild. In addition to insects and other invertebrates most species for which there is information eat some plant material, such as fruit, seeds, and shoots. Thus, Harrison (1955), analyzed the stomach contents of 15 specimens of *Tupaia montana* as consisting of 83% insects, 10% fruits and nuts, and 7% leaves and shoots. *Lyonogale tana* appears to be more carnivorous: stomach contents of a specimen consisted of terrestrial arthropods and earthworms (Davis, 1962), but in captivity it showed a preference, after insects, for mice and canned horsemeat (Sorenson and Conaway, 1964). *Urogale* also eats mice, lizards, and raw meat in captivity, in addition to insects and fruit (Wharton, 1950).

Le Gros Clark (1926) fed *Ptilocercus* in captivity on cockroaches and bananas, and stated that a stomach contained insect remains, including a large grasshopper. Lim (1967) found that nine stomach contents consisted almost entirely of insects, but one contained a young gecko. This does not support the view of Gregory (1910) that *Ptilocercus* is more omnivorous than *Tupaia*.

Kay (1975) found a relationship between molar structure and diet in a statistical study of *Tupaia glis* and 37 species of primates. After eliminating size-regression, he concluded that insectivorous primates (*Tarsius spectrum, Loris tardigradus, Arctocebus calabarensis,* and *Galago demidovii*) had longer shearing blades, a greater than average transverse (Phase I) shearing movement, a greater crushing and grinding area, and larger molar teeth. In a principal components analysis *Tupaia* grouped with these insectivorous forms. Kay's data show that *Tupaia* has a larger M_2 than in primates of similar size, with the exception of *Arctocebus*. It also has, for its size, a greater transverse chewing movement, its departure from the regression line for all primates being equalled only by *Alouatta* and the great apes.

6. Sequence of Tooth Development

Kindahl (1957), studying *Tupaia javanica*, found that germs of the deciduous incisors, canines, and third deciduous molars were present in the 11.5 mm fetus. From dP^3 and dP_3 backwards the teeth arise in serial order, M^2 and M_2 reaching the cap stage in the 35 mm fetus and M^3 and M_3 appearing after birth. dP^2 and dP_2 however arise only at the same time as the first permanent molars (23 mm). The order of eruption of the deciduous teeth seems to be the same as their order of development; among the deciduous molars this is 3>4>2 (Lyon, 1913).

Kindahl (1957) found the first germs of replacing teeth (I^1 and I^2) in the 30 mm fetus. Most replacing teeth made their appearance with the second permanent molars at the 35 mm stage, the fourth premolars being the best developed. P^2 and the canine were the last replacing teeth to arise in the upper jaw, developing with M^3 (Kindahl does not provide details of the lower dentition at this stage). Lyon (1913) stated that P^4 and P_4 usually erupt first, the second premolars at nearly the same time or just before, and the third premolars last. In the upper jaw P^3 is followed by $C>I^2>I^1$, but in the lower jaw the incisors erupt relatively earlier, overlapping with the premolars; their order is $I_3>I_1>I_2$, with I_1 erupting at about the same time as P_4. Shigehara (1975), from a radiographic examination of 23 specimens of *Tupaia glis*, obtained a slightly different result. In the upper jaw the sequence was $M^1>M^2>M^3>P^4=P^3>P^2>C>I^1>I^2$. In the lower jaw it was $M_1>M_2>M_3>I_3=P_2>P_4>I_1=C>P_3=I_2$. It should be noted that the order of eruption departs in a number of ways from the order of appearance in development, notably in the retardation of the upper incisors.

My own observations on museum specimens indicate that there is some variation in eruption sequence, both between and within species. In Tupaiinae M^3 and M_3 nearly always erupt before any antemolar is replaced, but there are a few specimens, such as *Tupaia javanica* BM 9.1.5.559, where P^2 and M^3 are erupting together. A specimen of *Ptilocercus* (BM 12.6.8.1) shows the tip of P^2 next to dP^2, and in the lower jaw P_2 erupted and P_3 in process of eruption, while M^2 and M_2 are incompletely erupted and the third molars are unerupted. This indicates that in *Ptilocercus* the premolars erupt earlier, in comparison with the molars, than in Tupaiinae. The commonest order in which the premolars of Tupaiinae erupt was found to be 2>4>3 in both jaws, in disagreement with the sequence reported by Shigehara (1975) but more in agreement with Lyon (1913). There are some specimens in which second and fourth premolars are erupting together, but no case was found in which fourth premolars were not in advance of the third premolars. The discrepancy with Shigehara's result may be due to the different method of observation.

From my observations the upper canine erupts after, together with, or before P^3 with about equal frequency; in one case it was erupting together with P^4. The sequence $C > I^1 > I^2$ seems to be invariable, in agreement with Shigehara but not with Lyon. In the lower jaw I_3 seems to be erupted later than P_2 and often later than P_4, but without the use of radiography it is not always possible to distinguish I_3 from dI_3. I_1 and the lower canine erupt approximately together, but either can be in advance. The most frequent order seems to be $P_4 > I_1 > C > P_3$, but there are several variants. I_2 is always erupted after I_1 and P_3.

Only three specimens of *Ptilocercus* in the British Museum show erupting teeth. In BM 12.6. 8.1 P_3 is erupting but dP_4 is still in place, an order not observed in Tupaiinae. Two older specimens show that the second incisors are the last teeth to erupt in both jaws, as in Tupaiinae.

In prosimian primates (Schwartz, 1975) the incisors and canines erupt earlier than in Tupaiidae, usually before the second molars and in some cases before the first molars. Upper incisors erupt before the canines in primates, after them in tupaiids. A premolar eruption sequence of 2>4>3, resembling that of Tupaiinae, occurs in all prosimians that have three premolars, except certain Lemurinae. At least one premolar erupts before the third molar as in *Ptilocercus*, again with the exception of some Lemurinae.

7. *Primitive and Advanced Characters of the Dentition*

In the absence of an early Tertiary fossil record, the evolutionary history of the Tupaiidae must be hypothetical. We know only the end-result, and can only speculate about the past evolutionary changes that produced it. Without knowing the intermediate steps we can however compare the end with the beginning, the modern tupaiid dentition with the dentition of the ancestral placental stock from which tupaiids, like other placentals, have been derived. Advances in knowledge of Cretaceous and Paleocene placentals are leading to a conception of the direction of dental evolution; though still hypothetical, a picture of the ancestral placental dentition is beginning to emerge. The cheek teeth of Cretaceous placentals have recently been reviewed by Butler (1977), who concludes that most features of the ancestral placental were retained by *Kennalestes* and Palaeoryctidae such as *Cimolestes*. By comparing Tupaiidae with these Cretaceous forms (Fig. 9) it is possible to form an opinion as to how far the character states observed in the Recent animals represent ancestral retentions (plesiomorphs) and how far they represent evolutionary changes that have taken place since the Tupaiidae separated from the basal stock (apomorphs). In making such a judgment it must be borne in mind that, in the long time between the Cretaceous and the present, evolutionary reversals may have occurred, so that characters which appear primitive may not be really so. Of the apomorphic characters of tupaiids, many have evolved in parallel in other groups of mammals of which the history is better known, and analogy with these groups may assist in the reconstruction of tupaiid evolution.

Turning first to the molars, the differences between Tupaiidae and primitive Cretaceous placentals (e.g., *Kennalestes, Cimolestes)* may be listed as follows.

1. Primitively M^1 and M^2 are much wider (buccolingually) than long. In Tupaiidae width and length are more equal, with the exception of *Anathana*,

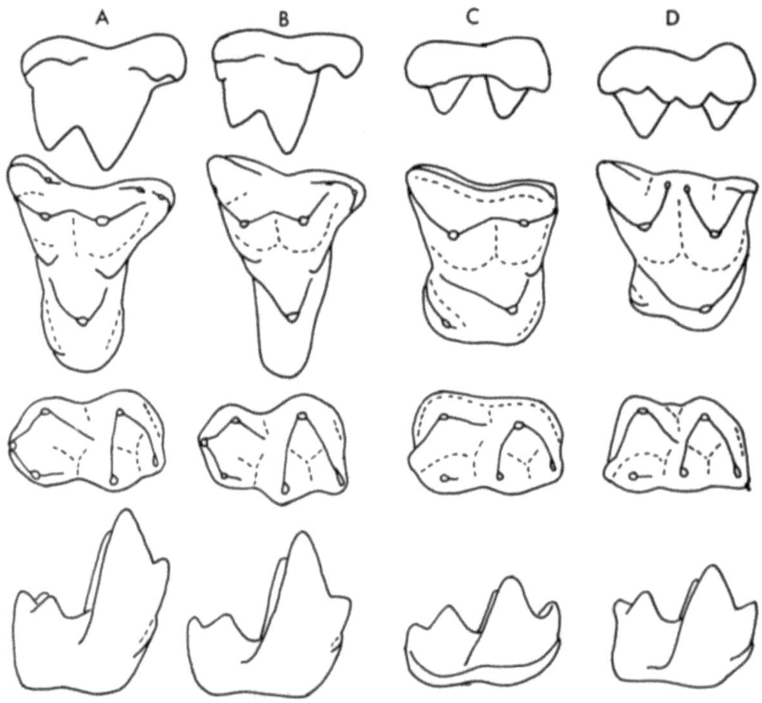

Fig 9. Comparison of tupaiid molars with those of two Cretaceous eutherians. A. *Kennalestes gobiensis;* B, *Cimolestes incisus;* C, *Ptilocercus lowii;* D, *Tupaia glis*. Not to scale.

which is specialized in other respects and may have shortened its molars secondarily. In *Ptilocercus*, however, the slightly transverse development of the molars may represent the retention of a more primitive state. Transverse molars are widespread in Paleocene mammals, including primates (Fig. 10), but there is a common tendency to equalize length and width. The markedly transverse molars of some primates such as *Tetonius* may be a secondary adaptation to shortening of the face.

2. Primitively the parastyle on M^1 and M^2 is a forwardly projecting hook, forming the buccal wall of a deep embrasure between the teeth into which the elevated trigonid of the lower molar penetrates. In Tupaiidae the parastyle is less prominent and the trigonid less elevated, a resemblance to many Tertiary mammals, including early primates.

3. The paracone and metacone were primitively much higher than in Tupaiidae, placed more closely together and joined at the base, so that the valley between them is at a much more ventral level than the buccal cingulum. The metacone is lower than the paracone. Reduction in height and equalization of the buccal cusps have occured in most groups of Tertiary placentals. Elevation of the metacone of M^1 in Tupaiinae to a greater height than the paracone is a specialization of that subfamily that is not found in *Ptilocercus*; a similar elevation of the metacone occurs in Soricidae, *Nesophontes*, and some Erinaceidae, among insectivores, as well as in didelphid marsupials.

4. Both the paracone and the metacone primitively had well developed transverse shearing crests (paracrista and metacrista) which functioned against the crests of the trigonid. These crests are retained in Tupaiinae, but in *Ptilocercus*, owing to the buccal position of the paracone, the paracrista of M^1 and M^2 is nearly longitudinal. A similar modification of the paracrista occurs in primates, leptictids, erinaceoids, condylarths, and other mammals with a bucally situated paracone; some of these, including erinaceoids and leptictids (Fig.10), resemble *Ptilocercus* in retaining the transverse metacrista.

5. The paracone is primitively directly connected with the metacone by a crest, and there is no mesostyle. In this respect *Ptilocercus* seems to retain the primitive condition; the dilambdodont condition of Tupaiinae, where the hypoconid works in a deep groove between the paracone and the metacone, extending to the mesostyle, is almost certainly derivative. It has evolved in many groups of mammals, including lipotyphlous insectivores (Soricidae, Talpidae, Nyctitheriidae, *Nesophontes*), insectivorous bats, Mixodectidae, Dermoptera, Microsyopidae, and some primates *(Plesiadapis, Notharctus, Propithecus, Alouatta)*. Intermediate stages of mesostyle evolution can be seen in a number of fossil forms (Fig. 10). Thus in some species of the nyctitheriid *Saturninia* (Sigé, 1976) and in *Nyctitherium velox* (Krishtalka, 1976b) the notch between the paracone and the metacone has deepened and extended more buccally. In *Mixodectes* and *Microsyops* (Szalay, 1969) the paracone-metacone crest at the bottom of the notch is connected by a short transverse crest to the buccal edge of the tooth where the mesostyle arises, and the tip of the hypoconid travels along this transverse crest. With further deepening the notch reaches the

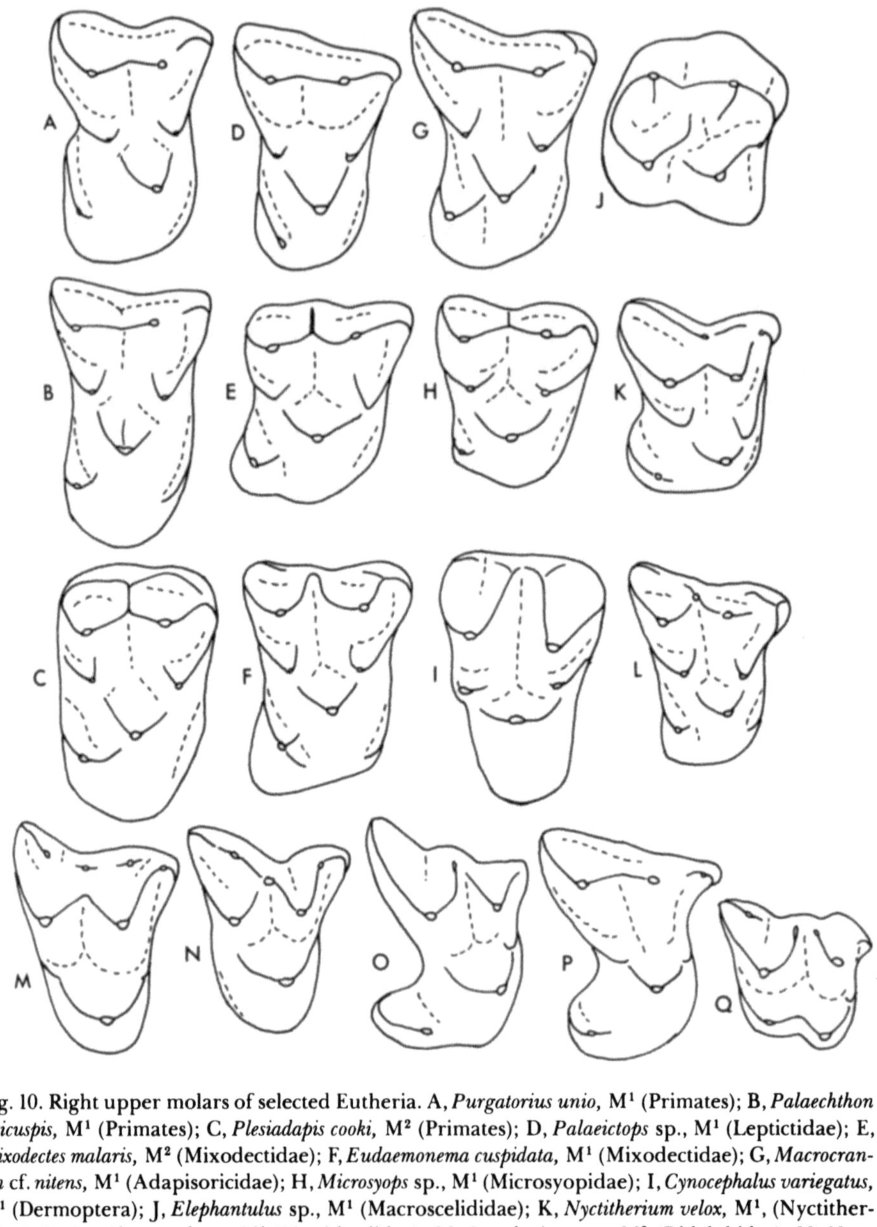

Fig. 10. Right upper molars of selected Eutheria. A, *Purgatorius unio*, M[1] (Primates); B, *Palaechthon alticuspis*, M[1] (Primates); C, *Plesiadapis cooki*, M[2] (Primates); D, *Palaeictops* sp., M[1] (Leptictidae); E, *Mixodectes malaris*, M[2] (Mixodectidae); F, *Eudaemonema cuspidata*, M[1] (Mixodectidae); G, *Macrocranion* cf. *nitens*, M[1] (Adapisoricidae); H, *Microsyops* sp., M[1] (Microsyopidae); I, *Cynocephalus variegatus*, M[1] (Dermoptera); J, *Elephantulus* sp., M[1] (Macroscelididae); K, *Nyctitherium velox*, M[1], (Nyctitheriidae); L, *Pontifactor celatus*, M[1] (Nyctitheriidae); M, *Peratherium* sp., M[2] (Didelphidae); N, *Nesophontes micrus;* M[1] (Nesophontidae); O, *Myosorex robinsoni*, M[1] (Soricidae); P, *Amphidozotherium cayluxi*, M[1] (Nyctitheriidae); Q, *Mesoscalops* sp., M[1] (Talpidae). Not to scale.

mesostyle (e.g., the mixodectid *Eudaemonema* and the nyctitheriid *Pontifactor*) and eventually divides the mesostyle into two cusps (e.g., *Elpidophorus*, the Dermoptera, and the microsyopid *Craseops*). Only the last two stages occur in living Tupaiinae.

6. Primitively the protocone shelf is transversely extended, occupying more than half the tooth width, but short mesiodistally; the protocone crests meet at an acute angle when seen in crown view, and they bear conules. In Tupaiidae the protocone shelf is narrower and longer, the crests of the protocone are more divergent, and the conules are reduced or lost. *Ptilocercus* is slightly less advanced in this respect than the Tupaiinae, for its protocone occupies a somewhat greater proportion of the crown width and the protoconule is more frequently present, but the difference is small. An analogous modification took place in the insectivore family Nyctitheriidae (Krishtalka, 1976b; Sigé, 1976). The Paleocene *Leptacodon tener* has a primitive, transversely extended protocone shelf, but in some later forms (*Nyctitherium velox, Saturninia gracilis, Amphidozotherium*; Fig. 10) the protocone occupies only half the tooth width, the angle between its crests increases, and the conules are reduced in size or lost. The microsyopid *Craseops* (Szalay, 1969) also has a longer than wide protocone shelf with divergent crests, though its conules are retained. It is possible that the tupaiid modification of the protocone shelf is associated with a reduction in the importance of the lingual phase of occlusion.

7. *Kennalestes* possesses cingula anterior and posterior to the protocone, the posterior (hypocone) cingulum making contact with the paraconid in occlusion. Such cingula are usually absent in *Cimolestes*, and when present in this genus they have no occlusal function. They occur however in nearly all Paleocene placentals, and it is very probable that their presence in *Ptilocercus* is an ancestral character of the Tupaiidae. In the development of the posterior cingulum and hypocone *Ptilocercus* closely resembles *Purgatorius*, the earliest known primate. The enlarged hypocone of *Anathana* and *Urogale* is certainly derivative, for increase of the hypocone is a common trend in placental evolution. At the same time the absence of cingula, including the hypocone cingulum, in *Dendrogale* and *Tupaia minor* is probably also derivative, the extreme of a process of reduction that can be seen in other species of *Tupaia* and in *Lyonogale*. The abbreviated, reduced hypocone cingulum of these, sometimes represented only by a prominence of the outline, does not resemble the hypocone cingulum of Cretaceous and Paleocene placentals. The hypocone cingulum is also absent, presumably secondarily, in some other dilambdodont forms, such as the Talpidae and Dermoptera. In *Craseops* the hypocone is smaller than in its ancestor *Microsyops*.

8. Primitively the talonid on M_1 and M_2 is narrower than the trigonid; its oblique crest joins the middle of the trigonid wall; the three talonid cusps are of comparable size and are equally spaced. In Tupaiidae the talonid is proportionately wider; the connection of the oblique crest with the trigonid is displaced buccally in *Ptilocercus*, while retaining its primitive position in Tupaiinae; the hypoconulid is much lower than the other two talonid cusps and it is displaced towards the entoconid. Widening of the talonid is a common trend: it occurs in primates, Microsyopidae, *Mixodectes* (where it is wider than the trigonid even on M_3), Dermoptera, and erinaceoids (Fig 11). Reduction in the size of the hypoconulid on M_1 and M_2 is also very widespread, but it usually remains in a median position. Its lingual displacement occurs in Microsyopidae,

Mixodectidae, and some Nyctitheriidae (e.g., *Nyctitherium velox*), which are like *Ptilocercus*, and more markedly in Soricidae, Talpidae, *Nesophontes*, and didelphid marsupials, which resemble Tupaiinae (Fig. 11). There is no doubt that the Tupaiinae are more advanced in this respect than *Ptilocercus*. Lingual displacement of the hypoconulid lengthens the posterior hypoconid crest and indicates an increased importance of the shearing function of the anterior surface of the enlarged metacone. Elevation of the entoconid in comparison with the hypoconid, which occurs in *Anathana*, represents another direction of advance, shared with Mixodectidae and erinaceoids. Its functional significance seems to be an increase in the entoconid-protocone contact: in Erinaceidae the entoconid occludes in a lingual groove between the protocone and the hypocone.

9. In the Cretaceous *Kennalestes, Asioryctes, Cimolestes*, and *Gypsonictops* the talonid of M_3 is similar in length to that of M_2; the hypoconulid of M_3 is moderately enlarged and prominent posteriorly. In these characters they resemble *Ptilocercus*. In Tupaiinae the hypoconulid of M_3 resembles that of M_2,

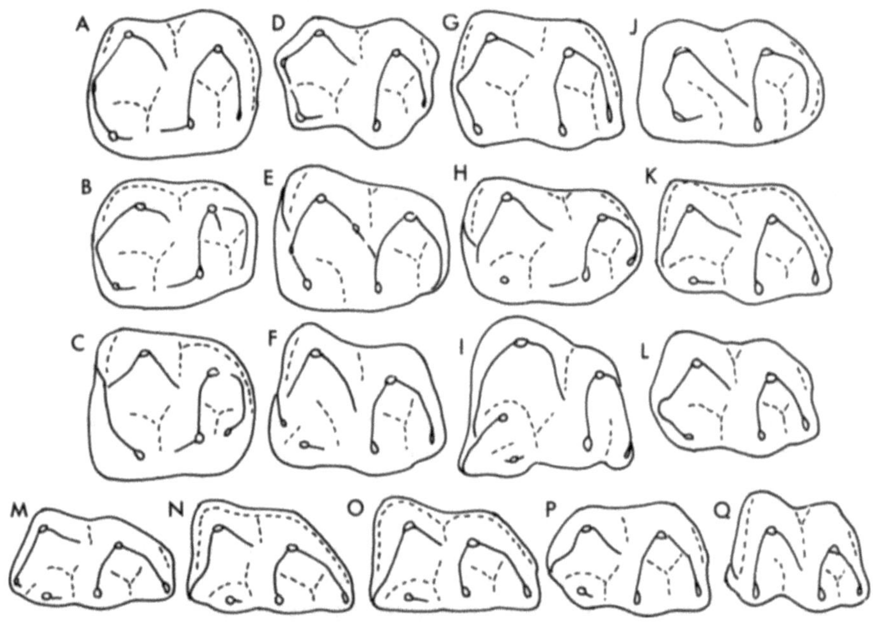

Fig. 11. Left lower molars. A, *Purgatorius unio*, M_2 (Primates); B, *Palaechthon* sp., M_2 (Primates); C, *Plesiadapis farisi*, M_2 (Primates); D, *Palaeictops* sp., M_2 (Leptictidae); E, *Mixodectes pungens*, M2 (Mixodectidae); F, *Eudaemonema cuspidata*, M_2 (Mixodectidae); G, *Macrocranion nitens*, M_2 (Adapisoricidae); H, *Microsyops elegans*, M_1 (Microsyopidae); I, *Cynocephalus variegatus*, *M_1 (Dermoptera); J, Elephantulus* sp., M_1 (Macroscelididae); K, *Nyctitherium velox*, M_2 (Nyctitheriidae); L, *Pontifactor celatus*, M_2 (Nyctitheriidae); M, *Peratherium* sp., M_2 (Didelphidae); N, *Nesophontes micrus*, M_1 (Nesophontidae); O, *Myosorex robinsoni*, M_1 (Soricidae); P, *Amphidozotherium*, M_2 (Nyctitheriidae); Q, *Proscalops* cf. *secundus*, M_2 (Talpidae). Not to scale.

and the talonid of M_3 is noticeably smaller than that of M_2. It would seem that the Tupaiinae have undergone some secondary reduction of the last molars. In many groups, including the primates, the hypoconulid of M_3 becomes more enlarged than in *Ptilocercus*.

10. Primitively the cingulum is developed on the lower molars only on the anterior face of the trigonid. *Ptilocercus* is therefore advanced in having a complete buccal cingulum. Such a cingulum has developed independently in many groups of mammals, including primates (e.g., Plesiadapidae, Notharctidae) and insectivores (e.g., *Nyctitherium velox*, Erinaceidae, Soricidae). On the other hand, many Tupaiinae have lost even the primitive anterior part of the buccal cingulum, paralleling for example the mixodectid *Eudaemonema*.

Primitive molar characters retained by all Tupaiidae include the metacrista of M^1 and M^2, and the triangular shape of the lower molar trigonids, in which a shearing paraconid projects forward away from the metaconid. Both these characters are modified in primates: the metacrista becomes reduced as the metacone takes up a more buccal position, and the paraconid becomes closely applied to the metaconid and eventually disappears.

Less can be said about the evolution of the anterior teeth, which are less often preserved in fossils than the molars. In *Kennalestes* the upper incisors are vertical and spaced, and the lower incisors are only moderately procumbent (Kielan-Jaworowska, 1968). The canines are larger than the incisors, but the upper canine has two roots. If this is taken as the primitive placental condition, Tupaiidae have reduced the number of upper incisors to two and have increased the procumbency of the lower incisors, which at least in Tupaiinae are used as a tooth-comb: they have also reduced the size of the canines, especially in the upper jaw. The two-rooted condition of the upper canine of *Ptilocercus* and *Dendrogale* may be a primitive retention, unless it is the secondary result of canine dedifferentiation. Enlargement of the incisors of *Ptilocercus* is presumably derivative, as may be the development of crests and accessory cusps on the anterior teeth. It is possible that the Tupaiinae have passed through a stage with enlarged incisors like *Ptilocercus* and their incisors have been secondarily reduced and simplified. The late eruption of the incisors might be taken as an indication of their secondary reduction, but it must be noted that *Ptilocercus* as well as the Tupaiinae has this character.

Enlargement of anterior incisors, accompanied by reduction or loss of I^3 and the canine (diprotodonty), is a common phenomenon in mammalian evolution. It is characteristic of the paromomyiform primates and of mixodectids. In at least some of these (*Plesiadapis*, *Elpidophorus*; Szalay, 1969) I^1 is complex, with additional cusps. In primates the third incisors are absent from both jaws, but this is true even of families in which diprotodonty did not develop, e.g., Notharctidae, and it must be due to another cause, perhaps shortening of the jaws. In Tupaiidae, though I^3 is lost, I_3 is retained, an unusual and as yet unexplained situation.

Enlargement of the lower canine of *Urogale* is most probably a secondary development, associated with the enlargement of I^2 and the reduction of I_3.

The Tupaiidae have lost the first premolars of the primitive placental dentition. A possible explanation is that at a previous stage in their evolution the jaws were short and a tooth was lost to avoid overcrowding. A similar process seems to have occurred in primates. Since there are four premolars in *Purgatorius* and in some later primates such as *Notharctus*, the first premolars must have been lost within the evolution of the order. First premolars are lacking in some insectivores, including *Erinaceus, Nesophontes, Solenodon*, Tenrecidae, and Chrysochloridae. All of these have a closed post-canine tooth row, except certain Tenrecidae such as *Tenrec* and *Hemicentetes* in which the jaws may well have elongated secondarily. If the closed tooth row of *Ptilocercus* is a retention from an earlier stage of tupaiid evolution, it would follow that the jaws of Tupaiinae have secondarily lengthened, least so in *Dendrogale* and most in *Lyonogale* and *Urogale*.

The remaining premolars are referred to as P^2 to P^4, following normal practice, though it is possible that an additional premolar was originally present between P^2 and P^3, as in *Gypsonictops* (Clemens, 1973). The Tupaiidae retain what is believed to be the primitive organization of P^2—P^4 (Butler, 1977): P^2 is a simple, two-rooted tooth (secondarily one-rooted in many Tupaiinae), P^3 is a triangular tooth, usually with three roots, at an intermediate stage of molarization, and P^4 is more transverse, with a well developed protocone shelf. The absence of the metacone from P^4 is probably primitive. This cusp is present in *Purgatorius* and other paromomyiform primates, as well as in the Cretaceous *Procerberus, Batodon,* and *Gypsonictops*, in Leptictidae, in Mixodectidae (but not *Mixodectes*), and in Nyctitheriidae. I believe that in all these P^4 is to some degree secondarily molarized (Butler, 1977).

Of the lower premolars, P_2 was probably primitively two-rooted, and its one-rooted condition in Tupaiidae is secondary. The simple talonids of P_3 and P_4 resemble those of Cretaceous placentals with non-molariform premolars, and may be considered primitive. In *Purgatorius* and other paromomyiform primates the talonid of P_4 is somewhat better developed, and this is also true of Paleocene erinaceoids *(Leipsanolestes)* and nyctitheriids *(Leptacodon).* There is no metaconid on P_4 in *Kennalestes, Asioryctes,* and *Cimolestes;* if this is the primitive condition, the presence of the cusp in Tupaiidae represents an advance. The P_4 metaconid is lacking in the earliest primates but develops in later forms. It is absent in *Mixodectes* but present in erinaceoids and nyctitheriids.

Reduction of P^3 and P_3 in *Ptilocercus* is a specialization possibly associated with an emphasis on the fourth premolars as puncturing teeth used when food is introduced through the side of the mouth. A similar adaptation is found in *Erinaceus,* where P^3 is very small and P_3 is absent. The third premolars are also reduced in the nyctitheriid *Amphidozotherium* (Sigé, 1976) and in *Nesophontes* (McDowell, 1958). It is not clear why P^3 erupts after P^2 and P^4, but as this happens in prosimian primates as well as in Tupaiinae it is possibly a primitive character.

8. Interrelationships of the Tupaiid Genera

In dental characters *Ptilocercus* stands apart from the other genera of Tupaiidae, justifying the separation of the two subfamilies, Ptilocercinae and Tupaiinae. Steele (1973), in his study of dental variability in Tupaiidae, found that of 43 traits used, 20 served to separate *Ptilocercus* from the remaining genera. Steele did not distinguish between primitive and advanced character traits. The analysis in the previous section indicates that in several respects *Ptilocercus* is more primitive, retaining ancestral characters modified in other tupaiids: the absence of the mesostyle, the less lingual position of the hypoconulid, the primitive arrangement of cingula around the protocone, the less reduced M_3, and the two-rooted upper canine. To these characters may perhaps be added the enlarged incisors, the development of crests and basal cusps on the upper teeth back to P^2, and the closed postcanine tooth row. On the other hand, *Ptilocercus* has advanced away from the Tupaiinae probably in the following respects: the conical, buccally placed paracone, with loss of the transverse paracrista shear and the associated more longitudinal orientation of the oblique crest; the buccal cingulum of the lower molars; and reduction of the third premolars.

The Tupaiinae advanced from the common ancestor in the development of the mesostyle, enlargement of the metacone, a further lingual displacement of the hypoconulid, some reduction in the size of M_3, loss of the preprotocone cingulum, elongation of the face with spacing of the postcanine teeth, and probably also reduction in size and simplification of the anterior teeth. These changes may be conceived as taking place in parallel along a number of different evolutionary lines, but at different rates, so that some species have reached a more advanced level than others. For example, the mesostyle may be single or divided, the snout may be lengthened to various degrees, and the canine may *(Dendrogale)* remain two-rooted.

Two genera of Tupaiinae show very distinctive specializations. In *Anathana* the upper molars are markedly transverse, the entoconid and metaconid are elevated, and P^3 is more like P^4 than in other Tupaiidae. In *Urogale* the lower canine is enlarged, together with I^2, while I_3 is reduced; moreover the face is more elongated than in other Tupaiidae except *Lyonogale*. *Anathana* and *Urogale* agree in their enlarged hypocones; in this they contrast with the remaining Tupaiinae in which the hypocone and its cingulum show various degrees of reduction. In view of their divergence in other respects it seems likely that *Anathana* and *Urogale* developed the enlarged hypocone independently, starting from a stage in which the hypocone resembled that of *Ptilocercus*.

Of the three remaining genera *Dendrogale* resembles *Ptilocercus* in the development of crests on the anterior teeth, in the two-rooted upper canine, and in the almost closed tooth row. It is however very different from *Ptilocercus* in its molars, on which the mesostyle is divided and the hypocone is completely absent. On the assumption that its resemblances to *Ptilocercus* are plesio-

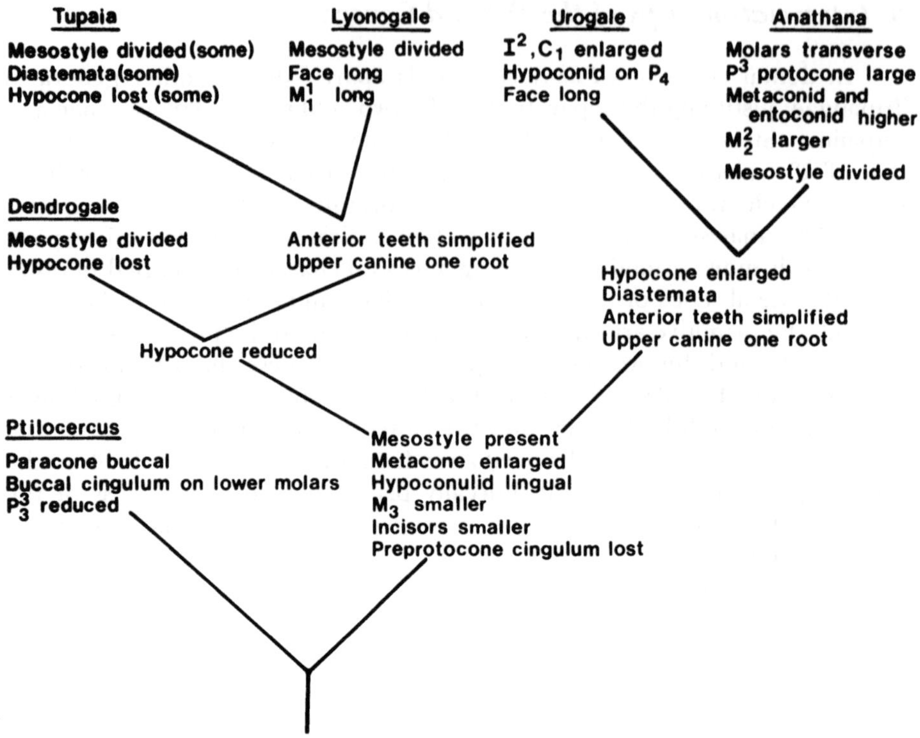

Fig. 12. Distribution of apomorphic dental characters in the genera of Tupaiidae. The dendrogram presupposes that hypocone reduction took place once, and that simplification of the anterior teeth occurred twice.

morphic, *Dendrogale* may be regarded as an offshoot from the *Tupaia* line in which reduction of the hypocone has taken place independently. Steele (1973) regards *Lyonogale* as a subgenus of *Tupaia,* and they are undoubtedly closely related. However, *Lyonogale* falls outside the range of the species of *Tupaia* in the elongation of the face, spacing of the teeth, and relative narrowness of the molars, as well as in its more carnivorous habits. Its skull is very similar to that of *Urogale,* probably by convergence.

A dendrogram to illustrate possible phyletic relations of the genera of Tupaiidae is given in figure 12. Other dendrograms could be constructed, depending upon which assumptions about parallel evolution are made. For example, if hypocone reduction occurred twice, *Dendrogale* would stand apart from the other Tupaiinae.

9. *Relationship of Tupaiidae to Other Mammals*

The fossil record of the order Primates goes back to the earliest Paleocene with *Purgatorius unio,* recently redescribed by Clemens (1974). A molar from the terminal Cretaceous has been placed in the same genus. *Purgatorius* is considered to be a member of the suborder Paromomyiformes, a group which diversified in the Paleocene, but is not necessarily ancestral to later primates.

If *Purgatorius* (Figs. 10, 11) is compared with *Ptilocercus,* several points of resemblance may be noted: the bluntness of the molar cusps is similar, the metacone is subequal to the paracone, the hypocone is developed in a similar manner, the talonid of M_1 is wider than the trigonid, the paracone on M^1 and M^2 is more buccally situated than the metacone, and the metacrista is better developed than the paracrista. *Purgatorius* is more primitive than *Ptilocercus,* as might be expected from its early date: the notch between paracone and metacone is at a higher level in relation to the buccal cingulum; the protocone shelf is more transversely developed and it bears well developed conules; the trigonid is higher in comparison with the talonid; the hypoconulid is less reduced and it is median in position; the buccal cingulum of the lower molars is incomplete. None of these characteracters would exclude a derivation of *Ptilocercus* from *Purgatorius.*

There are, however, other characters which show that *Purgatorius* is already evolving in a primate direction. The trigonid is compressed anteroposteriorly, and the paraconid is less divergent from the metaconid; there is a flange on the posterior surface of the metaconid; the hypoconulid of M_3 is more enlarged than in *Ptilocercus;* the buccal surface of the protocone is flattened and shows lingual-phase wear; the metacrista is shorter than in *Ptilocercus;* a metacone is present on P^4, and P_4 has a talonid basin. Because of these advances of *Purgatorius,* its common ancestor with *Ptilocercus* must be put back into the Cretaceous. Of the derived characters shared between the two genera, the buccal position of the paracone, associated with the more longitudinal direction of the paracrista, is not shared with the Tupaiinae, which apparently retain the primitively transverse paracrista. It is therefore likely that this resemblance is due to parallel evolution in *Ptilocercus* and *Purgatorius,* and not an indication of special relationship. The other similarities concern features that are common in other groups of placentals. A further resemblance of tupaiids to Paleocene primates is the loss of I^3 (unknown in *Purgatorius*), but this is associated in primates with the loss of I_3, a tooth retained in tupaiids. The dentition therefore provides no reason for including the Tupaiidae within the order Primates.

Turning now to other groups whose relationship to Tupaiidae has been suggested, the Leptictidae (Paleocene—Oligocene) are again too specialized in their own way to be ancestral to tupaiids, despite some resemblances in the ear region (McKenna, 1966). The paracone and metacone are more buccally situated and their shearing crests are reduced, though a short metacrista remains. The trigonid is compressed mesiodistally, the paraconid being reduced and, except in *Myrmecoboides,* median in postion. P^4 is almost fully molariform, and on P_4, which has a molariform talonid, the paraconid is enlarged and projects forward to occlude with the protocone of P^3. A metacone is present on P^3 and P^2. These advances were already present in the Paleocene (Novacek, 1977). Most of them are shared with the Cretaceous genus *Gypson-*

ictops, in which, however, the paraconid of P_4 is reduced, and the protocone of P^3 occludes with the semi-molariform talonid of P_3. In other respects the Leptictidae remain primitive: the upper molars are transverse, the metacone is lower than the paracone, the protocone shelf is transversely extended and bears well developed conules, the trigonid is much higher than the talonid, and the hypoconulid is median and comparatively large. As far as known, the incisors are never enlarged, but one incisor is lost in the Oligocene *Leptictis.* The upper canine is caniniform, with one root, but the lower canine is only moderately enlarged. The first premolars are retained. There is no indication from the dentition that Leptictidae and Tupaiidae are related.

Again, the African family Macroscelididae, formerly classified with the Tupaiidae in a suborder Menotyphla, shows very little resemblance to them in the teeth. Its paleontological history goes back only to the Oligocene, when many of the distinctive characters had already evolved. As in Leptictidae, P^4 is fully molariform, and P_4 is also molariform except that it has a greatly enlarged, forwardly projecting paraconid. At the same time the third molars have been lost in most genera, though sometimes present in a reduced form. P^4 and M^1 are quadrate teeth in which the hypocone resembles the protocone; the buccal cusps are conical, without shearing crests, and the buccal cingulum is rudimentary or absent. On the lower molars the talonid and trigonid are equal in height and the paraconid is reduced. On unworn teeth the metaconid and entoconid are higher than the buccal cusps. The chewing movement appears to be more horizontal than in *Tupaia.* The metacone is present on P^{2-4}. The teeth of Macroscelididae show some resemblances to those of Leptictidae, but McKenna (1975) has suggested that their relationships lie with the Asiatic Early Tertiary order Anagalida (Szalay and McKenna, 1971). *Anagale* was believed to be related to Tupaiidae by Simpson (1931), but McKenna (1963) showed that the ear region as well as the dentition excluded such a relationship.

Among the lipotyphlous Insectivora the erinaceoids (Adapisoricidae, Erinaceidae) can be traced back into the Paleocene (for reviews see Russell *et al.*, 1975; Krishtalka, 1976a). Two Eocene genera have been suggested as tupaiid relatives: *Entomolestes* by Matthew (1909) and *Macrocranion (=Messelina)* by Weizel (1949). The resemblances to tupaiids consist mostly of primitive characters, but on M^1 of erinaceoids the paracone stands nearer to the buccal edge than the metacone, so that the transverse metacrista is retained but the paracrista has become nearly longitudinal, as in *Ptilocercus* and *Purgatorius.* Most other advances of erinaceoids are in a direction away from Tupaiidae. The trigonid is compressed, especially on M_2 and M_3, the paraconid forming a transverse ridge that wears on its edge against the hypocone. This cusp enlarges early in the evolution of the group. On the lower molars the metaconid and entoconid stand higher than the buccal cusps. The hypoconid and entoconid are joined to form a transverse crest along the posterior edge of the talonid, similar in height to the paraconid of the following tooth. P_4 has a small but molariform talonid in the Paleocene genera *Leipsanolestes* and *McKennatherium;* in Eocene genera its talonid is shortened but usually retains the entoconid. In *Scenopagus* P^4 possesses a hypocone as in later Erinaceidae. The anterior teeth are not adequately known in Paleocene and Eocene erinaceoids. The canines are often small and premolariform, at least in the lower jaw, but the upper canine is large in *Macrocranion.* There is no evidence that the incisors are enlarged in the early forms, but such enlargement is characteristic of Oligocene and later Erinaceinae.

Another family of lipotyphlous insectivores, the Nyctitheriidae (reviewed by Krishtalka, 1976b, and Sigé, 1976) is interesting in that its molar evolution parallels the Tupaiidae in a number of ways. In various species the protocone becomes less transversely extended, and its crests more divergent; the paracone and metacone evolve towards the dilambdodont pattern, and the hypoconulid is displaced lingually. In *Saturninia* the canines are premolariform and only a little larger than the anterior premolars. However, there are also advanced features not shared by tupaiids. Even in the Paleocene the posterior premolars are more molariform, P^4 usually possessing a small metacone and P_4 a basined, molariform talonid. In *Saturninia* the lower incisors have additional posterior cusps or lobes, quite different from those of tupaiids. It is probable therefore that the development of tupaiid features on the molars is another example of parallel evolution.

If the dilambdodont upper molars of Tupaiinae represent a specialization that has developed within the family, and was avoided by *Ptilocercus,* fossil mammals with dilambdodont teeth cannot be ancestral to the Tupaiidae as a whole. This applies to the Paleocene family Mixodectidae, all members of which possess a mesostyle. Mixodectids also share with tupaiids a lingual displacement of the hypoconulid (to the same degree as *Ptilocercus*). In both jaws the first two incisors are enlarged, but the third incisor has been lost. Canines and first premolars are reduced or lost. The last premolars of *Mixodectes* are not molariform, and share primitive characters with tupaiids, but in the other genera they are molariform. The lower molars are more advanced than in tupaiids in that the paraconid is less salient, the metaconid and entoconid are higher than the buccal cusps (a resemblance only to *Anathana* where the character is surely secondary), and an additional cusp (mesoconid) is present on the oblique crest. Szalay (1969) considers that mixodectids may be related to tupaiids, but the advances shared between the two families seem more likely to be due to parallel evolution.

The same may apply to the Paleocene genus *Adapisoriculus,* which Van Valen (1965) put into the family Tupaiidae. This shares with all Tupaiidae the lingual displacement of the hypoconulid (as much as in *Ptilocercus*) and the single root of P_2; with the Tupaiinae it shares the mesostyle (not divided as in most Tupaiinae) and the reduction or loss of the hypocone cingulum. It also has some derived characters not found in Tupaiidae: M_3 is longer than M_2; the oblique crest of the lower molars continues up the posterior trigonid wall towards the metaconid, instead of ending near the middle of the trigonid wall as in Tupaiinae. The talonid of P_4 has two cusps (Russell, 1964), a resemblance to *Urogale.* In other ways *Adapisoriculus* is more primitive: the protocone shelf is more transversely developed, conules are less reduced, the parastyle is more prominent on the upper molars, the lower molar trigonids are taller, and P_1 is present. The canines and incisors are unknown. Until better material becomes available, the taxonomic position of *Adapisoriculus* must remain uncertain; its tupaiid affinities seem very doubtful.

Dental resemblances are difficult to interpret because of the frequency with which parallel evolution occurs. Nearly all the derived characters of Tupaiidae can be found in other groups of mammals: only in the combination of their characters are the Tupaiidae unique. Adding to the difficulty is the necessity of comparing living tupaiids with fossils separated from them by 50–60 million years. For these reasons any conclusions about tupaiid affinities based upon the dentition must be regarded as very tentative. As far as it goes, the dental evidence points to an isolated position for the Tupaiidae. They seem to have branched off at least as far back as the beginning of the Tertiary and possibly in the Cretaceous. They cannot be derived from Paleocene primates unless an improbable amount of evolutionary reversal is postulated; neither can early primates be derived from Tupaiidae. Inclusion of the Tupaiidae within the order Primates is therefore not justified. This is not to say that the Tupaiidae and the Primates do not have a common root, but the dentition provides no evidence in support of a special relationship. If the Tupaiidae are removed from the Primates, should they be returned to the Insectivora? The grounds for their relationship to any of the known insectivore families seem to be no stronger than for their relationship to the Primates. Moreover, the Insectivora is an undefinable, waste-basket order containing a variety of more or less primitive families that cannot be placed in other, better defined orders. I proposed (Butler, 1972) that the Tupaiidae should be given the rank of a separate order, Scandentia. If they have had such a long independent history as the dental evidence indicates, ordinal status seems to be the best way of expressing this.

10. References

Butler, P. M. 1939. Studies of the mammalian dentition. Differentiation of the post-canine dentition. *Proc. zool. Soc. Lond.* 109B:1–36.

Butler, P. M. 1948. On the evolution of the skull and teeth in the Erinaceidae, with special reference to fossil material in the British Museum. *Proc. zool. Soc. Lond.* 118:446–500.

Butler, P. M. 1972. The problem of insectivore classification, pp. 253–265. *In* K. A. Joysey and T. S. Kemp (eds.). *Studies in Vertebrate Evolution*. Oliver and Boyd Publishers, Edinburgh.

Butler, P. M. 1973. Molar wear facets of Tertiary North American primates. *Symp. 4th intern. Cong. Primatol.* 3:777–817. Karger, Basel.

Butler, P. M. 1977. Evolutionary radiation of the cheek teeth of Cretaceous placentals. *Acta palaeont. pol.* 22: 241–271.

Clemens, W. A. 1973. Fossil mammals of the type Lance Formation, Wyoming. Part III, Eutheria and Summary. *Univ. Calif. Publs. geol. Sci.* 94:1–102.

Clemens, W. A. 1974. *Purgatorius*, an early paromomyid primate (Mammalia). *Science.* 184:903–905.

Crompton, A. W., and Hiiemae, K. M. 1970. Functional occlusion and mandibular movements during occlusion in the American opossum, *Didelphis marsupialis* L. *J. Linn. Soc. (Zool.)* 49:21–47.

Davis, D. D. 1962. Mammals of the lowland rain-forest of North Borneo. *Bull. natn. Mus. St. Singapore* 31:1–129.

Gregory, W. K. 1910. The orders of mammals. *Bull. Am. Mus. nat. Hist.* 27:3–524.

Harrison, J. L. 1955. The natural food of some rats and other mammals. *Bull. Raffles Mus.* 25:157–165.

Hiiemae, K. M., and Kay, R. F. 1973. Evolutionary trends in the dynamics of primate mastication. *Symp. 4th intern. Cong. Primatol.* 3:24–68. Karger, Basel.

James, W. .W. 1960. *The Jaws and Teeth of Primates*. Pitman Medical Publ. Co. Ltd., London.

Kay, R. F. 1975. The functional adaptation of primate molar teeth. *Am. J. phys. Anthrop.* 43: 195–216.

Kay, R. F., and Hiiemae, K. M. 1974. Jaw movement and tooth use in Recent and fossil primates. *Am. J. phys. Anthrop.* 40:227–256.

Keilan-Jaworowska, Z. 1968. Results of the Polish-Mongolian palaeontological expeditions. Part I. Preliminary data on the Upper Cretaceous eutherian mammals from Bayn Dzak, Gobi Desert. *Palaeont. polon.* 19:171–191.

Kielan-Jaworowska, Z. 1975. Results of the Polish-Mongolian palaeontological expeditions. Part VI. Preliminary description of two new eutherian genera from the Late Cretaceous of Mongolia. *Palaeont. polon.* 33: 5–16.

Kindahl, M. 1957. On the development of the teeth in *Tupaia javanica*. *Ark. Zool.* 10:463–479.

Krishtalka, L. 1976a. Early Tertiary Adapisoricidae and Erinaceidae (Mammalia, Insectivora) of North America. *Bull. Carnegie Mus. nat. Hist.* 1:1–40.

Krishtalka, L. 1976b. North American Nyctitheriidae (Mammalia, Insectivora). *Ann. Carneg. Mus.* 46:7–28.

Le Gros Clark, W. E. 1926. On the anatomy of the pen-tailed tree-shrew *(Ptilocercus lowii). Proc. zool. Soc. Lond.* 1926:1179–1309.

Lim Boo Liat 1967. Note on the food habits of *Ptilocercus lowii* Gray (Pentail tree-shrew) and *Echinosorex gymnurus* Raffles (Moonrat) in Malava with remarks on "ecological labelling" by parasite patterns. *J. Zool. Lond.* 152:375–379.

Lyon, M. W., Jr. 1913. Tree shrews: An account of the mammalian family Tupaiidae. *Proc. U.S. natn. Mus.* 45:1–183.

McDowell, S. B. 1958. The Greater Antillean insectivores. *Bull. Am. Mus. nat. Hist.* 115:113–214.

McKenna, M. C. 1963. New evidence against tupaioid affinities of the mammalian family Anagalidae. *Am. Mus. Novit.* 2158:1–16.

McKenna, M. C. 1966. Paleontology and the origin of the primates. *Folia primat.* 4:1–25.

McKenna, M. C. 1975. Toward a phylogenetic classification of the Mammalia, pp. 21–46. *In* W. P. Luckett and F. S. Szalay (eds.). *Phylogeny of the Primates*. Plenum Press, New York.

Maier, W. 1977. Die Evolution der bilophodonten Molaren der Cercopithecoidea. Eine funktionsmorphologische Untersuchung. *Z. Morph. Anthrop.* 68:26–56.

Maier, W. 1979. A new dental formula for the Tupaiiformes. *J. human Evol.* 8: 319–321.

Martin, R. D. 1968. Reproduction and ontogeny in the tree-shrews *(Tupaia belangeri)* with reference to their general behaviour and taxonomic relationships. *Z. Tierpsychol.* 25:409–495, 505–532.

Matthew, W. D. 1909. The Carnivora and Insectivora of the Bridger Basin, Middle Eocene. *Mem. Am. Mus. nat. Hist.* 9:291–567.

Mills, J. R. E. 1955. Ideal dental occlusion in the Primates. *Dental Practnr., Bristol* 6:47–61.

Mills, J. R. E. 1963. Occlusion and malocclusion of the teeth of primates, pp. 29–51. *In* D. R. Brothwell (ed.). *Dental Anthropology*. Pergamon Press, London.

Mills, J. R. E. 1966. The functional occlusion of the teeth of Insectivora. *J. Linn. Soc. (Zool.)* 47:1–25.

Mivart, St. G. 1867. Notes on the osteology of the Insectivora. *J. Anat. Physiol., Lond.* 1:281–312.

Moseley, H. N., and Lankester, E. R. 1868. On the nomenclature of mammalian teeth, and on the dentition of the mole *(Talpa europaea)* and the badger *(Meles taxus)*. *J. Anat. Physiol., Lond.* 3:73–80.

Novacek, M.J. 1977. A review of Paleocene and Eocene Leptictidae (Eutheria: Mammalia) from North America. *PaleoBios* 24:1–42.

Russell, D. E. 1964. Les mammifères paléocènes d'Europe. *Mém. Mus. natn. Hist. nat., Paris. Ser. C,* 13:1–324.

Russell, D. E., Louis, P., and Savage, D.E. 1975. Les Adapisoricidae de l'éocène inférieur de France. Réévaluation des formes considerées affines. *Bull. Mus. natn. Hist. nat., Paris* 327:129–193.

Schwartz, J. H. 1975. Re-evaluation of the morphocline of molar appearance in the primates. *Folia primat.* 23:290–307.

Schwartz, J. H., and Krishtalka, L. 1976. The lower antemolar teeth of *Litolestes ignotus,* a late Paleocene erinaceid (Mammalia, Insectivora). *Ann. Carneg. Mus. 46:1–6.*

Shigehara, N. 1975. On tooth replacement in *Tupaia glis. Proc. 5th intern. Cong. Primatol.* pp. 20–24.

Sigé, B. 1976. Insectivores primitifs de l'Éocène supérieur et Oligocène inférieur d'Europe occidentale. Nyctitheriidés. *Mém. Mus. natn. Hist. nat., Paris* 34:1–140.

Simpson, G. G. 1931. A new insectivore from the Oligocene, Ulan Gochu horizon, of Mongolia. *Am. Mus. Novit.* 505:1–22.

Sorenson, M. W., and Conaway, C. H. 1964. Observations of tree shrews in captivity. *Sabah Soc. J.* 2:77–91.

Sorenson, M. W., and Conaway, C.H. 1966. Observations on the social behavior of tree shrews in captivity. *Folia primat.* 4:124–145.

Steele, D. G. 1973. Dental variability in the tree shrews (Tupaiidae). *Symp. 4th intern. Cong. Primatol.* 3:154–179.

Swindler, D. R. 1976. *Dentition of Living Primates*. Academic Press, London.

Szalay, F. S. 1969. Mixodectidae, Microsyopidae and the insectivore-primate transition. *Bull. Am. Mus. nat. Hist.* 140:193–330.

Szalay, F. S., and McKenna, M.C. 1971. Beginnings of the age of mammals in Asia. *Bull. Am. Mus. nat. Hist.* 144:269–318.

Van Valen, L. 1965. Treeshrews, primates and fossils. *Evolution* 19:137–151.

Weizel, K. 1949. Neue Wirbeltiere (Rodentia, Insectivora, Testudinata) aus dem Mitteleozän von Messel bei Darmstadt. *Abh. senckenb. naturforsch. Ges.* 480:1–24.

Wharton, C. H. 1950. Notes on the Philippine tree shrew, *Urogale everetti* Thomas. *J. Mammal.* 31:352–354.

Ziegler, A. C. 1971. Dental homologies and possible relationships of Recent Talpidae. *J. Mammal.* 52: 50–68.

Siwalik Fossil Tree Shrews 6

LOUIS L. JACOBS

1. Introduction

Fossils directly contribute to the construction of phylogenies in two ways. They provide additional samples (usually taxa) which are used in biological comparisons, so far as the completeness of the fossil material warrants, in order to determine morphocline polarity; that is, in forming a hypothesis of primitive and derived character states. In addition, fossils provide a temporal frame of reference (see Luckett, this volume) which makes the phylogeny more complete and practical in terms of geological applications. Although geologic age is not a biological attribute of a taxon, it follows that for a lineage with a reasonably complete fossil record, older taxa often (but not always) are on average more primitive than younger taxa, simply because ancestors must occur before descendants. Therefore, a sequence of fossils may reflect ancestor-descendant relationships, even if the fossils in hand are not themselves the actual ancestors or descendants.

Modern tree shrews have been intensively studied because of their possible relevance to the origin of primates. The lack of an adequate fossil record for tree shrews necessitates the use of modern taxa in comparisons made with fossil or living primates and insectivores (see McKenna, 1963a, 1966). Several Tertiary taxa have been considered at one time or another to be closely related to tree shrews (see below), but nevertheless the lack of a definitive fossil record for tree shrews, and hence knowledge of older, more primitive, tree shrews reduces the value of outgroup comparisons based solely on living tupaiids.

The purpose of this paper is to report fossil tupaiids from Miocene deposits of the Siwalik Group in Pakistan. Tree shrews presently do not occur as

LOUIS L. JACOBS • Department of Geology, Museum of Northern Arizona, Flagstaff, Arizona 86001 far west as Pakistan although they do occur in India and further east. The fossils comprise the anterior portion of a skull back to the level of the fourth premolar with the crowns of the teeth broken off, one complete lower molar, and the talonid of another molar. The specimens are from three separate localities. The configuration of the skull fragment allows certain identification as a tupaiid, the morphology of the complete lower molar is quite probably tupaiid, and the reference of the talonid to the Tupaiidae is probable but less so than for the other specimens. Similar statements concerning fossils supposedly related to tupaiids have been made in the past and have later been questioned. The reason the Siwalik specimens can be confidently referred to the Tupaiidae is because they are essentially modern in aspect, albeit distinct from modern genera. This being the case, they yield more information about the relationships of modern genera than they do about the early differentiation of tupaiids or the origin of primates.

Mammalian paleontologists have traditionally dealt with characters of the dentition and, to a lesser extent, the skull, mainly because that was the material at hand (but see Szalay, 1977). Here too some characters of the dentition and, so far as possible, the skull are the information to be gleaned from the Siwalik specimens. The most useful references with respect to these fossils and the tree shrew dentition in general are Lyon (1913), Steele (1973), and Butler (this volume). Dental terminology follows Butler (this volume).

For assitance and criticism, I gratefully acknowledge Everett H. Lindsay, David Pilbeam, W. Patrick Luckett, Michael J. Novácek, William R. Downs III, Timothy B. Rowe, and A. K. Dutta.

2. *Previous Reports of Fossil Tree Shrews*

Several early and middle Tertiary taxa have been considered closely related to, or members of, the Tupaiidae. Reference of fossil taxa to the Tupaiidae has been based almost exclusively on dental characters. Tree shrews have fairly primitive teeth (see the description in Butler, this volume), and therefore, other taxa more or less distantly related to tupaiids, but without specialized characters in the teeth, resemble tree shrews because of shared primitive characters. The most important references are discussed below, but this is not intended as an exhaustive review of the taxonomic history of fossils previously considered to have tupaiid affinities.

Matthew (1909) described the genus *Entomolestes* stating, with reference to dental characters, that it resembled *Tupaia* more than any other living genus. The type species is *Entomolestes grangeri* of middle Eocene (Bridgerian) age. Gregory (1913) considered *E. grangeri* a tupaiid. Krishtalka and West (1977) recently reviewed *E. grangeri* and considered it to be an erinaceid because of dental characters and size relationships of the teeth.

A second species of *Entomolestes, E. nitens*, from the early Eocene (Wasatchian) was described by Matthew (1918) and referred questionably to the Tupaiidae. Various authors have noticed the distinctions in the dentition of *E. nitens* relative to *E. grangeri*. Recently Krishtalka (1976a) reviewed the taxonomic history and dental characteristics of *E. nitens*. He considered it an adapisoricid and referred it to the genus *Macrocranion*.

Tupaiodon morrisi from the Oligocene Hsanda Gol Formation and Loh Red Beds, Mongolia, was described by Matthew and Granger (1924). Their diagnosis was based primarily on dental features, and *Tupaiodon* was provisionally referred to the Tupaiidae. Simpson (1931), after reviewing additional material, placed *Tupaiodon* in the Erinaceidae. This has been followed in most later work (for example, Butler, 1948; McKenna and Simpson, 1959; McKenna, 1960b), although Szalay (1977) suggested again that *Tupaiodon* may be a tupaiid.

Leipsanolestes siegfriedti was described as an adapisoricid by Simpson (1928), but he noted a morphological resemblance to tupaiids. After further study of the original material, the collection of better specimens, as well as additional documentation of European *Adapisorex*, Simpson (1929) considered *Leipsanolestes* a subgenus of *Leptacodon*, the type-species of which is *Leptacodon tener*. More recently, Krishtalka (1976a) transferred *Leipsanolestes* to the Erinaceidae because of its molar morphology; he (Krishtalka, 1976b) reviewed *Leptacodon* and referred the type-species to the Nyctitheriidae.

Simpson (1931) described *Anagale gobiensis*, represented by relatively complete skull and skeletal material from the Oligocene Ulan Gochu Formation of Mongolia, and erected the family Anagalidae for it. He considered the Anagalidae closely related to the Tupaiidae, and placed them both as Tupaioidea in the primate infraorder Lemuriformes (Simpson, 1945). Bohlin (1951) described *Anagalopsis kansuensis* from the middle Tertiary of Kansu, China, and noted the detailed resemblance between *Anagale* and *Anagalopsis*. Bohlin's interpretation of bullar structure in *Anagalopsis* did not suggest a close relationship with lower primates. The bulla of *Anagale* was at that time insufficiently prepared for detailed comparisons. McKenna (1963b) restudied *Anagale* after further preparation of the bulla and interpreted the homology of bullar elements. His interpretation suggested that the bullae of *Anagale* and *Anagalopsis* were similarly constructed, and that the construction of the bulla precluded a close relationship to tupaioids. McKenna considered anagalids Eutheria, *incertae sedis*. Van Valen (1966) noted similarities between the teeth of *Anagale* and *Ptolemaia* (from the Egyptian Fayum) in height of crown, thickness of enamel, and other characters. Later, Van Valen (1967) listed the Anagalidae under the superfamily Tupaioidea in his classification of insectivores. Szalay and McKenna (1971) considered affinity between *Anagale* and *Ptolemaia* improbable.

Evans (1942) noted similarities between macroscelidids and *Anagale*, as well as between *Anagale* and tupaiids. Patterson (1965), in his review of fossil elephant shrews, considered that "whatever the anagalids may prove to be, they are not related to the macroscelidids." McKenna (1975) has since revitalized the idea of such a relationship. The idea of a close relationship between anagalids and tupaiids (or lower primates) is not strongly supported by present evidence, regardless of what the relationship between anagalids and macroscelidids might be.

McKenna (1960a) referred a lower jaw fragment with the last two molars from the early Eocene (Wasatchian) to ?Tupaioidea. In the same paper he noted the resemblance of *Eudaemonema* to *Tupaia* and placed it in the Tupaioidea, *incertae sedis*. Szalay (1969) considered *Eudaemonema* an undoubted mixodectid and noted a possible mixodectid-tupaiid relationship (but see Butler, this volume). Van Valen (1965) did not consider *Eudaemonema* to have any particular similarity to tupaiids.

Van Valen (1965) reviewed the status of some taxa previously referred to tree shrews and evaluated the position of tree shrews relative to primates. He considered the Paleocene genus *Adapisoriculus* a tupaiid and placed it in the Tupaiidae as its own subfamily, the Adapisoriculinae (Van Valen, 1967). Szalay (1968, 1969) considered the Eocene genus *Messelina* similar to tupaiids and a likely candidate for that family. *Messelina* was referred to *Macrocranion*, an adapisoricid genus, by Russell *et al.* (1975; see also Krishtalka, 1976a; Weitzel, 1949).

Szalay (1977) suggested that the Paleocene genus *Litolestes* may be a tupaiid. Recent studies (Krishtalka, 1976a; Schwartz and Krishtalka, 1976) recognized *Litolestes ignotus* from the late Paleocene (Tiffanian) of Wyoming as an erinaceid.

The interest in tree shrews with respect to understanding the origin of primates has precipitated the search for early tupaiids. Similarity between the unspecialized teeth of tree shrews and several early Tertiary taxa has suggested to some authors an affinity with tupaiids; however, the resemblance may be due only to shared primitive characters. So far as is now known, there are no unequivocal early Tertiary tree shrews represented in fossil collections. From the late Tertiary, a rib cage, possibly *Tupaia*, has been reported from the upper Siwalik Group of India (Dutta, 1975). This specimen is difficult to evaluate because it has never been adequately described or figured. In addition, R.N. Vasishat, Panjab University, Chandigarh, India, has recently recovered tupaiid specimens from the middle Siwaliks of India (Vasishat, personal communication, 1979).[1]

3. The Siwalik Fossils

The Siwalik Group is a thick sequence of continental sediments deposited at the foot of the Himalaya in Pakistan and India. Fossils have been collected from the Siwalik Group for over a century, but efforts have recently been renewed because of the presence of several hominoid taxa in the Siwaliks (Pilbeam *et al.*, 1977a, b).

The biostratigraphy and biochronology of the Siwalik Group are presently

[1]A description of these tupaiine specimens from the Indian Siwaliks has appeared while the present paper was in press (Chopra and Vasishat, 1979; Chopra *et al.*, 1979).

being revised by David Pilbeam and others as part of the Yale-Geological Survey of Pakistan expedition (YGSP). The tree shrews reported here were collected as a part of that project.

Tree shrews have been collected from three localities in the Siwaliks of Pakistan. An anterior skull fragment was recovered from YGSP locality 450 in the middle Siwaliks. This is apparently the same locality (Yale Peabody Museum locality 94) from which Lewis (1934) recovered the hominoid primate he named *Sugrivapithecus*, as indicated by original field notes and maps. *Sugrivapithecus* is now considered a synonym of *Ramapithecus*. Specimens from this locality are middle Miocene (about 10 million years) in age.

An isolated lower molar was recovered from YGSP locality 182A in the middle Siwaliks, forty feet below a level yielding *Ramapithecus punjabicus* and *Sivapithecus indicus* (Pilbeam *et al.*, 1977b). Pilbeam *et al.* (in prep.) consider this locality younger than YGSP 450, but still Miocene.

The talonid of a lower molar, questionably referred to the Tupaiidae, was recovered from YGSP locality 259 in the lower Siwaliks. This locality is middle Miocene and is the oldest locality from which tupaiids have been recovered.

The specimens from localities YGSP 450 and 182A probably represent new taxa. They will not be named until continuing work in Pakistan and additional material representing these forms allow their thorough evaluation.

Dutta (1975) reported a rib cage possibly of *Tupaia* from the upper Siwaliks (Tatrot) of India. Tatrot beds are now generally considered Pliocene. If Dutta's referral of the rib cage to *Tupaia* is correct, and if my identification of the talonid from YGSP 259 is correct, then tupaiids range through the major part of the Siwalik Group.

YGSP 8089 (Figs. 1 and 2) is the anterior portion of a skull back to, and including, the level of the fourth premolars. The crowns of the teeth are broken off leaving only alveoli, roots, or the bases of teeth. Most sutures are indistinct, and there is some crushing. The snout is moderately elongate with diastemata separating the roots of first and second incisors, the second incisor and the canine, and a shorter diastema between the canine and the second

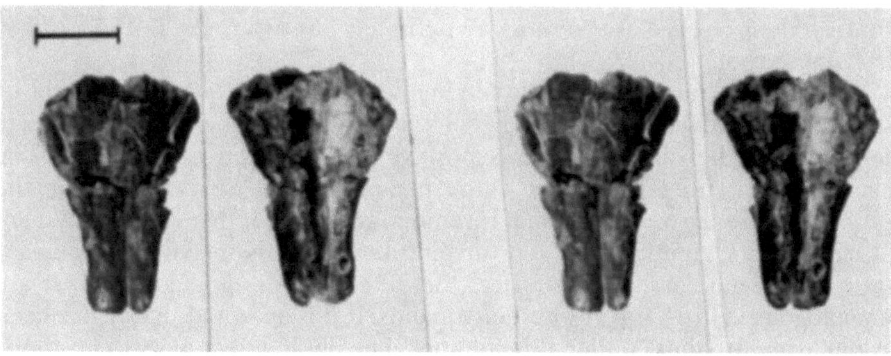

Fig. 1. Stereophotographs of YGSP 8089, tupaiid anterior skull fragment. Bar equals 10 mm.

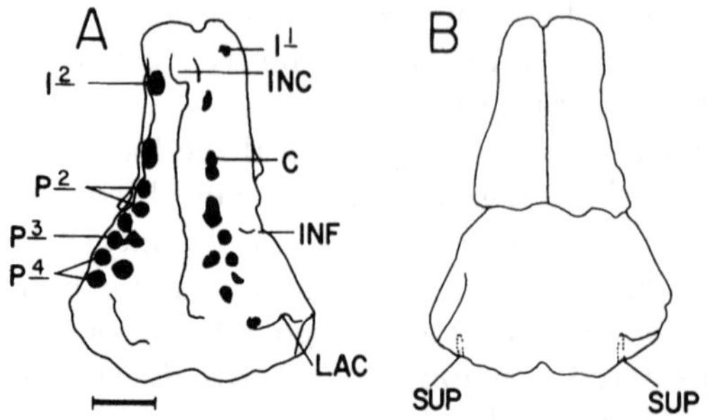

Fig. 2. Diagram of YGSP 8089, tupaiid anterior skull fragment. (A), palatal view; (B), dorsal view. Roots of teeth are labelled. Abbreviations: INC, incisive foramen; INF, infraorbital foramen; LAC, lacrimal foramen; SUP, grooves leading to supraorbital foramina. Bar equals 5 mm.

premolar. The posterior border of the incisive foramen is at the level of the transverse midline of the second incisor. A maxillopremaxillary suture is discernable extending from the posterior margin of the right incisive foramen, posterolaterally behind I^2, then dorsally up the side of the skull to the nasal. There are no palatine vacuities in this specimen and the palate, so far as demonstrated by this specimen, is fully ossified. There are numerous small foramina at about the level of the P^2. A small foramen is present posterior to the right I^2. The infraorbital foramen opens above the anterior root of the P^3. The lacrimal foramen is located in a distinct notch on the lateral surface of the skull. Grooves, apparently leading to supraorbital foramina, can be seen at the posterior edge of the dorsal surface of the specimen.

The anterior surface of the specimen is damaged. Crowns of the teeth are broken off leaving only alveoli, roots, or the bases of teeth. I^1 has a single root directed somewhat anteriorly. I^2 has a single root. The canine apparently has two roots. The outline of the left canine where it is broken off (probably just below the crown) is somewhat dumbell-shaped with a distinct groove on the lingual side, and a less distinct groove on the labial side. I interpret this as indicating the presence of two distinct roots distal to the crown. This interpretation is tentative as there is no definite proof such as X-ray photographs to confirm the presence of two roots. P^2 definitely has two roots. Both P^3 and P^4 have three roots.

This skull fragment is essentially modern in aspect, although, as will be shown, it is distinct from any living genus. The fact that it is essentially modern, so far as can be observed, allows positive identification as a tupaiid. The elongate snout, the arrangement of the infraorbital and lacrimal foramina, the probable presence of supraorbital foramina, and the number and deployment of teeth justify identification as a tupaiid. Assuming the maxillopremaxillary suture has been correctly identified, then the teeth designated I^1 and I^2 are homologous with I^1 and I^2 of other tupaiids, if it is accepted that I^3 has been lost in living tupaiids (Butler, this volume). The deployment of teeth posterior to the I^2 as preserved in this fossil is so similar to that of living tree shrews that there is hardly any doubt that they correspond to C, P^2, P^3, and P^4.

The characters listed above and the interpretation of the dentition serve to identify the specimen as a tree shrew and distinguish it from other taxa. The presence of supraorbital foramina (implied by the presence of grooves) in the Siwalik fossil is distinct from *Anagale* and erinaceids (so far as known), yet similar to tupaiine tree shrews. In addition, the presence of two upper incisors is consistent with the tree shrew dental formula, but distinct from that of most erinaceids which have three upper incisors (although some erinaceids have only two). Simpson's (1931) interpretation of *Anagale* included all upper incisors with the possible exception of I^1. The I^3 has been lost in living tree shrews, and I^1 and I^2 retained.

The Siwalik skull fragment is distinct from *Ptilocercus* in having supraorbital foramina, a lacrimal foramen in a distinct notch, and well separated lacrimal foramen and anterior opening of the infraorbital foramen (they are close together in *Ptilocercus*). The snout of the Siwalik specimen is more elongate than that of *Ptilocercus* with a relatively longer diastema between I^2 and the canine. In addition, the Siwalik specimen is larger than *Ptilocercus* and does not have a double-rooted I^2 as is sometimes present in *Ptilocercus*. It is similar to *Ptilocercus* in having double-rooted canines and in having the posterior border of the incisive foramina not extending posterior to the posterior margin of I^2.

The Siwalik skull fragment is similar to *Tupaia, Lyonogale, Dendrogale, Urogale*, and *Anathana* in having supraorbital foramina, in having the lacrimal foramen opening in a distinct notch, and in having the infraorbital foramen open above P^3. Further, it is similar to *Dendrogale* but distinct from *Tupaia, Lyonogale, Urogale*, and *Anathana* in having a double-rooted upper canine.

The specimen is comparable in size to *Tupaia glis*, but the snout appears more robust and the teeth relatively larger. The relatively large size of the teeth as indicated by the roots compared to other genera is a distinct feature of this specimen. The snout is only moderately elongate.

Incisive foramina are small and usually do not extend posterior to the I^2 in *Dendrogale* (a similarity to *Ptilocercus*), whereas they extend posterior to the I^2 in *Lyonogale, Anathana*, and some species of *Tupaia*. In *Urogale* the incisive foramina are large and do not extend posterior to I^2. The position of the posterior margin of the Siwalik specimen is more similar to *Ptilocercus, Dendrogale*, and some species of *Tupaia* than to *Lyonogale, Anathana*, or *Urogale*.

YGSP 8090 (Fig. 3) is a moderately worn and slightly corroded left lower molar, probably M_1, about the size of *Tupaia minor*. The cusp pattern is primitive with three cusps in the trigonid and three in the talonid. The talonid is wider than the trigonid. The cusps of the trigonid are distinct with the paraconid lower than the protoconid or metaconid. The hypoconid is the largest cusp on the talonid. The hypoconulid is small and located near the entoconid. The entoconid is well developed and distinctly separated from the hypoconulid. There is no entoconulid. The oblique crest extends from the hypoconid to the trigonid, joining the trigonid at the base of the protoconid, slightly buccal to the longitudinal midline of the tooth. There is no buccal cingulum. The base

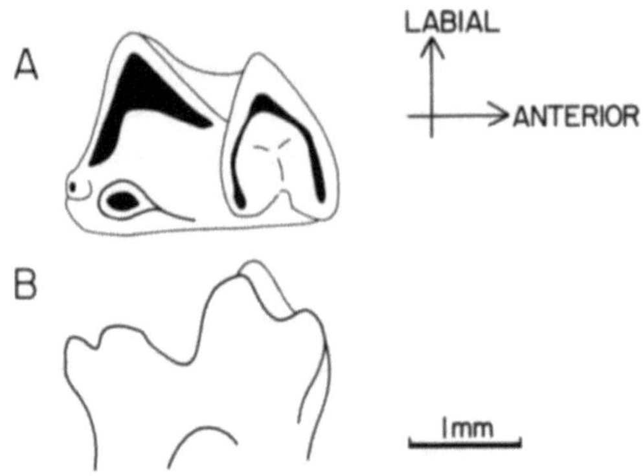

Fig. 3. Diagram of tupaiid lower molar (YGSP 8090). (A), occlusal view; (B), lingual view. Orientation arrows apply to occlusal view only.

of the notch separating the metaconid and paraconid is at a level higher than that separating the metaconid and entoconid.

This specimen is referred to the Tupaiidae because the hypoconulid is in a lingual position near the entoconid, there is no buccal cingulum, and the notch separating the paraconid and metaconid does not extend below the level of the notch separating the metaconid and entoconid. These characters are not unique to tree shrews (some are not present in all tupaiids), but nevertheless, the specimen compares more favorably with tupaiids than any other group, so far as comparisons can be made with this one specimen.

Shrews generally have a buccal cingulum and paraconid and metaconid more deeply separated than in tupaiids. Some shrews such as *Nectogale* and *Anourosorex* have reduced the buccal cingulum, but the separation of the paraconid and metaconid remains deep and thus distinguishes these genera from tupaiids (see Repenning, 1967). The deeper separation of the paraconid and metaconid plus the development of a buccal cingulum in most microchiropteran bats precludes the Siwalik specimen from being considered a bat. In addition, the cusps of the Siwalik fossil are not so steep walled and the cusps are less sharp than in shrews, moles, or bats.

Krishtalka (1976a) characterized adapisoricids as having (among other traits) the hypoconulid on lower molars median or barely lingual in position. The Siwalik specimen is like modern tree shrews rather than fossil adapisoricids in having the hypoconulid more strongly displaced lingually. In addition, Krishtalka (1976a) also characterized early Tertiary erinaceids as having the hypoconulid greatly reduced. In the Siwalik fossil, the hypoconulid is not reduced, and it is displaced strongly toward the entoconid, although the two cusps are distinctly separated. In addition, there is no posterior cingulum projecting from the hypoconulid as in some erinaceids. Finally, the less steep walls of the cusps, and the position of the hypoconulid next to the entoconid with the two apparently more distinctly separated suggest that this specimen is probably not a nyctitheriid. The size of the Siwalik lower molar relative to the skull fragment described above (YGSP 8089) indicates that these specimens represent two distinct species of tree shrews.

The Siwalik lower molar is distinct from *Ptilocercus* in lacking a buccal cingulum, in the oblique crest meeting the trigonid in a slightly more lingual position, and in the hypoconulid being placed slightly more lingually. It is quite similar to *Tupaia*, although the entoconid and hypoconulid may be a bit more distinct. The hypoconulid is more vertical than in *Dendrogale*. The metaconid and entoconid are not so well developed as in *Anathana*. The hypoconulid is more distinct than in *Urogale*. The talonid basin is better defined than in *Lyonogale*.

A third Siwalik specimen, a talonid from a molar tooth, does not warrant detailed consideration here. It is questionably referred to the Tupaiidae because of lingual position of the hypoconulid near the distinct entoconid. If it does in fact represent a tree shrew, the size of this talonid is similar to what would be expected to accompany the skull fragment described above, although they were found at separate localities and cannot represent the same individual.

4. Discussion

The division of the Tupaiidae into subfamilies has received an inordinate amount of attention considering that there are only six genera, and, as a whole, the tupaiids are structurally not very diverse. Each of the named genera has unique (apomorphous) characters which are the definitive aspects of the genus. *Ptilocercus* is usually considered to have a mosaic of primitive and derived characters, and it is usually placed in its own subfamily, the Ptilocercinae. The other five genera comprise the Tupaiinae.

Steele (1973), in his analysis of the tupaiid dentition, demonstrated the distinct nature of the genus *Ptilocercus* based on dental characters, but only by extension does his work demonstrate distinctness at a higher level. While there is no question as to whether *Ptilocercus* is a distinct genus, Davis (1938) has questioned the validity of the subfamily Ptilocercinae because *Dendrogale* exhibits characters which appear to bridge the gap between the two subfamilies. Luckett (this volume) pointed out that characters used by Davis were primitive and therefore provided no evidence for the pattern of branching among *Ptilocercus, Dendrogale,* and *Tupaia*. Davis's use of primitive characters was justifiable for his purpose, as he was attempting to demonstrate the primitive nature of *Dendrogale* among tree shrews and its similarity to *Ptilocercus* on the one hand and tupaiines on the other.

Ptilocercus is very probably derived in some characters, for instance cingula on teeth. Other characters are very probably primitive, such as double-rooted upper canines (although double-rooted second incisors might be advanced). Such characters as absence of supraorbital foramen, posterior position of infraorbital foramen, lacrimal foramen not in notch, are more difficult to evaluate. Because *Ptilocercus* is the only tree shrew that exhibits these characters, it is difficult to say with confidence whether these characters are primitive or derived.

Le Gros Clark (1925) suggested that the absence of a supraorbital foramen in *Ptilocercus* may be associated with the forward and inward rotation of the orbits. If so, then the absence of a supraorbital foramen is a derived character, unless the primitive morphotype for tree shrews had forward eyes which rotated outward in the Tupaiinae.

The Siwalik skull fragment shares the character of double-rooted canines with *Dendrogale* and *Ptilocercus*. This is probably primitive for the Tupaiidae. In addition, the position of the posterior border of the incisive foramina is similar between *Dendrogale, Ptilocercus*, and the Siwalik fossil (and some of the smaller species of *Tupaia*). The Siwalik skull fragment appears more primitive than other tree shrews in having a relatively robust snout and relatively larger teeth as indicated by the roots (except in comparison with the canine of *Urogale* which is obviously derived). Although the Siwalik specimen appears primitive relative to other tree shrews in these characters, it is similar to *Dendrogale* and *Ptilocercus* in others, and, in addition, it exhibits such tupaiine characters as supraorbital foramina, lacrimal foramen in notch, and infraorbital foramen forward. The age of the specimen is middle Miocene, roughly ten million years old.

Davis (1938) may have been correct that *Dendrogale* is more primitive than either *Ptilocercus* or *Tupaia*. It is possible that modern tree shrew genera (and subfamilies) differentiated within the last ten million years. These hypotheses may be tested in the Siwalik forms. Better knowledge of Siwalik fossil tree shrews hopefully may improve our knowledge of the relationships of modern forms and enhance the recognition of even earlier tupaiids.

5. References

Bohlin, B. 1951. Some mammalian remains from Shih-ehr-ma-ch'eng, Hui-hui-p'u area, Western Kansu, pp. 1–47. *In* S. Hedin (ed.). Reports from the scientific expedition to the northwestern provinces of China. Stockholm, vol. 6, Vertebrate Paleontology 5.

Butler, P.M. 1948. On the evolution of the skull and teeth in the Erinaceidae, with special reference to fossil material in the British Museum. *Proc. Zool. Soc. Lond.* 118:446–500.

Chopra, S. R. K., Kaul, S., and Vasishat, R. N. 1979. Miocene tree shrews from the Indian Śivaliks. *Nature* 281: 213–214.

Chopra, S. R. K., and Vasishat, R. N. 1979. Śivalik fossil tree shrew from Haritalyangar, India. *Nature* 281: 214–215.

Davis, D.D. 1938. Notes on the anatomy of the treeshrew *Dendrogale*. *Field Mus. Nat. Hist., Zool. Series* 20:383–404.

Dutta, A.K. 1975. Micromammals from Siwaliks. *Indian Minerals* 29:76–77.

Evans, F.G. 1942. The osteology and relationships of the elephant shrews (Macroscelididae). *Bull. Amer. Mus. Nat. Hist.* 80:85–125.

Gregory, W.K. 1913. Relationship of the Tupaiidae and of Eocene lemurs, especially *Notharctus*. *Bull. Geol. Soc. Amer.* 24:247–252.

Krishtalka, L. 1976a. Early Tertiary Adapisoricidae and Erinaceidae (Mammalia, Insectivora) of North America. *Bull Carnegie Mus. Nat. Hist.* 1:1–40.

Krishtalka, L. 1976b. North American Nyctitheriidae (Mammalia, Insectivora). *Ann. Carnegie Mus.* 46:7–28.

Krishtalka, L., and West, R.M. 1977. Paleontology and geology of the Bridger Formation, southern Green River Basin, southwestern Wyoming. Part 2. The Bridgerian insectivore *Entomolestes grangeri. Contr. Biol. Geol. Milwaukee Pub. Mus.* 14:1–11.

Le Gros Clark, W.E. 1925. On the skull of *Tupaia. Proc. Zool. Soc. Lond.* 1925:559–567.

Lewis, G.E. 1934. Preliminary notice of new man-like apes from India. *Amer. J. Sci.* 27:161–179.

Lyon, M.W., Jr. 1913. Tree shrews: An account of the mammalian family Tupaiidae. *Proc. U.S. Nat. Mus.* 45:1–188.

Matthew, W.D. 1909. The Carnivora and Insectivora of the Bridger Basin, middle Eocene. *Mem. Amer. Mus. Nat. Hist.* 9:291–567.

Matthew, W.D. 1918. Insectivora (continued), Glires, Edentata. *Bull. Amer. Mus. Nat. Hist.* 38: 565–657.

Matthew, W.D., and Granger, W. 1924. New insectivores and ruminants from the Tertiary of Mongolia, with remarks on the correlation. *Amer. Mus. Novitates* 105:1–7.

McKenna, M.C. 1960a. Fossil Mammalia from the early Wasatchian Four Mile fauna, Eocene of northwest Colorado. *Univ. Calif. Publ. Geol. Sci.* 37:1–130.

McKenna, M.C. 1960b. The Geolabidinae, a new subfamily of early Cenozoic erinaceoid insectivores. *Univ. Calif. Publ. Geol. Sci.* 37:131–164.

McKenna, M.C. 1963 a. The early Tertiary Primates and their ancestors. *Proc. XVI Int. Cong. Zool.* 4:69–74.

McKenna, M.C. 1963b. New evidence against tupaioid affinities of the mammalian family Anagalidae. *Amer. Mus. Novitates* 2158:1–16.

McKenna, M.C. 1966. Paleontology and the origin of the primates. *Folia primatol.* 4:1–25.

McKenna, M.C. 1975. Toward a phylogenetic classification of the Mammalia, pp. 21–46. *In* W.P. Luckett and F.S. Szalay (eds.). *Phylogeny of the Primates*. Plenum Press, New York.

McKenna, M.C., and Simpson, G.G. 1959. A new insectivore from the middle Eocene of Tabernacle Butte, Wyoming. *Amer. Mus. Novitates* 1952: 1–12.

Patterson, B. 1965. The fossil elephant shrews (family Macroscelididae). *Bull. Mus. Comp. Zool., Harvard Univ.* 133:205–335.

Pilbeam, D., Barry, J., Meyer, G.E., Shah, S.M.I., Pickford, M.H.L., Bishop, W. W., Thomas, H., and Jacobs. L.L. 1977a. Geology and paleontology of Neogene strata of Pakistan. *Nature* 270:684–689.

Pilbeam, D., Meyer, G.E., Badgley, C., Rose, M.D., Pickford, M.H.L., Behrensmeyer, A.K., and Shah, S.M.I. 1977b. New hominoid primates from the Siwaliks of Pakistan and their bearing on hominoid evolution. *Nature* 270:689–695.

Repenning, C.A. 1967. Subfamilies and genera of the Soricidae. *U.S. Geol. Surv. Prof. Pap.* 65: 1–74.

Russell, D.E., Louis, P., and Savage, D.E. 1975. Les Adapisoricidae de l'éocène inférieur de France. Réévaluation des formes considerées affines. *Bull. Mus. Nat. Hist. Nat., Paris* 327:129–193.

Schwartz, J.H., and Krishtalka, L. 1976. The lower antemolar teeth of *Litolestes ignotus*, a late Paleocene erinaceid (Mammalia, Insectivora). *Ann. Carnegie Mus.* 46:1–6.

Simpson, G.G. 1928. A new mammalian fauna from the Fort Union of southern Montana. *Amer. Mus. Novitates* 297:1–15.

Simpson, G.G. 1929. A collection of Paleocene mammals from Bear Creek, Montana. *Ann. Carnegie Mus.* 19:115–122.

Simpson, G.G. 1931. A new insectivore from the Oligocene, Ulan Gochu horizon, of Mongolia. *Amer. Mus. Novitates* 505:1–22.

Simpson, G.G. 1945. The principles of classification and a classification of mammals. *Bull. Amer. Mus. Nat. Hist.* 85:1–350.

Steele, D.G. 1973. Dental variability in the tree shrews (Tupaiidae). *Symp. 4th Intern. Cong. Primatol.* 3:154–179.

Szalay, F. S. 1968. The beginnings of primates. *Evolution* 22:19–36.

Szalay, F.S. 1969. Mixodectidae, Microsyopidae, and the insectivore-primate transition. *Bull. Amer. Mus. Nat. Hist.* 140:193–330.

Szalay, F.S. 1977. Phylogenetic relationships and a classification of the eutherian Mammalia, pp. 315–374. *In* M.K. Hecht, P.C. Goody and B.M. Hecht (eds.). *Major Patterns in Vertebrate Evolution*. Plenum Press, New York.

Szalay, F.S., and McKenna, M.C. 1971. Beginning of the age of mammals in Asia: The late Paleocene Gashato fauna, Mongolia. *Bull. Amer. Mus. Nat. Hist.* 144:269–318.

Van Valen, L. 1965. Tree shrews, primates, and fossils. *Evolution* 19:137–151.

Van Valen, L. 1966. Deltatheridia, a new order of mammals. *Bull. Amer. Mus. Nat. Hist.* 132:1–126.

Van Valen, L. 1967. New Paleocene insectivores and insectivore classification. *Bull. Amer. Mus. Nat. Hist.* 135:217–284.

Weitzel, K. 1949. Neue Wirbeltiere (Rodentia, Insectivora, Testudinata) aus dem Mitteleozän von Messel bei Darmstadt. *Abh. Senckenb. Nat. Ges.* 480:1–24.

Nervous System III

The Nervous System of the Tupaiidae: Its Bearing on Phyletic Relationships

7

C.B.G. CAMPBELL

1. Introduction

Evidence derived from studies of the brain of *Tupaia* by Sir Wilfrid Le Gros Clark played a major role in the acceptance of the classification of tupaiids as primates by many authorities (e.g. Simpson, 1945). Their current popularity as experimental subjects in neuroanatomy and neurophysiology has been a direct outgrowth of that status. Now that the view that they are primitive primates is less widely held, other considerations are being used to rationalize their study.

Le Gros Clark in 1924, upon returning from three years spent as Principal Medical Officer of Sarawak, took up the Chair of Anatomy at St. Bartholomew's Hospital Medical School in London. His prior anatomical experience consisted of one year working as a Demonstrator in Anatomy at St. Thomas's Hospital Medical School and the publication of one paper on the cranial characters of the Eskimo and another on the Pacchionian granulations. Consequently, he was appointed as Reader initially and elevated to the professorship three years later after the publication, in rapid succession, of a series of papers on the anatomy of insectivores and primates (Le Gros Clark, 1968).

As a medical officer in Sarawak he had the unique opportunity to collect tree shrew and tarsier material in the field and this formed the basis of much

C.B.G. CAMPBELL • Department of Medical Neurosciences, Division of Neuropsychiatry, Walter Reed Army Institute of Research, Washington, D.C. 20012

of his early work. Gregory (1910) and Carlson (1922) had both pointed out anatomical similarities between tupaiids and prosimians, the former proposing that primates were derived from a large-brained arboreal insectivore resembling extant tupaiids. Elliot Smith, Le Gros Clark's great mentor, had apparently accepted a tupaioid ancestry of the primates (Elliot Smith, 1924, p. 31) and Le Gros Clark himself went further, arguing that evidence drawn from the brain especially, but from other structures as well, supported the classification of tupaiids as prosimian primates. All of his arguments in this regard were gathered into a single volume (Le Gros Clark, 1934), whose publication coincided with his assumption of the Chair of Anatomy at Oxford University. The following year he was elected to Fellowship in the Royal Society of London. He conducted no further original work on tupaiids, but his establishment as an anatomist of note was to a large degree originally based on his studies of tupaiid anatomy and their implications for primate taxonomy.

This work was done in a fascinating era in the history of primatology and comparative neurology. Great figures occupied the stage of British anatomical science and the young Le Gros Clark moved among them and inherited their mantle. They included such men as Grafton Elliot Smith, Frederick Wood Jones, and Arthur Keith. Modern comparative neuroanatomy and primatology differ greatly from the science they knew. One of the ways in which comparative neuroanatomy has changed is that it has become largely an experimental science. Interestingly enough, one of the individuals most responsible for this change was Wilfrid Le Gros Clark.

It is my intent to review here much of what is currently known concerning the nervous system of the Tupaiidae. The vast majority of the work has involved a single species, *Tupaia glis*.

2. General Comments on the Tupaiid Brain

The brain of *Tupaia* is elongate, with an expanded occipito-temporal region and a somewhat constricted frontal area (Fig. 1). The olfactory bulbs are conspicuous; however, they appear somewhat smaller in relative size than those of extant insectivores such as *Tenrec* and larger than those of extant primates, especially the Haplorhini. A great deal of comparative data on the relative size of the olfactory bulbs of living mammals is available due to the work of Stephan (1972). It has been assumed that these data are consistent with Elliot Smith's (1924) thesis that a reduction in the importance of the sense of smell is a characteristic feature of primate brain evolution. Martin (1973) has indicated that although some data from fossil primates relative to olfactory bulb size exist (e.g. Radinsky, 1970), no thorough study has been made which clearly establishes the direction of trends in relative size in different mam-

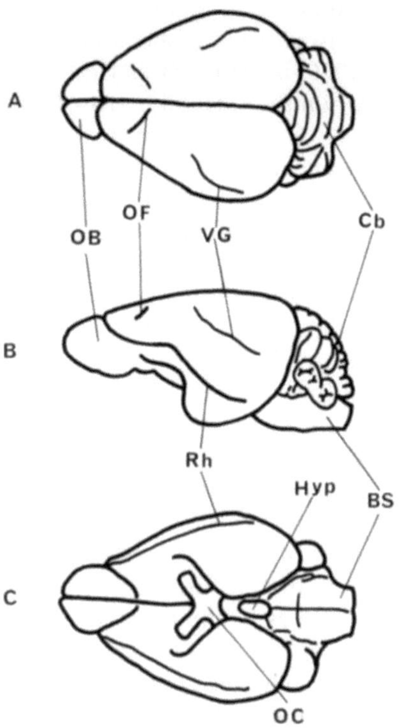

Fig. 1. Gross morphology of the brain of *Tupaia glis* in (A) dorsal view, (B) lateral view, and (C) ventral view.

malian lineages. The lateral ofactory tract is prominent and can be readily seen below the rhinal sulcus on the ventrolateral aspect of the brain.

The cerebral hemispheres of *Tupaia* abut posteriorly onto the cerebellum, completely obscuring the thalamus and midbrain when viewed from above. This is true in adult brains, as Tigges and Shantha (1969) noted that in the newborn the midbrain is exposed. The forebrain of *Tupaia* is lissencephalic, except for the rhinal sulcus which separates the neocortex from the pyriform cortex, and a rudimentary calcarine fissure on the medial aspect of the hemisphere (Fig. 1). Martin (1973) has noted that Elliot Smith considered a triradiate calcarine sulcus complex (calcarine, retrocalcarine, and paracalcarine sulci) to be characteristic of primates. Tree shrews do not possess such a sulcus complex. A shallow vascular groove forms an arc in the mid- and posterior portions of the dorsolateral hemisphere surface (Fig. 1B). An inconstant, shallow depression can be seen to lie between the rostral and middle thirds of the hemisphere in the superior view (Fig. 1a). It runs diagonally from near the sagittal fissure to near the rhinal fissure (Lende, 1970).

The vascular groove lying posteriorly in the dorsolateral hemisphere surface fairly consistently marks the lateral margin of cortical area 19, one of the so-called extrastriate visual areas (Kaas *et al.*, 1972b; Killackey, personal communication). Lende (1970) has shown that the diagonal indentation located anteriorly, and termed by him the "orbital furrow," marks the anterior border of the somatic sensory area. He considered this to be the homologue of the orbital sulcus, found in insectivores and the opossum, which similarly marks the rostral limit of the somatic sensory area. A central sulcus, characteristic of the primate brain, as depicted by Bernard Campbell (1966, p. 216) in *Tupaia* is chimerical.

The neocortex makes up a larger proportion of the cerebral hemisphere in *Tupaia* than it does in insectivores such as *Erinaceus* and *Tenrec*. This is indicated by the low position of the rhinal sulcus in *Tupaia* (Figs. 1b, c). This sulcus can be seen virtually in its entirety in a ventral view of the tupaiid brain. This contrasts with the brain of *Erinaceus*, for example, where the sulcus is found approximately halfway up the dorsolateral surface of the hemisphere and is not visible from the ventral aspect. This character was used by Le Gros Clark in his arguments in favor of classifying tupaiids as primates. A rhinal sulcus location similar to that of tupaiids can be found in a number of genera of the orders Marsupialia, Rodentia, and Lagomorpha (Campbell, 1966).

The optic nerves, optic chiasm, and optic tracts are quite prominent. The superior colliculus, a major optic relay center, is extremely large in relation to the inferior colliculus and the remainder of the brain when compared to many other mammals of equivalent size. The brain stem and spinal cord are not otherwise remarkable and resemble those of other small mammals.

3. Motor Systems

A. The Pyramidal System

Although hardly matching the visual system as an object of neuroscience interest, the tupaiid motor system has been extensively studied. The motor neurons of the spinal cord ventral horns are subject to the influences of a number of neuronal systems. The interaction of these systems determines the output of the motor neurons and, therefore, muscle activity. One such system, the pyramidal tract, is composed of neurons whose cell bodies lie in the cerebral cortex and send their axons out of the cortex via the internal capsule, cerebral peduncle, and medullary pyramid to terminate in various brain stem centers (so-called corticopontine and corticobulbar fibers) and in the gray matter of the spinal cord (corticospinal fibers). The vast majority of corticospinal fibers in most mammals cross to the opposite side in a decussation at the bulbospinal junction and travel in either the dorsal, lateral, or ventral funiculi before diverting into the gray matter to their site of termination. Some uncrossed tract fibers are usually also present. Different groups of mammals (usually at the ordinal level) have different patterns of disposition of corticospinal fibers in the funiculi of the spinal cord.

The cortical origin of the tupaiid pyramidal tract has been studied by anatomical methods (Jane *et al.*, 1969). This study showed that Le Gros Clark's (1924) Betz cell, insular, and anterior portions of the parietal area contain pyramidal tract neurons. The Betz cell area appears to be the primary source (Fig. 2). In addition, a cortical mapping study by Lende (1970), using macroelectrode stimulation to elicit motor movements, demonstrates an area of excitable cortex which essentially coincides with the anatomically derived map (Fig. 3). Jane *et al.* (1965, 1969) and Shriver and Noback (1967) have described the pyramidal tract pathways of *Tupaia*. Uncrossed axon projections to the periaqueductal gray matter of the midbrain in the vicinity of neurons of the mesencephalic nucleus of the trigeminal nerve were found, although termination on these neurons was not established. Uncrossed projections to the midbrain reticular formation, red nucleus, and pontine nuclei were also found. Bilateral projections to the lateral and medial reticular formation of the medulla oblongata were seen. These were predominantly to the ipsilateral medial reticular formation and to the contralateral lateral reticular formation. Corticobulbar fibers terminated bilaterally in the sensory nuclei of the trigeminal nucleus and in the ipsilateral inferior olivary nuclear complex. No direct projections to the motor nuclei of the cranial nerves were found. Pyramidal system axons terminate in the nuclei of the tractus solitarius and the contralateral nucleus gracilis and nucleus cuneatus.

At the pyramidal decussation the majority of corticospinal axons cross and descend in the contralateral dorsal funiculus. A small number of axons do not cross but descend in the ipsilateral dorsal funiculus. An even smaller number of axons cross and descend in the lateral and ventral funiculi but were not discernible below C6. The contralateral dorsal tract extends throughout cervical and thoracic levels. Ipsilateral dorsal tract axons, although few in

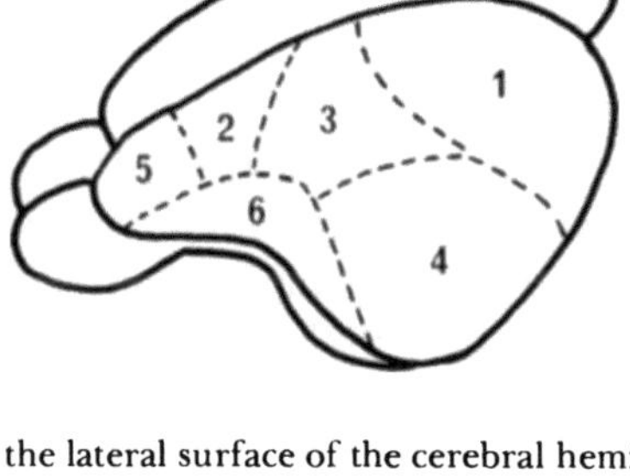

Fig. 2. The cortical areas on the lateral surface of the cerebral hemisphere (after Le Gros Clark, 1924).

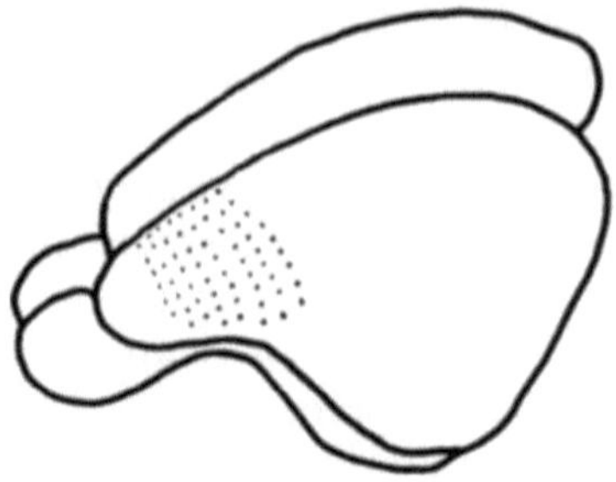

Fig. 3. Area of excitable cortex according to Lende (1970).

number, were seen at thoracic levels as well (Jane *et al.*, 1969). Axon terminations were found only in the dorsal horn and intermediate zone of the spinal gray matter. None was seen in the ventral horn where the spinal motor neuron cell somata lie. The dendrites of motor cells whose cell somata lie in the ventral horn extend into the intermediate zone and dorsal horn where axodendritic contacts with ventral horn motor neurons are probable. Valverde (1966) has shown such contacts in the rat, an animal with a corticospinal system organized in a similar manner.

Axosomatic corticospinal contacts with ventral horn motor neurons have been reported (Campbell *et al.*, 1966b; Goode, 1974; Goode and Haines, 1975; Harting and Noback, 1970; Kuypers, 1964) in all primates examined thus far except the marmoset, *Saguinus oedipus*. Shriver and Matzke (1965) deny their presence in this species in a short abstract. This study should be repeated with newer methods, as the suppressive Nauta-Gygax technique applied in that study can suppress sparse terminals if not employed with care. As has been pointed out elsewhere (Campbell, 1965; Jane *et al.*, 1965), the tupaiid pyramidal tract system conforms in every aspect to a nonprimate pattern of organization. Primate pyramidal tracts project to cranial nerve motor nuclei in the brain stem and the principal tracts extend throughout the cord in the lateral funiculi.

B. The Extrapyramidal System

The only portion of what has been termed the extrapyramidal motor system which has been examined in *Tupaia* is the rubrospinal tract (Murray *et al.*, 1976). This tract was found to correspond to that found in the marsupial and primate and nonprimate placental mammals examined previously. The cell bodies of origin form a conspicuous pair of nuclei on either side of the midline in the mesencephalon. Following lesions in one of the nuclei, rubrospinal axons were found to leave the nucleus, cross the midline in the ventral tegmental decussation and form a tract in the ventrolateral brain stem. In the spinal cord the tract shifts to the dorsal portion of the lateral funiculus and extends to the caudalmost sacral cord. Its precise position in the coccygeal cord was undetermined in this study. In the spinal gray matter rubrospinal axons were found to terminate in Rexed's (1954) laminae V-VII after leaving the tract. Medial and dorsomedial regions of the nucleus project primarily to cervical levels; lateral and ventrolateral regions of the nucleus project chiefly to lumbar and sacral levels of the cord. Murray *et al.* (1976) point out that since in *Tupaia* no corticospinal axons reach the lumbosacral cord, rubrospinal fibers may play a major role in the initiation and control of hindlimb movement, probably via last-order internuncial neurons as in many other therian mammals. Other descending tract systems, such as the vestibulospinal, interstitiospinal tracts, etc., have not yet been examined.

C. The Cerebellum

A series of studies of the tupaiid cerebellum has appeared (Haines, 1969, 1971a, 1971b, 1975; Haines and Culberson, 1974; Ware, 1973). The corpus cerebelli and flocculonodular lobe are described (Haines, 1969) in regard to their gross morphology and contrasted with the condition in *Galago*. The anterior lobe was found to be poorly developed in *Tupaia*, but well developed and with larger hemispheric portions in *Galago*. Haines made comparisons only among other tupaiids (*Tupaia minor* and *Ptilocercus lowii*), prosimians (*Microcebus murinus, Nycticebus coucang, Tarsius, Perodicticus potto*, and *Galago senegalensis*), and an anthropoid (*Macaca*). In general, the anterior lobe is large in *Macaca*, somewhat less well developed in *Microcebus, Nycticebus, Tarsius*, and *Galago*, and poorly developed in the tupaiids and *Perodicticus*.

Haines (1969) described the posterior lobe of *Tupaia* as better developed than that of *Galago*. He was particularly impressed with the seventh vermian lobule and part of its hemispheric extensions, crus I and crus II of the ansiform lobule. The ansiform lobule of *Tupaia* has a crus I divisible into three sublobules and a crus II divisible into two sublobules, while those structures are not subdivided in *Galago*, except for an inconstant shallow sulcus sometimes seen in crus I. This sulcus, when present, does not penetrate sufficiently deeply to subdivide crus I. The ansiform lobule of *Perodicticus potto* resembles that of *Tupaia*, while *Lemur, Tarsius*, and *Nycticebus* resemble that of *Galago* (Haines, 1969, based on Elliot Smith, 1903 and Bolk, 1906). Haines stated that the ansiform lobule is greatly expanded in *Macaca, Pan, Hylobates*, and *Homo*.

Another character Haines (1969) considered to be significant is the copula pyramidis, a band of cerebellar cortex connecting the paraflocculus to the pyramid of the vermis. He considered the copula to be unfissured in edentates, the opossum, the rat, and a variety of insectivores ("in most instances"). He further stated that it is moderately fissured in *Lemur* and is markedly so in the rhesus monkey. All of these comparisons were based on a literature survey. The copula pyramidis is longitudinally fissured in *Galago* and divided into medial and lateral portions in *Tupaia*. Haines suggested that the tupaiid copula pyramidis represents an "advance" over that of insectivores and other forms and is evidence of approximation of *Tupaia* to the primates. He considered the advanced differentiation of the ansiform lobule in *Tupaia* in a similar vein.

The ansiform lobule of *Scapanus townsendi*, a mole, is divided into five folia, two in crus I and three in crus II. The copula pyramidis is constricted into lateral and medial parts (Larsell, 1970). This resembles the condition in *Tupaia*. I have examined the cerebella of a number of sciurids and find them to have well developed, subdivided ansiform lobules as well. The elephant shrews, *Elephantulus* and *Rhynchocyon*, have large anterior lobes of the cerebellum (Le Gros Clark, 1932) and in this regard, therefore, appear more like *Galago*. These cerebellar features do not appear to me to be valid evidence of a tupaiid "approximation to the primates." Once again, comparisons have not been made sufficiently widely.

In another paper, Haines (1971a) found that much of the development of the cerebellum of *Galago* and *Tupaia* takes place in the prenatal period as occurs also in the rhesus monkey. He suggested that "such information may yet prove useful in ordering the ancestry of primates and non-primates." One can hardly deny that it "may" but he does not provide enough comparative information to make a valid suggestion that it will do so. Haines (1971b) has contrasted the cerebellar nuclei of *Galago* and *Tupaia*. The cerebellar nuclei of *Tupaia* resemble those of the rat, hedgehog, and mole in that there is a rostrocaudal orientation of the nucleus interpositus anterior and nucleus interpositus posterior, and incomplete separation of these nuclei from each other and the nucleus lateralis. *Galago* has the same nuclei, a nucleus medialis, interpositus anterior and posterior, and lateralis, but all are distinctly differentiated from each other. The nucleus lateralis has its lateral, rostral, and caudal sides thrown into an irregular contour suggestive of the undulations seen in the dentate nucleus of anthropoid primates.

An experimental study of the efferent projections of the cerebellar posterior lobe (Haines and Culberson, 1974) has shown that the vermis projects to caudal portions of the nucleus medialis and the rostral vermis projects to the interposed nuclei as well. The vermis also projects in a somatotopic manner to the vestibular nuclei. Paravermal cortex was found to have connections with the interposed nuclei, central areas of nucleus medialis, and medial areas of nucleus lateralis. The lateral cortex projects primarily to the nucleus lateralis and, to a lesser extent, the interposed nuclei.

Some connections of the anterior lobe with the vestibular nuclear complex have been described (Haines, 1975). The projections were found to be ipsilateral, axons from lobules V and IV traversing the lateral portions of the juxtarestiform body and those from lobules III and II occupying its more medial portion. Lobule V projects in a lesser degree to the superior vestibular nucleus and in large measure to dorsal, central, and dorsolateral areas of the lateral vestibular nucleus. Lobule V projects to the same sites, with the addition of some fibers to the spinal vestibular nucleus. Lobules II and III were found to connect massively with the dorsal half of the lateral vestibular nucleus and more sparsely to its ventral half. Sparse projections to the superior vestibular nucleus and the rostrodorsal spinal vestibular nucleus were also seen.

Ware (1973) has described the efferent projections of the deep cerebellar nuclei. Axons were found to ascend via the brachium conjunctivum (superior cerebellar peduncle) to the red nucleus of the midbrain, superior colliculus, oculomotor nuclei, nucleus of the posterior commissure, ventrolateral and ventral anterior nuclei of the thalamus, and recross through the reuniens nucleus to the ipsilateral ventral anterior nucleus. Projections to the contralateral pontine gray and ventromedial reticular formation via the descending limb of the superior cerebellar peduncle were also seen. The medial cerebellar nucleus connects with the vestibular nuclei, and, via the uncinate fasciculus, with the contralateral reticular formation. These findings are similar to those seen in other mammals.

4. Sensory Systems

A. The Somatosensory System

Afferent fibers in the spinal dorsal roots which project to the nucleus dorsalis of Clarke at hindlimb cord levels have been studied in *Tupaia glis* and *Galago senegalensis* (Albright and Haines, 1974). The nucleus dorsalis of Clarke is a column of conspicuous cells located at the base of the dorsal horn of the spinal cord. The column of cells is found at thoracic and lumbar levels. These cells receive proprioceptive and exteroceptive information and their axons, as the dorsal spinocerebellar tract, convey this information to the cerebellum.

Albright and Haines found a basic pattern in both animals fundamentally similar to that seen in the cat and rhesus macaque with similar methods of study. Incoming dorsal root fibers are distributed to cells of the nucleus dorsalis at multiple levels both above and below their level of entry at lower thoracic and upper lumbar levels, or entirely above their level of entry in the case of fibers entering at L5 or L6. They found a pattern of termination in the ventrolateral portion of the nucleus at caudal levels which shifts to the dorsal portion of the nucleus rostrally. In *Galago* fibers from L1 and L3 projected at least two to five segments further rostrally than in *Tupaia* or the rhesus macque (Carpenter *et al.*, 1968).

In *Galago* from T8–T11 the cells of Clarke's nucleus dorsalis are disposed in a circular manner when viewed in transverse sections (Albright and Haines, 1973). Albright and Haines (1974) suggest that the more rostral projection of dorsal root fibers to Clarke's column at some levels and the longitudinal orientation of fibers of this system within the medullary or center region of the circle of cells are related to the vertical clinging and leaping mode of locomotion in *Galago*. The lack of such an arrangement in the quadrupedal *Tupaia*, cat, and rhesus macaque led the authors to this interesting suggestion. The suggestion could be correct; however, not enough comparative data are presented to make the hypothesis a strong one. Data from *Galago crassicaudatus* would be particularly helpful, and of course data from a number of other primates and quadrupedal nonprimates would be useful as well.

B. The Auditory System

Recently a significant amount of work has been done on the secondary and higher order portions of the auditory system of *Tupaia*. Primary auditory cortex has been identified on the basis of its cytoarchitecture and its relations with the medial geniculate nucleus (Oliver and Hall, 1975). This area, where dense populations of small cells are so profuse in neocortical layers II, III, and IV that they blur the distinctions between the layers, is located posteriorly on the dorsal lip of the rhinal sulcus. It is surrounded by a cortical area whose appearance with the Nissl stain is somewhat different than the primary auditory cortex or "core" cortex. This is also auditory-related cortex and has been termed the "belt" by Casseday *et al.* (1976). It is akin to the extrastriate visual cortex which surrounds the primary visual cortical area, the so-called striate cortex.

The major auditory portion of the thalamus, the medial geniculate nucleus, has been subdivided on the basis of the Nissl stain into a number of subnuclei or divisions: the magnocellular, ventral, dorsal, and caudal divisions (Casseday *et al.*, 1976). The names for these nuclear divisions have been borrowed from terms applied to similar structures in the cat. The same authors have identified three divisions of the inferior colliculus, the principal midbrain auditory structure. The two inferior colliculi comprise the posterior pair of the corpora quadrigemina. The subdivisions of the inferior colliculus are: the central nucleus, an elliptical area of densely packed round cells in the center of the colliculus; the pericentral nucleus, which forms a dorsal cap over the central nucleus; and the external nucleus which lies lateral to the central nucleus. Ascending fibers from the auditory centers in the medulla oblongata travel in a prominent bundle, the lateral lemniscus, which terminates largely in the inferior colliculus. Intercalated in this bundle is a group of cells termed the nucleus of the lateral lemniscus. It has been subdivided into dorsal and ventral subdivisions.

Using modern tract tracing techniques, Diamond and his co-workers (Casseday *et al.* 1976; Jones *et al.* 1976) have studied the connectional relationships of these central nervous system auditory centers. They compared the auditory pathways of *Tupaia glis* and the more completely studied domestic cat and argued that the ascending auditory pathways of the two species closely correspond. Further, they suggested that multiple parallel pathways to different portions of auditory cortex are present in both species and this, therefore, probably represents a general mammalian plan. Figure 4 is intended to dia-

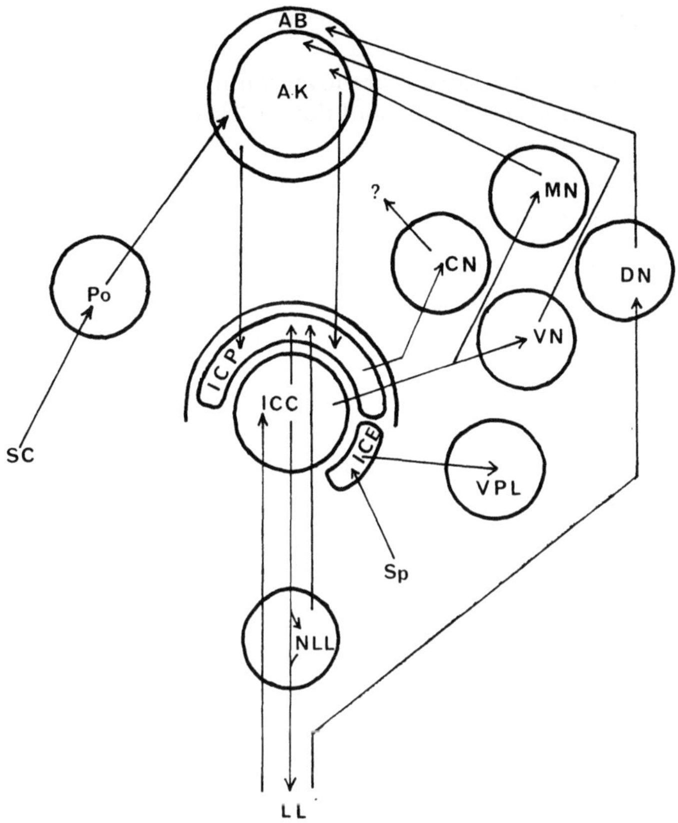

Fig. 4. Auditory connections of the inferior colliculus, medial geniculate body, and auditory cortex based on Casseday *et al.* (1976). Only ipsilateral connections are depicted.

grammatically depict the pathways described by Diamond and his co-workers.

One of the major points stressed by these workers is that there are at least two other parallel pathways to auditory cortex, in addition to the ascending auditory pathway in the lateral lemniscus which relays in the central nucleus of the inferior colliculus, projects to the ventral and to a lesser extent the *magnocellular* nucleus of the medial geniculate body and is relayed to the "core" of auditory cortex. One of these is via fibers in the lateral lemniscus and/or surrounding tegmentum which project to the *dorsal* nucleus of the medial geniculate body, without synapse in the inferior colliculus, where it is relayed to the "belt" of auditory cortex. It is not entirely clear from their description, but it appears that the described projection from the dorsal nucleus to the "belt" may be based on a single case. A third pathway from the deep portion of the superior colliculus projects to the posterior nucleus, another auditory-related thalamic nucleus, and is relayed to the "belt" of auditory cortex. The existence of this pathway is based on other work (Harting *et al.*, 1973; Diamond *et al.*, 1970).

Casseday *et al.* (1976) stated that the ascending auditory projections to the auditory thalamic nuclei in *Tupaia* and the cat "closely correspond." It is more difficult to demonstrate close correspondence of the cat and tree shrew in pathways from the auditory thalamus to the auditory cortex. The "belt" in the cat is clearly divisible into different areas, while no comparable subdivisions have been noted in *Tupaia*. Further, the described pathways in the cat from the subdivisions of the medial geniculate body to auditory cortex do not correspond to those described by Casseday *et al.* (1976) in *Tupaia*. Emphasis is placed on the fact that a distinction between "core" and "belt" auditory cortex is found in both animals.

The ascending projections to the auditory thalamus of the cat and tree shrew are very similar but are not precisely the same. The central nucleus of the inferior colliculus in the cat projects to the "core" of auditory cortex, area AI, via a relay in the ventral division of the medial geniculate body as in *Tupaia*. Also, the dorsal division of the medial geniculate body and the posterior nucleus do not receive projections from the inferior colliculus in the cat just as in *Tupaia*. These are important similarities.

There is disagreement in the cat literature in regard to projections from the inferior colliculus to the magnocellular division of the medial geniculate body. They are described by some workers (Moore and Goldberg, 1963) and denied by others (van Noort, 1969). Such a projection is also described in "the monkey," presumable *Macaca* (Moore and Goldberg, 1966), and in the rabbit a non-magnocellular, but topographically similar, nuclear group with similar connections has been described (Moore and Tarlov, 1963). The magnocellular division in the cat is also a target for fibers of the lateral lemniscus (Goldberg and Moore, 1967) and the lateral tegmental area (Morest, 1965). This is apparently not the case in *Tupaia* (Casseday *et al.*, 1976). Descending auditory projections to the pons have been described in the cat but were not seen in *Tupaia* (Casseday *et al.*, 1976).

Oliver *et al.* (1976) have mapped the tonotopic organization of the primary auditory cortex of *Tupaia glis* using both physiological and anatomical methods. They found that high frequency tones are represented rostrodorsally on the hemisphere surface in this cortical area and lower frequencies caudoventrally. The cortical area involved corresponds to the "core" area of Casseday *et al.* (1976). The rostral portions of this cortex receive projections from neurons in the rostral part of the ventral nucleus of the medial geniculate body, while caudal portions of the "core" cortex receive projections from the caudal part of the ventral nucleus. Figure 5 depicts the neocortical areas based on a reconstruction by Carey *et al.* (1979).

Two papers bearing on the central auditory pathways of *Tupaia* were published recently. Oliver and Hall (1978a, b), on the basis of careful cytoarchitectonic study and the use of the experimental silver impregnation, autoradiographic, and horseradish peroxidase tract tracing methods, described

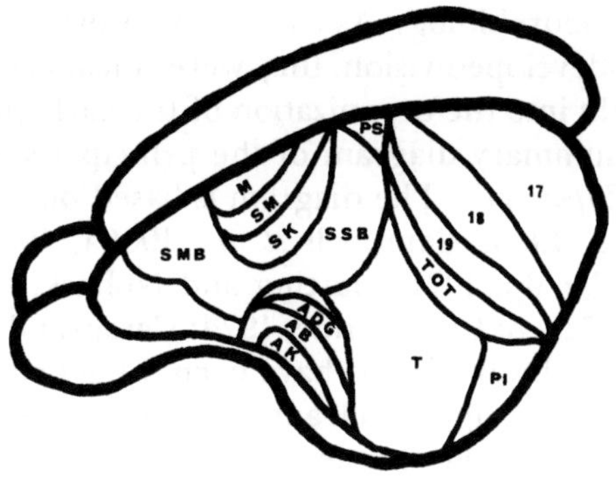

Fig. 5. The cortical areas on the lateral surface of the cerebral hemisphere (after Carey *et al.*, 1979). The reconstruction of the auditory areas is not in agreement with Oliver and Hall (1978).

somewhat different results than those obtained by Casseday *et al.* (1976). They parcellate the medial geniculate body and auditory cortex differently and use a different nomenclature than do Casseday *et al.* (1976).

Oliver and Hall's (1978a, b) dorsal division includes the posterior nucleus and dorsal division of Casseday *et al.* (1976). Their ventral division includes both the ventral and caudal division of Casseday *et al.*, and their medial division corresponds to the magnocellular division of Casseday *et al.* In addition, Oliver and Hall described a medial nucleus of the inferior colliculus which is not described by the other workers. Also, they parcellated the "belt" of auditory cortex into six areas, one divided into two subareas.

The divisions of the medial geniculate body are likewise divided into two or more nuclei each. Each nucleus projects to a different auditory cortical area. The ventral nucleus, equivalent to the ventral division of Casseday *et al.* (1976), is the major source of axons to the primary auditory cortex or "core." The remaining nuclei each project to a separate area or subarea within the auditory "belt," except for the caudal nucleus (a portion of the magnocellular nucleus of Casseday *et al.*). This nucleus may project to the entire region. Projections from the superior colliculus, midbrain tegmentum, and inferior colliculus to the auditory thalamus are similar in the two studies, but not precisely the same. The study of Oliver and Hall, because it made use of the horseradish peroxidase and autoradiographic methods, was able to avoid some of the difficulties encountered by Diamond and his co-workers who used electrolytic lesions which often involved more than one structure.

C. The Visual System

By far the greatest amount of work has been done on the visual system of *Tupaia*. In recent years vision has been one of the most active and productive areas of interest in neurobiology. As tree shrews were thought to be primitive primates with well-developed vision, they were considered ideal subjects from which to gain insight into the organization of the early primate visual system.

Figure 6 is a summary diagram of the principal known portions of the visual pathway of *Tupaia glis*. The diagram is based on the work of a number of workers (Tigges, 1966; Campbell *et al.*, 1966a, 1967; Glickstein, 1967; Laemle, 1968; Abplanalp, 1971; Harting and Noback, 1971; Diamond *et al.*, 1970; Kaas *et al.*, 1972a, b; Casagrande, 1974; Casagrande and Harting, 1975; Hubel, 1975). The pathways found have been, in general, similar to those of most mammals. Retinal ganglion cell axons form the optic nerve which undergoes partial decussation at the chiasm to distribute fibers via the optic tracts to the contralateral and ipsilateral sides of the brain stem. Some fibers leave at the chiasm and terminate in both the ipsilateral and contralateral suprachiasmatic nuclei of the hypothalamus. Retinal axons are distributed contralaterally via both a superior and an inferior accessory optic fasciculus to three accessory optic nuclei, the medial, lateral, and dorsal terminal nuclei in the brain stem.

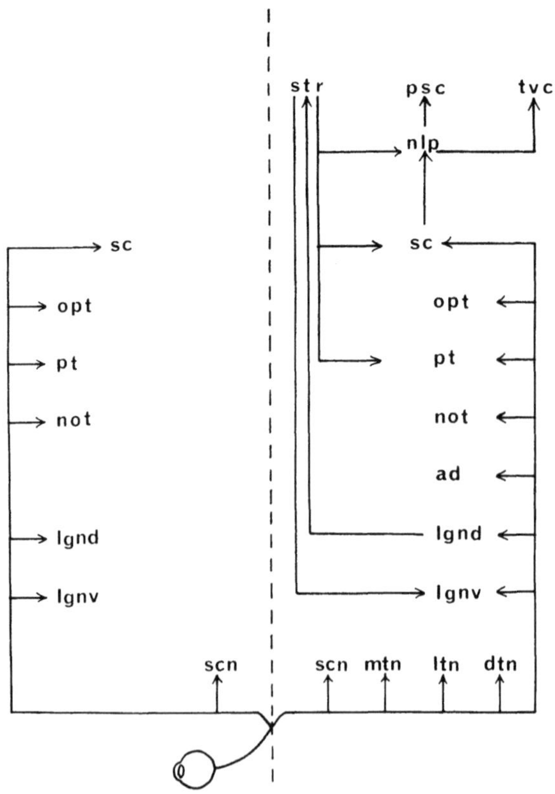

Fig. 6. Central visual pathways of *Tupaia glis*. The contralateral side is on the right, ipsilateral side on the left.

Other retinal fibers are distributed bilaterally to the lateral geniculate nucleus, pars dorsalis and ventralis, pretectal nuclear complex, and superior colliculus. The projection to the ipsilateral visual centers is not so extensive as that to the contralateral ones and is different in pattern.

The most prominent pattern difference between the ipsi- and contralateral sides is that of projections to the six layers of the lateral geniculate nucleus, pars dorsalis. There has been confusion about these patterns. Firstly, Campbell *et al.* (1967) noted a layer of cells adjacent to the optic tract which could have been interpreted as a sixth lamina but mistakenly chose not to report it. Subsequent workers have accepted its existence and it is present. Laemle (1968) described it and contrasted her findings with those of others. Unfortunately, her account of the data of others is in error. Another unnecessary complication is that while most of the earlier workers numbered the laminae from lateral to medial, later workers have reversed the sequence. The more recent studies have used the autoradiographic protein transport method and have numbered the laminae from medial to lateral (Casagrande, 1974; Casagrande and Harting, 1975; Hubel, 1975). On this basis, the pattern of projection is the following: the contralateral eye projects to laminae 2, 3, 4, and 6, while the ipsilateral eye projects to laminae 1 and 5.

It has been customary to consider the basic number of laminae in the primate dorsal lateral geniculate nucleus to be six. The pattern of inputs in anthropoid primates is described as laminae 1, 4, and 6 receiving input from the contralateral eye and 2, 3, and 5 from the ipsilateral eye. In prosimians other than *Tarsius*, laminae 1, 5, and 6 receive input from the contralateral eye and 2, 3, and 4 from the ipsilateral eye. This numbering scheme counts the lamina nearest to the optic tract as number 1. To complicate matters, an additional lamina is sometimes seen intercalated between lamina 1 and the optic tract (lamina S), and also occasionally thin layers are seen between the parvocellular laminae.

Kaas *et al.* (1972a, 1978) have proposed that the basic number of classical laminae in haplorhine primates is really four, two magnocellular layers adjacent to the optic tract and two more deeply-lying parvocellular layers. In portions of the nucleus the parvocellular laminae split each into two leaflets which makes the total number of laminae appear to be six. They state that the external magnocellular (closest to the superficial optic tract) lamina and the external parvocellular (deepest) lamina receive input from the contralateral eye. The remaining, more central laminae, receive input from the ipsilateral eye. The more centrally located leaflet of the external parvocellular lamina is believed to be interposed between the leaflets of the adjacent parvocellular lamina in anthropoids, resulting in the 1, 4, and 6 contralateral pattern. The basic number of classical relay laminae in strepsirhines appears to be six. In addition, one or two small "s" layers are found between the optic tract and the external magnocellular layer in most, possibly all, primates. Kaas *et al.* (1972a, 1978) concluded, as did Campbell (1966), that the layers of the tupaiid dorsal lateral geniculate nucleus are not easily homologized with those of primates.

Conrad and Stumpf (1975), using the autoradiographic protein transport method of tracing pathways, described a projection of retinal fibers from the optic tract in the environs of the rostral pole of the lateral geniculate nucleus which terminates in the anterodorsal nucleus of the thalamus. No such visual pathway has ever been described previously. Other studies using similar methods (Casagrande, 1974; Casagrande and Harting, 1975; Hubel, 1975) do not report such a finding. A recent study, however, utilizing the anterograde transport of horseradish peroxidase to trace the pathways from the retina (Carey and Conley, 1978) again demonstrates a projection to the anterodorsal nucleus of the thalamus in *Tupaia*. It is unclear as yet whether this finding is unique to *Tupaia* or is the result of advances in technology coupled with a prepared mind. Abplanalp (1971) described a projection from the superior colliculus to the nucleus lateralis posterior (usually called pulvinar by Diamond and his co-workers) and the ventral nucleus of the lateral geniculate body. It also sends axons to the pretectal complex of nuclei. He described topographic projections from the striate or primary visual cortex (area 17) to the superior colliculus and the pretectal complex. Harting and Noback (1971) reported similar findings. They pointed out that the rostral visual cortex projects to ventral portions of the ventral nucleus of the lateral geniculate body, to the rostral pretectal area, and to the lateral portions of the superior colliculus. The caudal visual cortex projects to dorsal regions of the ventral nucleus of the lateral geniculate body, to the caudal pretectal area, and to the medial superior colliculus. Medial cortex projects to caudal superior colliculus, and lateral visual cortex projects to rostral superior colliculus. Harting and Noback (1971) noted that no projection from the visual cortex upon the dorsal lateral geniculate nucleus was found. This contrasts with findings in the cat and many non-primate mammals (Garey *et al.*, 1968; Myers, 1962; Altman, 1962; Harting and Martin, 1970; Martin, 1968; Nauta and Bucher, 1954). It also contrasts with the situation in primates (e.g. Benevento and Fallon, 1975; Raczkowski and Diamond, 1978). If Harting and Noback are correct, *Tupaia* appears to be unique in this regard.

Although orientation columns have been demonstrated physiologically and anatomically (Humphrey *et al.*, 1977; Skeen *et al.*, 1977), ocular dominance columns have not been found in the striate cortex of *Tupaia glis* (Harting *et al.*, 1973, Casagrande and Harting, 1975; Hubel, 1975). These ocular dominance columns result from the more or less columnar aggregation of neurons through the full thickness of striate cortex which shows a preference for responding to stimulation of either the left or right eye. Some of the cells in the column are binocular and others monocular. Ocular dominance columns have been demonstrated in the domestic cat, *Galago*, and *Macaca*, but are absent in the mouse and rat (Hubel and Wiesel, 1977). Hubel and Wiesel noted that they appear to be lacking in *Saimiri* on anatomical grounds. These authors believe that further work is required to confirm this finding in *Saimiri*, as there were sources of error in the anatomical study and physiological confirmation had not yet been sought at that time.

Allman (1977) has pointed out that in non-primate mammals there is a complete representation of the visual field viewed by the contralateral retina in the superior colliculus. In the one strepsirhine and four haplorhine primates examined thus far, however, the superior colliculus has a representation restricted to the contralateral *half* of the visual field. Quite a number of mammals, including carnivores, rodents, lagomorphs, insectivores, and marsupials, as well as birds, reptiles, amphibians, and fishes have been shown to conform to the non-primate pattern. *Tupaia* also conforms to the non-primate pattern (Lane *et al.*, 1971). In addition, primates characteristically possess a large projection from the retina to the ipsilateral superior colliculus. This is characteristically small in non-primate mammals and is also small in *Tupaia* (Allman, 1977).

Allman (1977) noted another distinctive feature of the primate visual system in regard to VII, the second visual area adjacent to the striate cortex (VI) and included in Diamond's visual "belt." In *Galago, Aotus, Saimiri*, and *Macaca* the representation of the horizontal meridian is split apart except near the center of the visual field. This is not found in *Tupaia*, and is found in only one other non-primate mammal examined thus far, the domestic cat. Allman attributed this finding in the cat to the fact that it, like primates, has an expanded cortical representation of the center of the visual field and a large binocular field of vision.

5. Discussion

A very large amount of experimental neuroanatomical work has been performed on *Tupaia*, far more than one might expect in the case of what neurobiologists would ordinarily consider to be a "funny animal." Neurobiologists have traditionally considered any animal other than the rat, cat, or rhesus monkey to be a "funny animal" and, therefore, not worthy of serious study. Indeed, it is not uncommon to hear prominent neurobiologists say that animals other than the famous laboratory triumvirate are not "real animals." This would be merely an amusing attitude except for the fact that it has been so destructive for truly biological study of the nervous system. Fortunately, a trend away from this kind of thinking has appeared among some younger workers.

The reason for all of this interest in *Tupaia* has been, of course, that these animals were considered by Le Gros Clark, and accepted by Simpson (1945), to be primitive primates. In the past evolutionary inferences were often made by comparing the nervous systems of an "advanced" mammal with those of a "primitive" mammal. This most often meant comparing the cat, or the rhesus macaque if such were available, with the laboratory rat. A very active group of investigators began in the 1960's to make comparisons between *Erinaceus europaeus*, the European hedgehog, *Tupaia glis*, and the strepsirhine *Galago* in order to infer historical changes in the primate lineage.

I admit to making contributions to this approach early on, primarily to combat the rat-cat-monkey-man approach then prevalent. It has always been true that more extensive comparisons than the hedgehog-tree shrew-bushbaby one are required. Unfortunately, the hedgehog began to be thought of as a prototype placental mammal. This "primitive survivor" approach has been effectively criticized by Martin (1973) and McKenna (1976). Both of these authors have strongly urged comparative neurobiologists to broaden the base of their comparisons.

Tupaia was initially used as "the primitive primate" representing a stage just after the transition from insectivore to primate. As evidence began to accumulate that tree shrews might not actually be primates, many workers merely blurred the boundaries of their transitional position. Their study was sometimes rationalized on the basis of a presumed exclusivity in their transitional role (Abplanalp, 1971; Goode and Haines, 1975). Goode and Haines (1975), for example, stated that they are the only extant forms thought to be close to the ancestral stock from which primates evolved.

Although tupaiids have always been considered to be specialized in regard to their visual system, they have usually been regarded as generalized in other respects. Casseday *et al.* (1976) suggested that since they have no obvious auditory specializations their central auditory system pathways probably represent a general mammalian plan. Further, since the cat and tree shrew are phyletically quite divergent and are so similar in many respects, the pattern of organization discussed must represent the general mammalian plan. These authors are possibly correct, but such an inference based on two species does not stand on firm ground.

The more rigorous theoretical approach advocated by Martin (1973) and embodied in the Hennigian method (Hennig, 1966) is much more demanding of effort than the limited comparisons that have been made thus far. The time-consuming and often difficult experimental methods required to establish any finding in neurobiology militates against any extended comparison method. Nevertheless, more and more work is being done on "funny animals" and perhaps before very long some valid inferences can be drawn. Indeed, enough work has been done on various primates to allow some reasonable inferences regarding the evolution of the primate visual system to be made (e.g. Allman, 1977).

It has been pointed out elsewhere that the list of characters presented by Le Gros Clark (1959) in which the brain of *Tupaia* contrasts with that of insectivores is really only a reflection of a well-developed visual system, and that these characters can almost all be found in other mammals with good vision (Campbell, 1966, 1974). At the risk of being repetitive, the list is: (1) the relative size of the brain as a whole; (2) the expansion of the neopallium, accompanied by a displacement downwards of the rhinal sulcus; (3) the formation of a distinct temporal pole of the neopallium; (4) the backward projection of the occipital pole; (5) the presence of a calcarine sulcus; (6) the well-marked lamination and cellular richness of the neopallial cortex in general, and the degree of differentiation of the several cortical areas; (7) the pronounced elaboration of the visual apparatus of the brain, particularly at the cortical level; (8) the advanced degree of differentiation of the nuclear elements of the thalamus; and (9) the well-defined cellular lamination of the lateral geniculate nucleus.

What is perhaps more significant is that *Tupaia* possesses none of the findings characteristic of all primate visual systems examined thus far and not found in any or most non-primate mammals (Allman, 1977). These include: (1) a representation of the contralateral *half* of the visual field in the superior colliculus; (2) the second visual area, VII or area 18, representing the contralateral *hemifield* and with the representation of the horizontal meridian split apart except near the center of the visual field; (3) the lack of a medial terminal nucleus of the accessory optic system; and (4) a large ipsilateral retino-superior colliculus projection. The fact that *Tupaia* does not share these synapomorphies argues against their inclusion in the same taxon. It should be noted that these anatomic features are of a more detailed and specific nature than the rather general features used by Le Gros Clark.

The significance of the lack of ocular dominance columns in the striate cortex of *Tupaia* requires further examination. If *Saimiri* truly also lacks this feature then certainly many more species of primates and other mammals must be examined to infer whether the presence of ocular dominance columns is apomorphous or plesiomorphous in the primate lineage. Of course, broader comparisons are desirable in any case.

As I (Campbell, 1965; Jane *et al.*, 1965; Campbell, 1974) and others (Haines and Swindler, 1972) have pointed out, some of the tree shrew nervous system characters, e.g. the position of the pyramidal tract in the dorsal funiculus of the spinal cord and the pattern of inputs to the dorsal lateral geniculate nucleus, are divergent from both lipotyphlous insectivores and primates. The presence of a third visual area in the occipital cortex of *Tupaia* can possibly be added to this list, as it is lacking in *Erinaceus* (Kaas *et al.*, 1970). These characters support the initial suggestion of Straus (1949) that the tree shrews are ordinally distinct.

I have avoided using this evidence in isolation to suggest that they are in fact ordinally distinct on the grounds that evidence drawn from a single system is less desirable than multiple lines of evidence. I believe now that multiple lines of evidence do exist to support such a classification (Butler, 1972; McKenna, 1975; Luckett, this volume).

Interestingly, as more detailed information on the central nervous system of tupaiids, primates, sciurids, and other mammals has accumulated in recent years, the resemblances to primates can be seen to be very general and largely of a "trends" and "tendencies" nature. Although not precisely indentical, of course, when the location of the pyramidal tracts and many aspects of the visual system are considered, tree shrews resemble arboreal squirrels more than they resemble primates. Indeed, tupaiids occupy an ecological niche similar to that of sciurids. The Malays, when they used the word "tupai" for both tree shrews and squirrels, came closer to the truth than many who followed them.

ABBREVIATIONS

AB	auditory belt	OB	olfactory bulb
AD	anterodorsal nucleus	OC	optic chiasm
ADG	auditory dysgranular area	OF	orbital furrow
AK	auditory koniocortex or core	opt	olivary pretectal nucleus
AUD	auditory cortex	PI	proisocortex
BS	brain stem	Po	posterior nucleus
Cb	cerebellum	PS	pseudostriate cortex
CN	caudal nucleus of medial geniculate body	psc	peristriate cortex
DN	dorsal nucleus of medial geniculate body	pt	pretectal nucleus complex
dtn	dorsal terminal nucleus of accessory optic system	Rh	rhinal sulcus
Hyp	hypophysis	SC, sc	superior colliculus
ICC	central nucleus of inferior colliculus	scn	suprachiasmatic nucleus
ICE	external nucleus of inferior colliculus	SK	somatic koniocortex
ICP	pericentral nucleus of inferior colliculus	SM	sensory-motor belt
lgnd	dorsal nucleus of lateral geniculate body	SMB	somatic motor belt
lgnv	ventral nucleus of lateral geniculate body	Sp	spinal cord
LL	lateral lemniscus	SSB	somatic sensory belt
ltn	lateral terminal nucleus of accessory optic system	str	striate cortex
M	motor area	T	temporal cortex
MN	magnocellular nucleus of medial geniculate body	TOT	temporo-occipital transition zone
mtn	medial terminal nucleus of accessory optic system	tvc	temporal visual cortex
NLL	nn. of lateral lemniscus	VG	vascular groove
nlp	nucleus lateralis posterior	VN	ventral nucleus of medial geniculate body
not	nucleus of optic tract	VPL	ventral posterolateral nucleus
		1	striate area
		2	Betz cell area
		3	parietal area
		4	temporal area
		5	frontal area
		6	insular area
		17	area 17 of Brodmann
		18	area 18 of Brodmann
		19	area 19 of Brodmann

6. References

Abplanalp, P. 1971. The neuroanatomical organization of the visual system in the tree shrew. *Folia Primat.* 16: 1–34.

Albright, B.C., and Haines, D.E. 1973. The morphology of Clarke's column in the lesser bushbaby *(Galago senegalensis). Brain Behav. Evol.* 7: 165–190.

Albright, B.C., and Haines, D.E. 1974. Dorsal root afferents to Clarke's column from hindlimb cord levels in *Tupaia* and *Galago. Brain Behav. Evol.* 10: 274–289.

Allman, J. 1977. Evolution of the visual system in the early primates, pp. 1–53. *In* J. Sprague and A. Epstein (eds.). *Progress in Psychobiology and Physiological Psychology*, Vol. 7. Academic Press, San Francisco.

Altman, J. 1962. Some fiber projections to the superior colliculus in the cat. *J. Comp. Neur.* 119: 77–95.

Benevento, L. A., and Fallon, J.H. 1975. The projection of occipital cortex to the dorsal lateral geniculate nucleus in the rhesus monkey *(Macaca mulatta)*. *Exp. Neur.* 46: 409–417.

Bolk, L. 1906. *Das Cerebellum der Säugetiere. Eine vergleichende anatomische Untersuchung.* DeErven F. Bohn, Haarlem.

Butler, P.M. 1972. The problem of insectivore classification, pp. 253–265. *In* K.A. Joysey and T.S. Kemp (eds.) *Studies in Vertebrate Evolution.* Oliver and Boyd, Edinburgh.

Campbell, B.G. 1966. *Human Evolution: An Introduction to Man's Adaptations.* Aldine, Chicago.

Campbell, C.B.G. 1965. Pyramidal tracts in primate taxonomy. Unpublished Ph.D. Thesis, University of Illinois, Chicago, Illinois.

Campbell, C.B.G. 1966. The relationships of the tree shrews: The evidence of the nervous system. *Evolution* 20: 276–281.

Campbell, C.B.G. 1974. On the phyletic relationships of the tree shrews. *Mammal Rev.* 4: 125–143.

Campbell, C.B.G., Jane, J.A., and Yashon, D. 1966a. The retinal projections of tree shrew and hedgehog. *Anat. Rec.* 154: 326.

Campbell, C.B.G., Yashon, D., and Jane, J.A. 1966b. The origin, course and termination of corticospinal fibers in the slow loris, *Nycticebus coucang* (Boddaert). *J. Comp. Neur.* 127; 101–112.

Campbell, C.B.G., Jane, J.A., and Yashon, D. 1967. The retinal projections of the tree shrew and hedgehog. *Brain Res.* 5: 406–418.

Carey, R.G., and Conley, M. 1978. Anterograde transport of horseradish peroxidase in the visual system of the tree shrew. *Soc. for Neurosc., Abstr.* 4: 32.

Carey, R.G., Fitzpatrick, D., and Diamond, I.T. 1979. Thalamic projections to layer I of striate cortex shown by retrograde transport of horseradish peroxidase. *Science* 203: 556–559.

Carlsson, A. 1922. Uber die Tupaiidae und ihre Beziehungen zu den Insectivora und den Prosimiae. *Acta Zool.* 3: 227–270.

Carpenter, M.B., Stein, B.M., and Shriver, J.E. 1968. Central projections of spinal dorsal roots in the monkey. II. Lower thoracic, lumbosacral and coccygeal dorsal roots. *Amer. J. Anat.* 123: 75–117.

Casagrande, V.A. 1974. The laminar organization and connections of the lateral geniculate nucleus in the tree shrew *(Tupaia glis)*. *Anat. Rec.* 178: 323.

Casagrande, V.A., and Harting, J.K. 1975. Transneuronal transport of tritiated fucose and proline in the visual pathways of tree shrew *Tupaia glis*. *Brain Res.* 96: 367–372.

Casseday, J.H., Diamond, I.T., and Harting, J.K. 1976. Auditory pathways to the cortex in *Tupaia glis*. *J. Comp. Neur.* 166: 303–340.

Conrad, C.D., and Stumpf, W.E. 1975. Direct visual input to the limbic system: Crossed retinal projections to the nucleus anterodorsalis thalami in the tree shrew. *Exp. Brain Res.* 23: 141–149.

Diamond, I.T., Snyder, M., Killackey, H., Jane, J., and Hall, W.C. 1970. Thalamocortical projections in the tree shrew *(Tupaia glis)*. *J. Comp. Neur.* 139: 273–306.

Elliot Smith, G. 1903. On the morphology of the brain in the mammals, with special reference to that of the lemurs recent and extinct. *Trans. Linn. Soc. Lond., ser.* 2 8: 319–432.

Elliot Smith, G. 1924. *The Evolution of Man: Essays.* Oxford, London.

Garey, L.J., Jones, E.G., and Powell, T.P.S. 1968. Interrelationship of striate and extrastriate cortex with the primary relay sites of the visual pathway. *J. Neurol. Neurosurg. Psychiat.* 31: 135–157.

Glickstein, M. 1967. Laminar structure of the dorsal lateral geniculate nucleus in the tree shrew *(Tupaia glis)*. *J. Comp. Neur.* 131: 93–102.

Goldberg, J.M., and Moore, R.Y. 1967. Ascending projections of the lateral lemniscus in the cat and monkey. *J. Comp. Neur.* 129:143–156.

Goode, G.E. 1974. The intramedullary course and terminal distribution of cortical projections to the spinal cord of *Galago*. *Anat. Rec.* 178: 364.

Goode, G.E., and Haines, D.E. 1975. Origin, course and termination of corticospinal fibers in a prosimian primate, *Galago*. *Brain Behav. Evol.* 12: 334–360.

Gregory, W.K. 1910. The orders of mammals. *Bull. Amer. Mus. Nat. Hist.* 27: 1–524.

Haines, D.E. 1969. The cerebellum of *Galago* and *Tupaia*. I. Corpus cerebelli and flocculonodular lobe. *Brain Behav. Evol.* 2: 377–414.

Haines, D.E. 1971a. The cerebellum of *Galago* and *Tupaia*. II. The early postnatal development. *Brain Behav. Evol.* 4: 97–113.

Haines, D.E. 1971b. The morphology of the cerebellar nuclei of *Galago* and *Tupaia*. *Am. J. Phys. Anthrop.* 35: 27–42.

Haines. D.E. 1975. Cerebellar corticovestibular fibers of the anterior lobe in *Galago* and *Tupaia*. *Anat. Rec.* 181: 369–370.

Haines, D.E., and Culberson, J.L. 1974. Cerebellar corticonuclear projections of the posterior lobe in *Galago* and *Tupaia*. *Anat. Rec.* 178:368.

Haines, D.E., and Swindler, D.R. 1972. Comparative neuroanatomical evidence and the taxonomy of the tree shrews *(Tupaia)*. *J. Human Evol.* 1: 407–420.

Harting, J.K., and Martin, G.F. 1970. Neocortical projections to the mesencephalon of the armadillo *(Dasypus novemcinctus)*. *Brain Res.* 17: 447–462.

Harting, J.K., and Noback, C.R. 1970. Corticospinal projections from the pre- and postcentral gyri in the squirrel monkey *(Saimiri sciureus)*. *Brain Res.* 24: 322–328.

Harting, J.K., and Noback, C.R. 1971. Subcortical projections from the visual cortex in the tree shrew *(Tupaia glis)*. *Brain Res.* 25: 21–33.

Harting, J.K., Diamond, I.T., and Hall, W.C. 1973. Anterograde degeneration study of the cortical projections of the lateral geniculate and pulvinar nuclei in the tree shrew *(Tupaia glis)*. *J. Comp. Neur.* 150: 393–440.

Hennig, W. 1966. *Phylogenetic Systematics.* Univ. of Illinois Press, Urbana.

Hubel, D.H. 1975. An autoradiographic study of the retino-cortical projections in the tree shrew *(Tupaia glis)*. *Brain Res.* 96: 41–50.

Hubel, D.H., and Wiesel, T.N. 1977. Functional architecture of macaque monkey visual cortex. *Proc. Roy Soc. Lond. B.* 198: 1–59.

Humphrey, A.L., Norton, T.T., and Albano, J.E. 1977. Microelectrode analysis of the orientation column system in the striate cortex of the tree shrew *(Tupaia glis)*. *Soc. for Neurosc. Abstr.* 3: 563.

Jane, J.A., Campbell, C.B.G, and Yashon, D. 1965. Pyramidal tract: A comparison of two prosimian primates. *Science* 147: 153–155.

Jane, J.A., Campbell, C.B.G., and Yashon, D. 1969. The origin of the corticospinal tract of the tree shrew *(Tupaia glis)* with observations on its brain stem and spinal terminations. *Brain Behav. Evol.* 1: 160–182.

Jones, D.R., Casseday, J.H., and Diamond. I.T. 1976. Further study of parallel auditory pathways in the tree shrew, *Tupaia glis*. *Anat. Rec.* 184: 438–439.

Kaas, J., Hall, W.C., and Diamond, I.T. 1970. Cortical visual areas I and II in the hedgehog: Relation between evoked potential maps and architectonic subdivisions. *J. Neurophys.* 33: 595–615.

Kaas, J.H., Guillery, R.W., and Allman, J.M. 1972a. Some principles of organization in the dorsal lateral geniculate nucleus. *Brain Behav. Evol.* 6: 253–299.

Kaas, J., Hall, W.C., Killackey, H., and Diamond, I.T. 1972b. Visual cortex of the tree shrew *(Tupaia glis)*: Architectonic subdivisions and representations of the visual field. *Brain Res.* 42: 491–496.

Kaas, J.H., Huerta, M.F., Weber, J.T., and Harting, J.K. 1978. Patterns of retinal terminations and laminar organization of the lateral geniculate nucleus of primates. *J. Comp. Neur.* 182: 517–554.

Kuypers, H.G.J.M. 1964. The descending pathways to the spinal cord, their anatomy and function, pp. 178–202. *In* J.C. Eccles and J.P. Schadé (eds.). *Progress in Brain Research,* vol. 11. *Organization of the Spinal Cord.* Elsevier, Amsterdam.

Laemle, L.K. 1968. Retinal projections of *Tupaia glis*. *Brain Behav. Evol.* 1: 473–499.

Lane, R.H., Allman, J.M., and Kaas, J.H. 1971. Representation of the visual field in the superior colliculus of the grey squirrel *(Sciurus carolinensis)* and the tree shrew *(Tupaia glis)*. *Brain Res.* 26: 277–292.

Larsell, O. 1970. *The Comparative Anatomy and Histology of the Cerebellum from Monotremes through Apes.* University of Minnesota Press, Minneapolis.

Le Gros Clark, W.E. 1924. On the brain of the tree shrew *(Tupaia minor). Proc. Zool. Soc. Lond.* 1924: 1053–1074.

LeGros Clark, W.E. 1932. The brain of the Insectivora. *Proc. Zool Soc. Land.* 1932: 975–1013.

Le Gros Clark, W.E. 1934. *Early Forerunners of Man.* Bailliere, Tyndall & Cox, London.

Le Gros Clark, W.E. 1959. *The Antecedents of Man.* Edinburgh University Press, Edinburgh.

Le Gros Clark, W.E. 1968. *Chant of Pleasant Exploration.* E. & S. Livingstone Ltd., Edinburgh and London.

Lende, R.A. 1970. Cortical localization in the tree shrew. *Brain Res.* 18: 61–75.

Martin, G.F. 1968. The pattern of neocortical projections to the mesencephalon of the opossum, *Didelphis virginiana. Brain Res.* 11: 593–610.

Martin, R.D. 1973. Comparative anatomy and primate systematics. *Symp. Zool. Soc. Lond.* 33: 301–337.

McKenna, M.C. 1975. Toward a phylogenetic classification of the Mammalia, pp. 21–46. *In* W.P. Luckett and F.S. Szalay (eds.). *Phylogeny of the Primates.* Plenum Press, New York.

McKenna, M.C. 1976. Comments on Radinsky's "Later mammal radiations," pp. 245–250. *In* R.B. Masterton, M.E. Bitterman, C.B.G. Campbell, and N. Hotton (eds.). *Evolution of Brain and Behavior in Vertebrates.* Lawrence Erlbaum Associates, Hillsdale, N.J.

Moore, R.Y., and Goldberg, J.M 1963. Ascending projections of the inferior colliculus in the cat. *J. Comp. Neur.* 121: 109–136.

Moore, R.Y., and Goldberg, J.M. 1966. Projections of the inferior colliculus in the monkey. *Exp. Neur.* 14: 429–438.

Moore, R.Y., and Tarlov, E.C. 1963. Ascending projections of the inferior colliculus in the rabbit. *Anat. Rec.* 145: 262.

Morest, D.K. 1965. The lateral tegmental system of the midbrain and the medial geniculate body: Study with Golgi and Nauta methods in cat. *J. Anat.* 99: 611–634.

Murray, H.M., Haines, D.E., and Cote, I. 1976. The rubrospinal tract of the tree shrew *(Tupaia glis). Brain Res.* 116: 317–322.

Myers, R.F. 1962. Striate cortex connections in the monkey. *Fed. Proc.* 21: 352.

Nauta, W.J. H., and Bucher, V.M. 1954. Efferent connections of the striate cortex in the albino rat. *J. Comp. Neur.* 100: 257–286.

Oliver, D.L., and Hall, W.C. 1975. Subdivisions of the medial geniculate body in the tree shrew *(Tupaia glis). Brain Res.* 86: 217–227.

Oliver, D.L., and Hall, W.C. 1978a. The medial geniculate body of the tree shrew, *Tupaia glis.* I. Cytoarchitecture and midbrain connections. *J. Comp. Neur.* 182: 423–458.

Oliver, D.L., and Hall, W.C. 1978b. The medial geniculate body of the tree shrew, *Tupaia glis.* II. Connections with the neocortex. *J. Comp. Neur.* 182: 459–494.

Oliver, D.L., Merzenich, M.M., Roth, G.L., Hall, W.C., and Kaas, J.H. 1976. Tonotopic organization and connections of primary auditory cortex in the tree shrew, *Tupaia glis. Anat. Rec.* 184: 491.

Raczkowski, D., and Diamond, I.T. 1978. Connections of the striate cortex in *Galago senegalensis. Brain Res.* 144: 383–388.

Radinsky, L.B. 1970. The fossil evidence of prosimian brain evolution, pp. 209–224. In C.R. Noback and W. Montagna (eds.). *The Primate Brain,* Appleton-Century-Crofts, New York.

Rexed, B. 1954. A cytoarchitectonic atlas of the spinal cord in the cat. *J. Comp. Neur.* 100: 297–379.

Shriver, J.E., and Matzke, H.A. 1965. Corticobulbar and cortiscospinal tracts in the marmoset monkey *(Oedipomidas oedipus). Anat. Rec.* 151: 416.

Shriver, J.E., and Noback, C.R. 1967. Cortical projections to the lower brain stem and spinal cord in the tree shrew *(Tupaia glis). J. Comp. Neur.* 130: 25–54.

Simpson, G.G. 1945. The principles of classification and a classification of mammals. *Bull. Amer. Mus. Nat. Hist.* 85: 1–350.

Skeen, L.C., Humphrey, A.L., Norton, T.T., and Hall, W.C. 1977. Deoxyglucose mapping of the orientation column system in the striate cortex of the tree shrew *(Tupaia glis). Soc. for Neuroscience Abstr.* 3: 577.

Stephan, H. 1972. Evolution of primate brains: A comparative anatomical investigation, pp. 155–174. *In* R. Tuttle (ed.). *The Functional and Evolutionary Biology of Primates.* Aldine-Atherton, Chicago.

Straus, W.L., Jr. 1949. The riddle of man's ancestry. *Quart. Rev. Biol.* 24: 200–223.
Tigges, J. 1966. Ein experimenteller Beitrag zum subkortikalen optischen System von *Tupaia glis*. *Folia Primat.* 4: 103–123.
Tigges, J., and Shantha, T.R. 1969. *A Stereotaxic Brain Atlas of the Tree Shrew (Tupaia glis)*. Williams and Wilkins, Baltimore.
Valverde, F. 1966. The pyramidal tract in rodents. A study of its relations with the posterior column nuclei, dorsolateral reticular formation of the medulla oblongata, and cervical spinal cord. *Zeit. Zellforsch.* 71: 297–363.
Van Noort, J. 1969. *The Structure and Connections of the Inferior Colliculus*. Van Gorcum, Assen.
Ware, C.B. 1973. Efferent projections of the deep cerebellar nuclei in the tree shrew *(Tupaia glis)*. *Anat. Rec.* 175: 463.

Reproductive System IV

The Use of Reproductive and Developmental Features in Assessing Tupaiid Affinities

8

W. PATRICK LUCKETT

1. Introduction

The lack of unquestioned early Tertiary fossil tupaiids has stimulated a search for morphological, developmental, molecular, and behavioral features of extant tree shrews which might provide insight into their evolutionary relationships with primates or other eutherians. Reproductive features were among the earliest to be cited as evidence of a special tupaiid-primate relationship. Kaudern (1911) was one of the first investigators to suggest close affinities between tupaiids and primates as a result of his comparative studies on the male reproductive system in Tupaiidae, Macroscelididae, Lipotyphla, and Lemuroidea. He concluded that the male reproductive system provided no evidence of special affinities between tupaiids and macroscelidids; instead, he emphasized the occurrence of scrotal testes and seminal vesicles as shared similarities which supported the classification of tupaiids with Primates rather than Insectivora. These reproductive similarities were also cited as evidence of a close tupaiid-primate relationship by Carlsson (1922) in her extensive analysis of cranioskeletal and soft anatomical features in insectivores and

W. PATRICK LUCKETT • Department of Anatomy, Creighton University, Omaha. Nebraska 68178 USA

prosimians. Subsequently, developmental characters of the fetal membranes and placenta, particularly the nature of the definitive placenta and allantois, were also utilized to support the hypothesis of primate affinities for tree shrews (Meister and Davis, 1956, 1958; Le Gros Clark, 1959, 1971).

As emphasized by numerous investigators (Van Valen, 1965; McKenna, 1966; Martin, 1968, 1969; Luckett, 1969; Campbell, 1974), there is no *a priori* reason for assuming that the phyletic relationships of tupaiids must be either with Primates or Insectivora; instead, assessment of tupaiid affinities should entail comparison of features among a wide range of eutherian mammals. Such analyses are essential in order to determine the relatively primitive and derived nature of character states for Eutheria, Tupaiidae, Primates, Macroscelididae, and other eutherian taxa. Methodologies of character analysis and phylogenetic reconstruction are discussed in detail elsewhere in this volume (Luckett, Chapter 1; Novacek, Chapter 2) and will not be repeated here.

The aim of the present paper is to review briefly the major features of reproduction and placentation in tupaiids in order to determine whether there are derived character states in these organ systems which are shared uniquely with other eutherian taxa. Particular emphasis will be given to those features which have been considered to be indicative of tupaiid-primate affinities. These are: (1) scrotal testes and a pendulous penis; (2) presence of seminal vesicles; (3) occurrence of a peniform clitoris in the female, similar to the condition in many strepsirhines; (4) the possible occurrence of menstruation in *Tupaia,* resembling the condition in higher primates; (5) small number of mammae, correlated with a reduced litter size; (6) a relatively long gestation period, intermediate in length between that of Lipotyphla and Strepsirhini; and (7) several features of the fetal membranes and placenta which resemble those of higher primates. Syntheses of many aspects of tupaiid reproduction and placentation have been published recently (Martin, 1968, 1969, 1975; Luckett, 1969, 1974; Butler, 1974) and should be consulted for further details.

2. *Male Reproductive System*

A. External Genitalia

The slender, elongate penis of *Tupaia* is relatively pendulous (Jones, 1917; Martin, 1968), similar to the condition in Primates, Chiroptera, and Dermoptera. Smith and Madkour (1980) considered the pendulous penis to be a shared, derived character state which supports the inclusion of these taxa in the Archonta (*sensu* McKenna, 1975). However, the penis of *Ptilocercus* is short, stout, and less pendulous than that of *Tupaia* (Le Gros Clark, 1926), and this trait has not been described adequately in other tupaiid genera. Given the lack of data for most tupaiids and the considerable intergeneric variability in the gross and microscopic anatomy of the penis in primates, lipotyphlans, bats, rodents, and other eutherians (Kaudern, 1911; Hill, 1958; Raynaud, 1969), it is questionable whether much phyletic valence can be given to the common occurrence of a pendulous penis as an archontan synapomorph.

Considerable emphasis has been given to the presence of scrotal testes in tupaiids as indicators of their primate relationships (Kaudern, 1911; Carlsson, 1922; Le Gros Clark, 1959, 1971), because this trait is lacking in all macroscelidids and lipotyphlans. As emphasized by Martin (1968, 1969), however, the wide distribution of this character state in eutherians and metatherians suggests that scrotal testes may have characterized the ancestral therian stock. If this is true, such a shared primitive retention would provide no evidence of close tupaiid-primate affinities. Even if complete testicular descent into a scrotum did not occur in the ancestran therians or eutherians, the widespread distribution of this trait in therians would support an alternative hypothesis of extensive convergent evolution.

The prepenial position of the tupaiid scrotum differs considerably from the relationship in most primates and other eutherians (Kaudern, 1911; Jones, 1917; Le Gros Clark, 1926, 1959; Martin, 1968, 1969); instead, it resembles the typical marsupial condition. The disposition of the testes has been described as permanently scrotal in *Tupaia* (Jones, 1917; Le Gros Clark, 1959), seasonally scrotal in *Ptilocercus* (Le Gros Clark, 1926, 1959), and abdominal in *Anathana* (Verma, 1965). However, morphological studies by Martin (1968, 1969) demonstrated that the inguinal canal remains patent throughout life in *Tupaia belangeri*, similar to the condition in *Ptilocercus* (and presumably also in *Anathana* and other tupaiines). Retraction of the testes from the scrotum into the abdominal cavity can occur through the patent inguinal canal in *Tupaia* under experimental conditions of stress or fright (Martin, 1968; Von Holst, 1974), and this provides a likely explanation for the occurrence of abdominal testes in preserved specimens of *Ptilocercus* and *Anathana*, as suggested by Martin.

Other features of the external genitalia provide little evidence for the eutherian affinities of Tupaiidae at the present time. Accessory erectile tissue occurs in the glans penis and prepuce of microchiropterans and some lipotyphlans (Centetidae, Erinaceidae, Soricidae, Talpidae), but it is absent in tupaiids, dermopterans, macroscelidids, megachiropterans, and most primates (Kaudern, 1911; Wimsatt and Kallen, 1952; Smith and Madkour, 1980). Smith and Madkour (1980) have suggested that the various accessory erectile tissues of different eutherians are homologous, and that the presence of this tissue is a primitive mammalian or eutherian trait. Consequently, these authors considered the absence of accessory erectile tissue to be a shared, derived character state that supports a monophyletic relationship among Tupaiidae, Primates, Dermoptera, and Megachiroptera. Available descriptions of the distribution and morphology of this accessory tissue in eutherians raise grave doubts concerning its homology and primitiveness, however, and it seems premature to utilize this feature in phylogenetic analyses.

The presence of an os penis is widespread among eutherians, including primates, bats, dermopterans, and some lipotyphlans. Its absence in tupaiids is doubtlessly an autapomorphous feature; as such, it provides no evidence for the phylogenetic relationships of tupaiids to other archontans or eutherians.

B. Testes and Accessory Sex Glands

Histological descriptions of testicular morphology are lacking for tupaiids. However, consideration of the extremely conservative nature of testicular morphology in other eutherians suggests that it is unlikely that such observations will provide useful features for assessing the affinities of tree shrews.

The male accessory sex glands of *Tupaia* were described independently by Kaudern (1911) and Jones (1917), and data on *Ptilocercus* were provided by Le Gros Clark (1926). More recently, Alcalá and Conaway (1968) summarized the comparative histology of the accessory glands in several species of *Tupaia, Lyonogale,* and *Urogale.* Paired seminal vesicles (Fig. 1) develop as outgrowths of the ductus deferens, and their histological appearance is comparable to that of other eutherians. Several authors (Kaudern, 1911; Carlsson, 1922) cited the joint possession of seminal vesicles in tupaiids and primates as further evidence of their close relationship, because these glands are absent in macroscelidids and most lipotyphlans (excluding Erinaceidae). As with many shared tupaiid-primate similarities, the presence of seminal vesicles is widespread and appears to be a primitive eutherian characteristic that provides no evidence of tree shrew affinities.

Paired dorsolateral prostate glands are located immediately inferior (caudal) to the seminal vesicles in *Tupaia, Lyonogale,* and *Urogale* (Fig. 1), and the two organs share a common connective tissue sheath (Kaudern, 1911; Jones, 1917; Alcalá and Conaway, 1968). These tubuloalveolar glands are lined by simple columnar epithelium and empty into the urethra by a pair of ducts which lie lateral to the openings of the ejaculatory ducts. A separate medial lobe of the prostate has also been identified in these genera; it is situated dorsal to the uterus masculinus (when this structure occurs) and opens by a separate duct into the urethra. The histological structure of the lateral and medial lobes of the prostate is similar. The intimate relationship between the seminal vesicles and dorsolateral lobes of the prostate has led to confusion over the identification and homology of sex accessory glands in *Tupaia* and other tupaiids (see below).

Le Gros Clark (1926) described several apparent differences between the accessory sex glands of *Ptilocercus* and *Tupaia.* Most significant was his observation that the prostate gland of *Ptilocercus* is a compact, glandular mass which encircles the urethra near the neck of the bladder. Glandular acini completely surround the urethra, and multiple excretory ducts drain into the entire circumference of the urethra. In addition, there is a small, dorsomedial lobe of the prostate which is closely associated with the uterus masculinus and opens independently into the urethra. Le Gros Clark (1926, 1959, 1971) emphasized that the relative size and morphology of the prostate in *Ptilocercus* closely

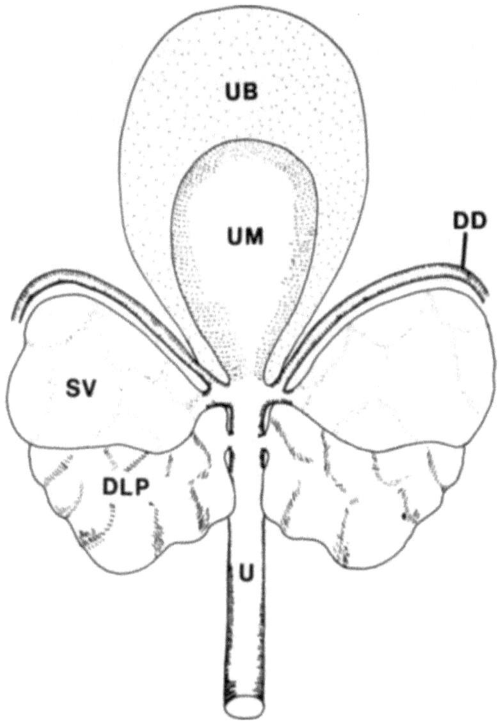

Fig. 1. Diagram (dorsal view) of the male accessory sex glands (excluding the bulbourethral glands) in *Lyonogale* and *Tupaia* (Adapted from Alcalá and Conaway, 1968). The medial lobe of the prostate, which lies dorsal to the uterus masculinus, is not illustrated. Some species of *Tupaia* lack a uterus masculinus. Abbreviations: DD, ductus deferens, DLP, dorsolateral prostate lobe; SV, seminal vesicle; U, urethra; UB, urinary bladder; UM, uterus masculinus.

resemble the human condition. In contrast, the glands identified as dorsolateral lobes of the prostate in *Tupaia* were designated as vesicular glands in *Ptilocercus* and considered to be specialized portions of the complex and lobulated seminal vesicles. The more superior portion of the seminal vesicle was called the "vesicular diverticulum" of the ductus deferens, and Le Gros Clark (1959, 1971) asserted that it functions as a receptacle for the temporary storage of spermatozoa.

Le Gros Clark (1926) acknowledged the probable homology between the dorsolateral prostate lobe of *Tupaia* and the "vesicular gland" of *Ptilocercus*, but he did not consider the latter organ to be a component of the prostate gland. No function was suggested for the vesicular gland, and Le Gros Clark (1959, 1971) admitted that there is no well-defined homologue of this gland in primates. He believed instead that the vesicular gland and vesicular diverticulum of *Ptilocercus* became incorporated into a common seminal vesicle in primates. A prostate gland which encircles the urethra was also described in *Dendrogale*

(Davis, 1938) and *Anathana* (Verma, 1965). It is unclear at present which pattern of prostate gland morphology is primitive for tupaiids. Available descriptions and illustrations of the accessory glands in *Ptilocercus, Dendrogale,* and *Anathana* suggest the probability that a dorsomedial and paired dorsolateral prostatic lobes characterize all genera of tupaiids, despite the fact that the dorsolateral lobes ("vesicular glands") have been considered to be related to the seminal vesicles in *Ptilocercus* and *Anathana.*

Considerable variability exists in prostate morphology among mammals, and two basic patterns are evident (Price and Williams-Ashman, 1961). Prostatic tissue may be disseminate or diffuse and remain confined to the lamina propria of the urethra, as in marsupials, edentates, and some artiodactyls. In other taxa, such as rodents, lagomorphs, insectivores, and many primates, the prostate forms one or more discrete extrinsic glands which lie outside the muscular coat of the urethra. Some mammals exhibit a combination of both these patterns, and this appears to be the case in tree shrews. Diffuse or disseminate prostatic tissue is the only type that occurs in monotremes and marsupials (Rodger, 1976), and its widespread distribution in eutherians suggests that this is the primitive mammalian and eutherian condition. Solid, glandular prostatic buds also develop in a diffuse manner in *Tupaia, Lyonogale,* and *Urogale* embryos (Alcalá and Conaway, 1968), in contrast to their more restricted distribution in the adult. It is possible that this ontogenetic reduction or loss of diffuse prostatic tissue may be a shared, derived attribute of these genera. Further embryonic and microscopic studies of the sex glands in all genera of tupaiids are required to corroborate this hypothesis.

Attempts to homologize the prostate glands of Tupaiidae, Primates, Lipotyphla, Chiroptera, and Macroscelididae are hindered by the scarcity of data on their embryology and morphology, as well as by the fact that prostatic lobules may develop from multiple sites on the circumference of the urethra. At present, the morphological pattern of the prostate appears to be of little value in assessing phylogenetic relationships among higher categories of mammals.

A prominent uterus masculinus (Fig. 1), derived developmentally from the fused caudal portions of the Müllerian ducts, has been found in *Tupaia javanica* (Kaudern, 1911), *T. minor, T. montana, T. palawanensis, Lyonogale tana, L. dorsalis* (Alcalá and Conaway, 1968), and *Ptilocercus lowii* (Le Gros Clark, 1926). Developmental, histological, and experimental studies by Alcalá and Conaway (1968) demonstrated that this organ has undergone a morphological, and presumably functional, differentiation into an accessory sex gland. The functional significance of this organ is unknown, but it undergoes a considerable increase in size and glandular proliferation under the influence of exogenous estrogenic stimulation. A persisting uterus masculinus is absent in *Tupaia glis* (Jones, 1917), *T. chinensis, T. longipes, T. gracilis, Urogale everetti* (Alacalá and Conaway, 1968), *Dendrogale frenata* (Davis, 1938), and *Anathana wroughtoni* (Verma, 1965).

The presence or absence of a large, glandular uterus masculinus was considered to be of possible taxonomic significance by Alcalá and Conaway (1968), although they did not assess its probable primitive or derived nature within Tupaiidae or Eutheria. Consideration of its relatively rare distribution among eutherian mammals suggests that its occurrence is a derived character state which results from pedomorphic retention and further differentiation of a portion of the Müllerian duct primordia. A glandular, enlarged uterus masculinus occurs only rarely in Primates, whereas it has been found in Macroscelididae, Erinaceidae, and Chrysochloridae. Its distribution suggests independent evolution of this trait within Tupaiidae, and therefore its presence appears to be of limited use in evaluating the eutherian affinities of tree shrews. However, further study of its presence or absence within Tupaiidae may provide additional data for assessing interspecific relationships.

C. Biology of Spermatozoa

In recent years the ultrastructural and biochemical attributes of mammalian spermatozoa have been investigated extensively in order to assess the structural and functional interrelationships which occur during the process of sperm maturation and fertilization. Accumulation of such data from a variety of mammals provides an additional form-function complex that can be evaluated phylogenetically. Spermatozoa of *Tupaia glis* have been examined by both transmission and scanning electron microscopy (Bedford, 1974; Matano *et al.*, 1976), and selected attributes have been compared with those of primates and other eutherians. Several morphological traits of tupaiid sperm, including the flattened, paddle-shaped head and absence of a pronounced rostral projection of the mature acrosome, bear a resemblance to those of Anthropoidea. Similarities are also evident between the sperm heads of tupaiids and strepsirhines (*Lemur, Galago, Nycticebus*) in some biochemical attributes of the cell membrane, such as the tendency for head-head autoagglutination in physiological solutions and the distribution of negative surface charge. Evolutionary analysis of these and additional features of sperm biology in primates, rodents, lagomorphs, and artiodactyls suggested, however, that shared tupaiid-primate similarities are all primitive eutherian features which provide no evidence of a close phylogenetic relationship between these taxa (Bedford, 1974).

3. Female Reproductive System

The female reproductive system and reproductive cycle of tupaiids have been investigated in greater detail than those of the male, although much of our knowledge remains limited to the genus *Tupaia*. However, available data on these features in other tupaiid genera indicate a relative uniformity within the family, in agreement with Mossman's (1953) observation that the female reproductive system of eutherians is very conservative at the family level.

A. Internal Reproductive Organs

The uterus of all tupaiid genera is bicornuate, with relatively long uterine horns and a short median corpus (Jones, 1917; Le Gros Clark, 1959; personal observations). A similar relationship is found in Cheirogaleinae and Lemurinae (Petter-Rousseaux, 1964) and probably represents the morphotype of strepsirhine primates. Comparative and ontogenetic studies of all mammalian orders suggest that a uterus duplex, characterized by the separation of uterine horns throughout their length and by their separate cervical openings into the vagina, is the primitive eutherian condition. This pattern occurs in dermopterans, megachiropterans, lagomorphs, and some rodents. Varying degrees of fusion of the uterine horns during ontogeny have led to the development of a longer uterine corpus and shorter uterine horns during the evolution of several eutherian orders, including Primates and Chiroptera. The most derived character state is the completely fused simplex uterus of Anthropoidea and phyllostomatid bats. A bicornuate uterus also characterizes the morphotype condition in Lipotyphla and Macroscelididae, and it is a widespread intermediately derived feature among eutherians. Its common occurrence in tupaiids, strepsirhines, microchiropterans, lipotyphlans, and macroscelidids does not clarity phylogenetic relationships among these taxa.

An unusual and autapomorphic feature of the tupaiid uterus is the occurrence of gland-free endometrial pads or cushions which occupy the lateral walls of the uterine horns in both juveniles and adults (Hubrecht, 1899; Luckett, 1963, 1968). These pads are actually bilateral ridges which extend the entire length of the uterine horn, and they represent specialized sites for implantation and placental disc formation during pregnancy (Figs. 2b, 3b, 4b).

As reported previously by Jones (1917) and Martin (1968), tupaiids exhibit an elongate urogenital sinus and relatively short vagina. This relationship is similar to that of many lipotyphlans and contrasts with the primate condition in which there is a long vagina and relatively inconspicuous urogenital canal. It is likely that the tupaiid-lipotyphlan pattern represents the primitive eutherian condition (Martin, 1968).

Histological features of the ovary during pregnancy and nonpregnant cycles have been examined in detail in several species of *Tupaia, Lyonogale, Urogale,* and *Dendrogale* (Luckett, 1963, 1966; Duke and Luckett, 1965; Kreisell, 1977), and comparative aspects of tupaiid and primate ovarian morphology were summarized by Koering (1974). Most ovarian features of tupaiids appear to be retentions of the primitive eutherian condition; these include the relatively complete ovarian bursa, persistence of presumably functional corpora lutea throughout pregnancy, and the occurrence of a rete ovarii and epoophoron tubules in the hilus and mesovarium. Available ovarian data provide no evidence for intergeneric relationships along tupaiids, and this organ appears to be of limited value for assessing the higher taxonomic affinities of the family.

B. External Genitalia

The clitoris is greatly elongate and grooved on its ventral surface in *Tupaia, Lyonogale, Urogale,* and *Dendrogale* (Pehrson, 1914; Jones, 1917; Hill, 1958; Luckett, 1963). The ventral groove is deepened posteriorly (caudally) to form the external opening of the urogenital sinus. Le Gros Clark (1959, 1971) described and illustrated the clitoris of *Ptilocercus* as a small structure occurring at the anterior end of the labia minora, but his diagram is very unclear. Curiously, he did not discuss the apparent differences between *Ptilocercus* and *Tupaia* in morphology of the external genitalia, and Le Gros Clark's observations on *Ptilocercus* require further confirmation.

The urethra enters the clitoris and extends throughout its length as a clitoral urethra in neonatal *Tupaia* (Luckett, 1963), whereas in the adult the urethra opens in common with the vagina as a urogenital sinus at the base of the clitoris. This difference appears to be functionally correlated with the fact that the external "vaginal" orifice is sealed in neonatal and juvenile tupaiids. The deep ventral groove of the adult clitoris probably develops as the result of secondary union between the clitoral urethra and the shallow ventral groove of the juvenile clitoris, but intermediate stages have not been examined to corroborate this hypothesis.

All strepsirhine primates are characterized by hypertrophied external genitalia in the female (Hill, 1958; Petter-Rousseaux, 1964), whereas those of haplorhines are usually small and inconspicuous, with the notable exception of some cebids. Two basic patterns are evident in strepsirhines. (1) A peniform clitoris tunneled by the urethra characterizes all lorisiforms and cheirogaleines, and this condition has also been described in the lemurid *Hapalemur.* (2) In indriids, *Daubentonia,* and some lemurids the elongate clitoris is deeply grooved ventrally, and the urethra opens at the posterior end of the groove near the base of the clitoris. In cheirogaleines and galagids, the peniform clitoris appears to be functionally correlated with the periodic closure of the external vaginal orifice during periods of sexual inactivity (Petter-Rousseaux, 1964; Butler, 1974), but such a relationship is less clear for lorisids.

The relative rarity of hypertrophied female external genitalia suggests that this condition is derived within Eutheria, and the shared similarity of this feature in tupaiines and lemuriforms might be considered as evidence of their common ancestry. However, the possible differences in clitoral morphology between *Ptilocercus* and tupaiines, and the known differences between lemuriforms and lorisiforms, impede the morphotype reconstruction of external genitalia traits in both taxa. An enlarged, peniform clitoris associated with seasonal closure of the external vaginal orifice also occurs in talpids (Matthews,

1935), and these functional relationships closely resemble the condition in cheirogaleines and galagids. In the absence of more extensive comparative and ontogenetic studies on the female external genitalia of tupaiids and strepsirhines, it is premature to suggest that shared similarities of this form-function complex may be the result of homology rather than convergence.

C. Reproductive Cycle

It has been suggested that an 8–12 day estrous cycle occurs in several species of *Tupaia* maintained in captivity (Sorenson and Conaway, 1964, 1968; Conaway and Sorenson, 1966; Hasler and Sorenson, 1974); this was based on the length of the interval between observed copulation periods. The basic estrous cycle was considered to be either anovulatory or one in which a functional corpus luteum is not formed. The hypothesis that such a cycle is anovulatory is supported by histological studies of the ovary which indicate the presence of large, preovulatory vesicular follicles but the absence of corpora lutea in adult, nonpregnant female tupaiines (Stratz, 1898; Luckett, 1963; Conaway and Sorenson, 1966). In addition to the basic estrous cycle, Conaway and Sorenson (1966) reported the frequent occurrence of a 20–22 day cycle that they considered to represent pseudopregnancy. Experimental induction of similar pseudopregnant cycles was also accomplished using matings with vasectomized males, and subsequent histological examination of the reproductive organs confirmed that ovulation and corpus luteum formation occurred during pseudopregnancy.

Conaway and Sorenson (1966) presented histological evidence that slight uterine bleeding and concomitant epithelial desquamation occur in *Tupaia* at the end of pseudopregnancy and under experimental conditions in which uterine regression was induced by estrogen and progesterone withdrawal. They considered this slight endometrial loss to be evidence of menstruation in tupaiids, and emphasized that "it occurs at the end of a non-pregnant luteal phase as in higher primates." Previously, Stratz (1898) described the occurrence of menstruation in *Tupaia* and reported that its onset was synchronized with follicular maturation. However, several investigators (Van Herwerden, 1908; Luckett, 1963; Martin, 1968) have suggested that Stratz's description of uterine and ovarian morphology associated with menstruation in *Tupaia* was probably indicative instead of postpartum recovery of the uterus in animals exhibiting postpartum estrus.

Conaway and Sorenson (1966) suggested the possible origin of the higher primate (Anthropoidea) menstrual cycle by the occurrence of two modifications in the tupaiid cycle. These are: (1) the loss of postpartum (and postpseudopregnancy) estrus; and (2) the acquisition of spontaneous ovulation and corpus luteum formation, leading to the development of spontaneous pseudopregnancy in the nonpregnant cycle. Although interesting from a theoretical viewpoint, this hypothesis does not account for the lack of evidence for a menstrual cycle in any strepsirhine primate, despite the fact that their reproductive cycles are more similar to anthropoids than are those of tupaiids. Strepsirhines undergo lengthy estrous cycles (30–55 days), associated with spontaneous ovulation and corpus luteum formation, and elevated progesterone levels are secreted during the long luteal phase of the cycle (Eaton *et al.*, 1973; Butler, 1974; Bogart *et al.*, 1977). A major difference between the estrous cycle of strepsirhines and the menstrual cycle of anthropoids is the relatively low levels of circulating estrogens during the follicular phase of the cycle in the former, and Bogart *et al.* (1977) have suggested that minimal estrogenic stimulation of the uterine endometrium may obviate the need for overt menses in strepsirhines. This interesting hypothesis warrants further analysis.

Despite their suggestion for the possible derivation of the higher primate menstrual cycle from the tupaiid pattern, Conaway and Sorenson (1966) acknowledged that the occurrence of menstruation in tree shrews is not necessarily evidence for their primate affinities. Subsequently, Sorenson and Conaway (1968) reported that pseudopregnancy was not detected in tupaiids collected in the field, and they suggested that pseudopregnant and nonpregnant cycles should be considered as an artifact among animals maintained under abnormal conditions of captivity. They also observed that adaptation to captivity was reflected by an increased number of births and postpartum pregnancies, and this was correlated with a decrease in the number of pseudopregnancies and nonpregnant estrous cycles. This observation is consistent with the hypothesis that the normal reproductive cycle of mammals is one of pregnancy rather than recurring nonpregnant estrous cycles (Conaway, 1971; Weir and Rowlands, 1973). Repeated pregnancy cycles characterized a highly successful breeding colony of *Tupaia belangeri*, and female receptivity and copulation were limited to postpartum estrus (Martin, 1968).

The occurrence or significance of menstruation in natural populations of tupaiids remain unclear. Menstruation has been reported also in phyllostomatid bats (Hamlett, 1934; Rasweiler 1972) and macroscelidids (Van der Horst and Gillman, 1941). Some features of the reproductive cycle in these two taxa are more similar to those of higher primates than are those of tupaiids; these include the occurrence of spontaneous ovulation and a functional luteal phase in nonpregnant cycles. Evidently, menstruation has evolved convergently in Anthropoidea, Macroscelididae, and Phyllostomatidae, and the reproductive cycle of tupaiids is dissimilar in its form-function relationships to the menstrual cycle of these taxa.

4. Pregnancy and Placentation

The gestation periods of tupaiines born in captivity range from 41–50 days for several species of *Tupaia* and *Lynogale* (Conaway and Sorenson, 1966; Martin, 1968), and 54 days for a single recorded birth of *Urogale* (Wharton,

1950). This is longer than the interval of soricids and talpids (18–30 days), but is comparable to the 49–64 day gestation period estimated for macroscelidids (Tripp, 1972). Strepsirhines have considerably longer gestation periods, ranging from 120–193 days (Petter-Rousseaux, 1964; Manley, 1966). A notable exception occurs among small-bodied cheirogaleines; they exhibit relatively short gestation periods of 59–70 days (Petter-Rousseaux, 1964). It is conceivable that short gestation periods, associated with twinning and small body size, have developed secondarily as the result of evolutionary miniaturization in cheirogaleines, instead of representing the primitive strepsirhine condition. A similar hypothesis has been presented for these relationships in callitrichids (Leutenegger, 1973).

There appears to be no seasonality of breeding or births in laboratory colonies of tree shrews, but it is uncertain whether similar conditions prevail among natural populations in the field. There are no published reports of tree shrews collected from the same locality throughout the year. Pregnancies are distributed throughout the 8-month collection period for *Tupaia javanica* specimens housed at the Hubrecht Laboratory, Utrecht, the Netherlands, and Zuckerman (1932) concluded that this species probably breeds throughout the year. Preliminary and incomplete data presented by Harrison (1955) and Wade (1958) suggest the possibility of a restricted breeding season in Malaysia and Borneo, and Sorenson and Conaway (1968) cited unpublished observations by Negus which indicate that seasonal breeding occurs in *Tupaia glis*. Unfortunately, corroborating evidence for this seasonality has not been published.

Seasonal breeding and birth peaks also characterize natural populations of lipotyphlans, macroscelidids, and strepsirhines, and neither the seasonality nor length of gestation periods provide insight into the mammalian affinities of Tupaiidae.

A. Placentation

Comparative features of the morphogenesis of the fetal membranes and placenta have provided more extensive evidence for evolutionary relationships among higher categories of eutherians than any other aspect of the reproductive systems (Mossman, 1937, 1953; Luckett, 1977). It was anticipated, therefore, that analysis of this developmentally and functionally complex system might prove valuable in assessing the phylogenetic affinities of tupaiids. Several investigators (Hubrecht, 1894, 1899; Van der Horst, 1949; Meister and Davis, 1956, 1958; Hill, 1965; Luckett, 1968, 1969; Martin, 1968, 1969; Kuhn and Schwaier, 1973) have described in varying detail the ontogenetic and definitive aspects of placentation in tupaiines, and different interpretations were presented for some morphological features. In turn, different and sometimes contrasting viewpoints have been espoused concerning the phylogenetic significance of similarities and differences in placentation among tupaiids, primates, lipotyphlans, and macroscelidids. An evolutionary analysis of fetal

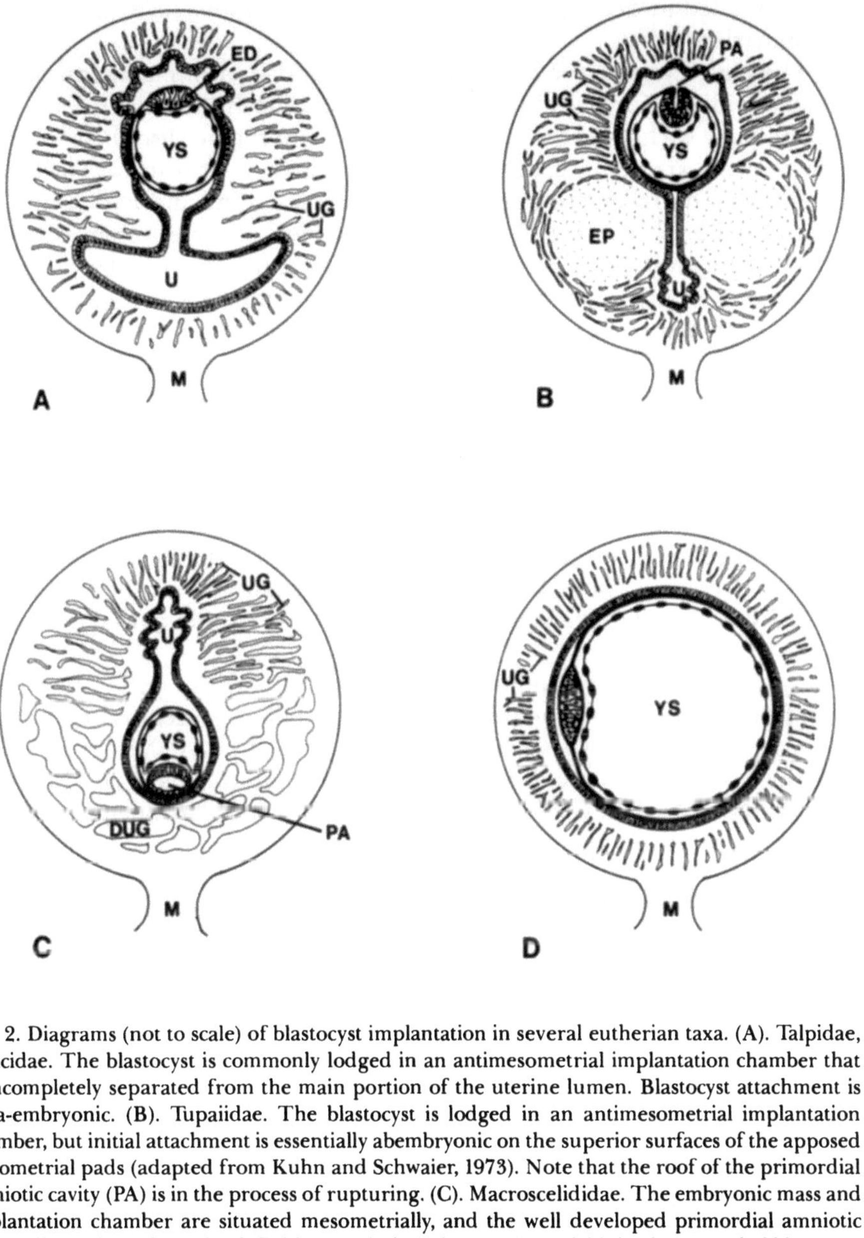

Fig. 2. Diagrams (not to scale) of blastocyst implantation in several eutherian taxa. (A). Talpidae, Soricidae. The blastocyst is commonly lodged in an antimesometrial implantation chamber that is incompletely separated from the main portion of the uterine lumen. Blastocyst attachment is para-embryonic. (B). Tupaiidae. The blastocyst is lodged in an antimesometrial implantation chamber, but initial attachment is essentially abembryonic on the superior surfaces of the apposed endometrial pads (adapted from Kuhn and Schwaier, 1973). Note that the roof of the primordial amniotic cavity (PA) is in the process of rupturing. (C). Macroscelididae. The embryonic mass and implantation chamber are situated mesometrially, and the well developed primordial amniotic cavity will persist to form the definitive amniotic cavity. (D). Strepsirhini. The expanded blastocyst is attached circumferentially, and the exposed embryonic disc is orientated orthomesometrially. The mesometrium is oriented toward the bottom in this and subsequent figures. Abbreviations: DUG, dilated uterine glands; ED, embryonic disc; EP, endometrial pad; M, mesometrium; PA, primordial amniotic cavity; U, uterine lumen; UG, uterine glands; YS, yolk sac cavity.

membrane morphogenesis in *Tupaia, Lyonogale, Urogale,* and *Dendrogale* has been presented in detail previously (Luckett, 1968, 1969, 1974) and will only be summarized here.

The blastocyst-uterine relationships at the time of implantation provide valuable evidence for assessing evolutionary relationships. Certain features of implantation, such as the orientation of the embryonic disc to the uterine endometrium, are highly conservative and rarely vary at the ordinal level. The disc is oriented orthomesometrially in both strepsirhine and haplorhine primates, despite the considerable differences which occur in other aspects of their implantation. In contrast, orientation is antimesometrial in tupaiids, dermopterans, and most lipotyphlans, and mesometrial in macroscelidids (Fig. 2). Both orthomesometrial and mesometrial arrangements are relatively rare among eutherians, while the antimesometrial pattern is common and may represent the primitive eutherian condition. Other features, including the site and degree of invasiveness of the attaching trophoblast, vary considerably among tupaiids, strepsirhines, lipotyphlans, and macroscelidids. An antimesometrial implantation chamber of varying proportions develops in tupaiids, soricids, talpids, and erinaceids, but consideration of the entire implantation process suggests convergence of this feature between tupaiids and lipotyphlans. An autapomorphous feature of tupaiid implantation is the secondarily bilateral attachment of the blastocyst to the specialized endometrial pads (Figs. 2b, 3b).

Following implantation, the bilateral endometrial pads become the sites of differentiation of the transitory choriovitelline placentae. This primitive eutherian feature is shared with strepsirhines, lipotyphlans, macroscelidids, dermopterans, and chiropterans. The origin and expansion of the vascular allantoic vesicle result in displacement of the choriovitelline placenta and the subsequent differentiation of bilateral chorioallantoic placental discs in all tupaiids (Fig. 4b). Expansion of the exocoelom and allantoic vesicle does not extend beyond the inferior border of the placental discs; consequently, the yolk sac persists mesometrially and forms a smooth choriovitelline membrane with the adjacent trophoblast (Fig. 4b). Continued growth of the embryo and amnion results in compression of the tupaiid yolk sac in later stages of pregnancy, but it remains moderately large, elongate, and trilaminar. This relationship is similar to that of carnivores, but contrasts with the reduced, free yolk sac of strepsirhines, macroscelidids, dermopterans, many chiropterans, and some lipotyphlans.

A primordial amniotic cavity develops within the embryonic mass during the peri-implantation period of tupaiids, macroscelidids (Fig. 2c), dermopterans, chiropterans, haplorhines, and most lipotyphlans. The root of the primordial cavity becomes disrupted at the time of implantation in tupaiids (Fig. 2b), so that the flattened embryonic disc becomes exposed secondarily to the uterine lumen during later stages of implantation (Fig. 3b). The definitive amnion develops in later stages by somatopleuric folding. Strepsirhines, which lack any trace of a primordial cavity (Fig. 2d), also exhibit amniogenesis by

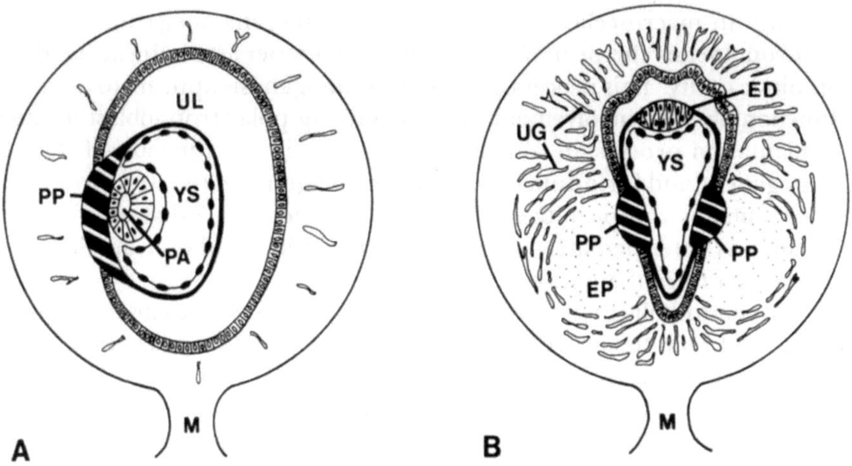

Fig. 3. Comparative aspects of late implantation phases in Anthropoidea and Tupaiidae (not to scale). (A). Implanted cercopithecid blastocyst, with orthomesometrial attachment and disc orientation. (B). Implanted tupaiid blastocyst, with secondarily bilateral attachment, and antimesometrial orientation of exposed embryonic disc. Abbreviations: ED, embryonic disc; EP, endometrial pad; M, mesometrium; PA, primordial amniotic cavity; PP, preplacenta; UG, uterine glands; UL, uterine lumen; YS, yolk sac cavity.

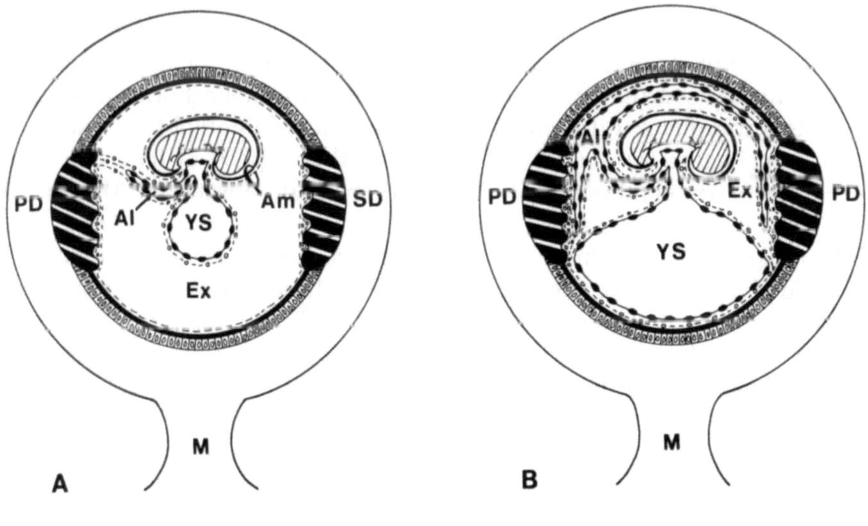

Fig. 4. Comparison of double discoidal placentation in Cercopithecidae and Tupaiidae (highly schematic and not to scale). (A). Cercopithecidae. Only the primary placental disc receives its blood supply from the umbilical cord (containing the allantoic duct). The allantoic duct is vestigial, and the yolk sac is free and reduced. (B). Tupaiidae. Both placental discs are primary and develop at the sites of bilateral implantation. Each disc receives a primary blood supply from allantoic vessels which follow the large vesicular allantois. The large yolk sac persists and forms an attached, mesometrial choriovitelline membrane. Abbreviations: Al, allantoic vesicle; Am, amnion; Ex, exocoelom; M, mesometrium; PD, primary placental disc; SD, secondary placental disc; YS, yolk sac cavity.

folding. In macroscelidids, tenrecids, dermopterans, megachiropterans, and anthropoideans the primordial amniotic cavity persists to form the definitive amniotic cavity. This process of cavitate amniogenesis appears to be correlated functionally with involvement of the overlying polar trophoblast in the initial implantation process, in contrast to the condition in tupaiids (cf. Fig. 3a, b). Ontogenetic and form-functional analyses suggest that the strepsirhine pattern approximates the primitive eutherian condition, and that amniogenesis in tupaiids is intermediately derived between this condition and the derived cavitate pattern of amniogenesis (Luckett, 1977).

An endotheliochorial relationship exists between the closely apposed maternal and fetal tissues of the definitive chorioallantoic placenta in all genera of tupaiines (Van der Horst, 1949; Hill, 1965; Verma, 1965; Luckett, 1968, 1969). Earlier, Hubrecht (1899) had reported the loss of maternal capillary endothelium within the placenta of *Tupaia javanica,* and subsequent studies of *Tupaia minor* and *Lyonogale tana* (Meister and Davis, 1956, 1958) were interpreted as showing a transition from endotheliochorial to hemochorial placentation during ontogeny. However, examination of numerous specimens of these species, as well as the illustrations of supposed hemochorial placentation by Meister and Davis, have convinced me that maternal endothelium persists in the placental labyrinth throughout gestation in these and all tupaiine species studied to date.

Differing viewpoints have been presented by several investigators with regard to the phylogenetic implications of tupaiid placentation. Van der Horst (1949) emphasized that the double discoidal, endotheliochorial placenta of tupaiids is completely different from the placenta of lipotyphlans, macroscelidids, or primates, but he did not compare other aspects of fetal membrane development among these taxa. On the other hand, Meister and Davis (1956, 1958) examined three stages of chorioallantoic placentation in *Tupaia minor* and *Lyonogale tana,* and they asserted (1956) that "the placenta and fetal membranes of *Tupaia* are extraordinarily similar to those of the more generalized members of the Cebidae and Cercopithecidae," and that the "agreement among *Tupaia, Tarsius,* and the lower Anthropoidea is so broadly based and detailed that it is unlikely to be due to parallelism or convergence."

The phylogenetic conclusions of Meister and Davis (1956, 1958) were utilized by Le Gros Clark (1959, 1971) to support his hypothesis of tupaiid—primate affinities. Supposed similarities shared by tupaiids and anthropoids included: (1) absence of a choriovitelline placenta; (2) allantoic vesicle small or vestigial, does not reach surface of placenta; and (3) definitive placenta double discoidal and hemochorial, with villus-like processes. Each of these character states is apparently derived within Eutheria, although none occurs in strepsirhines. However, both choriovitelline placentation and a large, permanent allantoic vesicle do develop in all tupaiines (Hill, 1965; Luckett, 1968), contrary to the assertions of Meister and Davis. They examined no early stages which would show evidence of a transitory choriovitelline placenta, and they simply ignored earlier reports of its occurrence (as well as the presence of a large allantoic vesicle).

Contrary to the claim by Meister and Davis (1956), the development of double discoidal placentation in tupaiids and anthropoideans is one of the clearest and best documented cases of convergent evolution (Martin, 1969; Luckett, 1974). Both placental discs of tupaiids are primary structures that develop at the bilateral sites of early implantation (Figs. 3b, 4b), and each receives an independent blood supply from the umbilical cord. In contrast, implantation in all anthropoideans is restricted initially to the embryonic pole of the blastocyst (Fig. 3a). The primary placental disc of ceboids and cercopithecoids differentiates at the initial implantation site, and subsequent attachment and invasive activity at the abembryonic pole of the blastocyst lead to development of a *secondary* placental disc (Fig. 4a). The secondary nature of this placental disc is reflected in the origin of its fetal blood supply solely from the primary disc, and also in the loss of the secondary disc in several genera (*Alouatta, Papio*). Developmental and definitive relationships of the placental discs, as well as an assessment of their associated fetal membrane features (cf. Fig. 4a, b), corroborate the convergent evolution of these features in tupaiids and anthropoideans.

As emphasized elsewhere (Luckett, 1974, 1975), the developmental pattern of the tupaiid fetal membranes and placenta is a combination of: (1) primitive eutherian retentions (amniogenesis by folding, large yolk sac and allantoic vesicle, transitory choriovitelline placenta); (2) widespread eutherian apomorphous features (invasive, para-embryonic implantation, antimesometrial implantation chamber, development of a transitory primordial amniotic cavity); and (3) tupaiid autapomorphs that do not occur in other eutherians (specialized bilateral endometrial pads, decidual knot formation, development of a double discoid, endotheliochorial placenta). There is no special resemblance in fetal membrane development between tupaiids and primates; indeed, the only two character states shared by both strepsirhine and haplorhine primates (orthomesometrial orientation of the embryonic disc, and a reduced, free yolk sac in late stages) do not occur in tupaiids.

5. Litter Size and Maternal Care

Evolutionary analysis of reproduction patterns in primates, insectivores, and other eutherians suggests that birth of precocial young, a small litter size (1–2 young), concomitant reduction in the number of mammae and nipples, and an extended and elaborate period of maternal care characterized the last common ancestor of living primates (Martin, 1969, 1975). The possible phylogenetic significance of these features in tupaiids has been discussed extensively by Martin, and only a few pertinent points will be considered here.

A. Litter Size and Nipple Count

The nipple count for all species of tupaiids falls within the range of 1–3 pair (Lyon, 1913), and three pair probably represents the morphotype condition for extant tupaiids. Both field and laboratory studies indicate that the number of young per litter is 1–4 (Lyon, 1913; Conaway and Sorenson, 1966; Martin, 1968). Strepsirhines also possess 1–3 pair of mammary glands (Schultz, 1948) and bear 1–2 young, whereas insectivores are considered genrally to have larger litter sizes. However, considerable variation in these traits occurs among lipotyphlans and other eutherians, and the causative factors involved are not clear. Macroscelidids also give birth to 1–2 precocial young (Tripp, 1972), and a small litter size characterizes arboreal, gliding, and flying eutherians in general.

B. Maternal Behavior

An unusual system of maternal behavior was described for *Tupaia belangeri* (Martin, 1968); a separate nest is built for the altricial young, and the mother suckles them only at 48 hour intervals. Martin also suggested that this absentee system of maternal care might characterize all tupaiines, but this hypothesis was disputed by Sorenson (1970) on the basis of his observations of maternal behavior in *Tupaia longipes* and *Lyonogale tana.* As emphasized by D'Souza and Martin (1974), however, cannibalism of the newborn young by the mother or other adults occurred in the majority of births observed by Sorenson (1970), and it is likely that stressful laboratory conditions which contributed to this high mortality may also have led to modifications of maternal suckling behavior. This suggestion is supported by Von Holst's (1974) report that increased stress modifies the suckling intervals and leads to increased cannibalism of infants of *Tupaia belangeri* under experimental conditions. D'Souza and Martin (1974) have demonstrated that a 48-hour suckling rhythm occurs also in *Tupaia minor* and *Lyonogale tana* under laboratory conditions of minimal stress and high infant survival rates. This tupaiine pattern of minimal mother-infant contact is strikingly different from the characteristic primate pattern of elaborate maternal care, juvenile dependence, and enhanced social organization (Martin, 1975).

6. Conclusions

The data analyzed during the present survey provide no evidence for clarifying the mammalian affinities of the family Tupaiidae (see Luckett, this volume, Chapter 1, Table 6 and Fig. 7). The negative aspects of this assessment can be taken as further indication of the probable early Tertiary separation of the order Scandentia from other eutherians, but insight into the possible sister group relationships of Scandentia remains elusive. A modified archontan hypothesis (*sensu* Goodman, 1975) of cladistic relationships among Scandentia, Primates, and Dermoptera remains somewhat more viable than other hypotheses of tupaiid affinities (see Cronin and Sarich, this volume; Dene *et al.*, this volume; Szalay and Drawhorn, this volume), but reproductive and placental traits provide virtually no corroboration for these suggested relationships. Tupaiids have retained numerous primitive eutherian features in these organ systems, and most of the relatively derived attributes that occur in tupaiids (os penis absent, bicornuate uterus, hypertrophied clitoris, transitory primordial amniotic cavity, endotheliochorial placenta) are distributed widely among other eutherians. Excluding autapomorphies, the only derived trait of the tupaiid reproductive system that is relatively rare is the presence of a pendulous penis. This feature occurs in Tupaiidae, Primates, Dermoptera, and Chiroptera, and it was utilized by Smith and Madkour (1980) as a synapomorphy of the Archonta (*sensu* McKenna, 1975). As noted earlier, the phylogenetic significance of this character is questionable.

It is likely that further study of the morphology and development of the reproductive systems in different genera and species of tupaiids will provide useful data for assessing evolutionary relationships within the family. This is evident particularly for various aspects of the male accessory sex glands.

Use of reproductive features for assessing hypotheses of archontan affinities is hindered by the lack of information on many facets of dermopteran reproduction. With the exception of fetal membrane development, there have been few microscopic studies of the reproductive system, and virtually nothing is known about dermopteran reproductive cycles. In light of the suggestion by Cronin and Sarich (this volume) that Dermoptera may be the sister group of Primates, it would be interesting to investigate whether the derived primate pattern of spontaneous ovulation, long estrous cycle, relatively long gestation period, and elaborate system of maternal care occurs in dermopterans.

ACKNOWLEDGMENT

The writer wishes to thank Dr. Nancy Hong for the illustrations. This research was supported in part by NSF grant DEB-7823906.

7. References

Alcalá, J. R., and Conaway, C. H. 1968. The gross and microscopic anatomy of the uterus masculinus of tree shrews. *Folia Primat.* 9: 216–245.

Bedford, J. M. 1974. Biology of primate spermatozoa. *Contrib. Primat.* 3: 97–139.

Bogart, M. H., Kumamoto, A. T., and Lasley, B. L. 1977. A comparison of the reproductive cycle of three species of *Lemur. Folia Primat.* 28: 134–143.
Butler, H. 1974. Evolutionary trends in primate sex cycles. *Contrib. Primat.* 3: 2–35.
Campbell, C. B. G. 1974. On the phyletic relationships of the tree shrews. *Mammal Rev.* 4: 125–143.
Carlsson, A. 1922. Uber die Tupaiidae und ihre Beziehungen zu den Insectivora und den Prosimiae. *Acta Zool.* 3: 227–270.
Conaway, C. H. 1971. Ecological adaptation and mammalian reproduction. *Biol. Reprod.* 4: 239–247.
Conaway, C. H., and Sorenson, M. W. 1966. Reproduction in tree shrews, pp. 471–492. *In* I. W. Rowlands (ed.) *Comparative Biology of Reproduction in Mammals*. Academic Press, London.
Davis, D. D. 1938. Notes on the anatomy of the treeshrew *Dendrogale. Field. Mus. Nat. Hist., Zool. Series* 20: 383–404.
D'Souza, F., and Martin, R. D. 1974. Maternal behaviour and the effects of stress in tree shrews. *Nature* 251: 309–311.
Duke, K. L., and Luckett, W. P. 1965. Histological observations on the ovary of several species of tree shrews (family Tupaiidae). *Anat. Rec.* 151: 450.
Eaton, G. G., Slob, A., and Resko, J. A. 1973. Cycles of mating behaviour, oestrogen and progesterone in the thick-tailed bushbaby *(Galago crassicaudatus crassicaudatus)* under laboratory conditions. *Anim. Behav.* 21: 309–315.
Goodman, M. 1975. Protein sequence and immunological specificity: Their role in phylogenetic studies of primates, pp. 219–248. *In* W. P. Luckett and F. S. Szalay (eds). *Phylogeny of the Primates*. Plenum Press, New York.
Hamlett, G. W. D. 1934. Uterine bleeding in a bat, *Glossophaga soricina. Anat. Rec.* 60: 9–17.
Harrison. J. L. 1955. Data on the reproduction of some Malayan mammals. *Proc. Zool. Soc. Lond.* 125: 445–460.
Hasler, J. F., and Sorenson, M. W. 1974. Behavior of the tree shrew, *Tupaia chinensis,* in captivity. *Am. Mid. Nat.* 91: 294–314.
Hill, J. P. 1965. On the placentation of *Tupaia. J. Zool.* 146: 278–304.
Hill, W. C. O. 1958. External genitalia, pp. 630–704. *In* H. Hofer, A. H. Schultz, and D. Starck (eds.). *Primatologia*, Vol. 3, Part 1. S. Karger, Basel.
Hubrecht, A. A. W. 1894. Spolia nemoris. *Quart. J. Micr. Sci.* 36: 77–125.
Hubrecht, A. A. W. 1899. Ueber die Entwickelung der Placenta von *Tarsius* und *Tupaja,* nebst Bemerkungen ueber deren Bedeutung als haematopoietische Organe. *Proc. 4th Internat. Congr. Zool.,* Cambridge, pp. 343–412.
Jones, F. W. 1917. The genitalia of *Tupaia. J. Anat.* 51: 118–126.
Kaudern, W. 1911. Studien über die männlichen Geschlechtsorgane von Insectivoren und Lemuriden. *Zool. Jahrb.* 31: 1–106.
Koering, M. J. 1974. Comparative morphology of the primate ovary. *Contrib. Primat.* 3: 38–81.
Kriesell, W. 1977. Das Ovar von *Tupaia.* Inaugural-Dissertation aus dem Anatomischen Institut der Georg-August-Universität zu Göttingen.
Kuhn, H.-J., and Schwaier, A. 1973. Implantation, early placentation, and the chronology of embryogenesis in *Tupaia belangeri. Z. Anat. Entw.* 142: 315–340.
Le Gros Clark, W. E. 1926. On the anatomy of the pen-tailed tree-shrew *(Ptilocercus lowii). Proc. Zool. Soc. Lond.* 1926: 1179–1309.
Le Gros Clark, W. E. 1959. *The Antecedents of Man*. Edinburgh University Press, Edinburgh.
Le Gros Clark, W. E. 1971. *The Antecedents of Man*, 3rd ed. Edinburgh University Press, Edinburgh.
Leutenegger, W. 1973. Maternal-fetal weight relationships in primates. *Folia Primat.* 20: 280–293.
Luckett, W. P. 1963. A histological examination of the female reproductive anatomy of tree shrews. Unpublished M. A. Thesis, University of Missouri, Columbia.
Luckett, W. P. 1966. The ovarian cycle of the tree shrews (family Tupaiidae). *Am. Zool.* 6: 574.
Luckett, W. P. 1968. Morphogenesis of the placenta and fetal membranes of the tree shrews (family Tupaiidae). *Am. J. Anat.* 123: 385–428.
Luckett, W. P. 1969. Evidence for the phylogenetic relationships of the tree shrews (family Tupaiidae) based on the placenta and foetal membranes. *J. Reprod. Fert., Suppl.* 6: 419–433.

Luckett, W. P. 1974. The comparative development and evolution of the placenta in primates. *Contrib. Primat.* 3: 142–234.

Luckett, W. P. 1975. Ontogeny of the fetal membranes and placenta: Their bearing on primate phylogeny, pp. 157–182. *In* W. P. Luckett and F. S. Szalay (eds.). *Phylogeny of the Primates*. Plenum Press, New York.

Luckett, W. P. 1977. Ontogeny of amniote fetal membranes and their application to phylogeny, pp. 439–516. *In* M. K. Hecht, P. C. Goody, and B. M. Hecht (eds). *Major Patterns in Vertebrate Evolution*. Plenum Press, New York.

Lyon, M. W., Jr. 1913. Treeshrews: An account of the mammalian family Tupaiidae. *Proc. U.S. Nat. Mus.* 45: 1–188.

Manley, G. H. 1966. Reproduction in lorisoid primates, pp. 493–509. *In* I. W. Rowlands (ed.). *Comparative Biology of Reproduction in Mammals*. Academic Press, London.

Martin, R. D. 1968. Reproduction and ontogeny in tree-shrews *(Tupaia belangeri)*, with reference to their general behaviour and taxonomic relationships. *Z. Tierpsychol.* 25: 409–532.

Martin, R. D 1969. Evolution of reproductive mechanisms in primates. *J. Reprod. Fert., Suppl.* 6: 49–66.

Martin, R. D. 1975. The bearing of reproductive behavior and ontogeny on strepsirhine phylogeny, pp. 265–297. *In* W. P. Luckett and F. S. Szalay (eds.). *Phylogeny of the Primates*. Plenum Press, New York.

Matano, Y., Matsubayashi, K., Omichi, A., and Ohtomo, K. 1976. Scanning electron microscopy of mammalian spermatozoa. *Gunma Symp. Endocrinology* 13: 27–48.

Matthews, L. H. 1935. The oestrous cycle and intersexuality in the female mole *(Talpa europaea* Linn.). *Proc. Zool. Soc. Lond.* 1935: 347–382.

McKenna, M. C. 1966. Paleontology and the origin of Primates. *Folia Primat.* 4: 1–25.

McKenna, M. C. 1975. Toward a phylogenetic classification of the Mammalia, pp. 21–46. *In* W. P. Luckett and F. S. Szalay (eds.). *Phylogeny of the Primates*. Plenum Press, New York.

Meister, W., and Davis, D. D. 1956. Placentation of the pigmy treeshrew *Tupaia minor. Fieldiana: Zool.* 35: 73–84.

Meister, W., and Davis, D. D. 1958. Placentation of the terrestrial treeshrew *(Tupaia tana). Anat. Rec.* 132: 541–553.

Mossman, H. W. 1937. Comparative morphogenesis of the fetal membranes and accessory uterine structures. *Contrib. Embryol. Carneg. Inst. Wash.* 26: 129–246.

Mossman, H. W. 1953. The genital system and the fetal membranes as criteria for mammalian phylogeny and taxonomy. *J. Mammal.* 34: 289–298.

Pehrson, T. 1914. Beiträge zur Kenntnis der äusseren weiblichen Genitalien bei Affen, Halbaffen, und Insectivoren. *Anat. Anz.* 46: 161–179.

Petter-Rousseaux, A. 1964. Reproductive physiology and behavior of the Lemuroidea, pp. 91–132. *In* J. Buettner-Janusch (ed.). *Evolutionary and Genetic Biology of Primates*, Vol. 2. Academic Press, New York.

Price, D., and Williams-Ashman, H. G. 1961. The accessory reproductive glands of mammals, pp. 366–448. *In* W. C. Young (ed.). *Sex and Internal Secretions*, Vol. 1, 3rd ed. The Williams and Wilkins Co., Baltimore.

Rasweiler, J. J. IV 1972. Reproduction in the long-tongued bat, *Glossophaga soricina.* I. Preimplantation development and histology of the oviduct. *J. Reprod. Fert.* 31: 249–262.

Raynaud, A. 1969. Les organes génitaux des Mammiferès, pp. 149–636. *In* P.-P. Grassé (ed.). *Traité de Zoologie*, Vol. 16. Masson et Cie, Paris.

Rodger, J. C. 1976. Comparative aspects of the accessory sex glands and seminal biochemistry of mammals. *Comp. Biochem. Physiol.* 55B: 1–8.

Schultz, A. H. 1948. The number of young at a birth and the number of nipples in primates. *Am. J. Phys. Anthrop.* 6: 1–24.

Smith, J. D., and Madkour, G. 1980. Penial morphology and the question of chiropteran phylogeny. *In* D. E. Wilson and A. L. Gardner (eds.) *Proceedings of the Fifth International Bat Research Conference.* Texas Tech. Univ. Press. Lubbock.

Sorenson, M. W. 1970. Behavior of tree shrews, pp. 141–194. *In* L. A. Rosenblum (ed.). *Primate Behavior*, Vol. 1. Academic Press, New York.

Sorenson, M. W., and Conaway, C. H. 1964. Observations of tree shrews in captivity. *Sabah Soc. J.* 2: 77–91.

Sorenson, M. W., and Conaway, C. H. 1968. The social and reproductive behavior of *Tupaia montana* in captivity. *J. Mammal.* 49: 502–512.

Stratz, C. H. 1898. *Der geschlechtsreife Säugetiereierstock.* Martinus Nijhoff, Den Haag.

Tripp, H. R. H. 1972. Capture, laboratory care and breeding of elephant-shrews (Macroscelididae). *Lab. Anim.* 6: 213–224.

Van der Horst, C. J. 1949. The placentation of *Tupaia javanica. Proc. Kon. Ned. Akad. Wet., Amsterdam* 52: 1205–1213.

Van der Horst, C. J., and Gillman, G. 1941. The menstrual cycle in *Elephantulus. S. Afr. J. Med. Sci.* 6: 27–47.

Van Herwerden, M. 1908. Bijdrage tot de Kennis van den Menstrueelen Cyclus. *Tijdsch. Nederl. Ver.* 10: 65–73.

Van Valen, L. 1965. Treeshrews, primates, and fossils. *Evolution* 19: 137–151.

Verma, K. 1965. Notes on the biology and anatomy of the Indian tree-shrew, *Anathana wroughtoni. Mammalia* 29: 289–330.

Von Holst, D. 1974. Social stress in the tree-shrew: Its causes and physiological and ethological consequences, pp. 389–411. *In* R. D. Martin, G. A. Doyle, and A. C. Walker (eds). *Prosimian Biology*. Duckworth, London.

Wade, P. 1958. Breeding season among mammals in the lowland rain-forest of North Borneo. *J. Mammal.* 39:429–433.

Weir, B. J., and Rowlands, I. W. 1973. Reproductive strategies of mammals. *Ann. Rev. Ecol. Syst.* 4: 139–163.

Wharton, C. H. 1950. Notes on the Philippine tree shrew, *Urogale everetti* Thomas. *J. Mammal.* 31: 352–354.

Wimsatt, W. A., and Kallen, F. C. 1952. Anatomy and histophysiology of the penis of a vespertilionid bat, *Myotis lucifugus lucifugus,* with particular reference to its vascular organization. *J. Morph.* 90: 415–466.

Zuckerman, S. 1932. The menstrual cycle of the primates. Part VI. Further observations on the breeding of primates, with special reference to the suborders Lemuroidea and Tarsioidea. *Proc. Zool. Soc. Lond.* 1932: 1059–1075.

Molecular Evolution V

Molecular Evidence for the Affinities of Tupaiidae

9

HOWARD DENE, MORRIS GOODMAN, WILLIAM PRYCHODKO, and GENJI MATSUDA

1. Introduction

The basic method for deducing relationships among living organisms of any type is by assessing the degree of difference and similarity in various characters common to the groups under consideration. Traditionally, these characters have been morphological in nature. As a result, most students of systematic biology and evolution have been anatomists and paleontologists. It was largely through the efforts of these investigators that modern taxonomy has been brought to its present state. However, as is so often the case among investigators in any science, there have been disagreements both in conclusions derived from comparative work and in taxonomies developed from this work.

Recently, an important stimulus to work on and discussion of systematic problems has been the growing recognition of the efforts of Willi Hennig of Germany in this field, largely as a result of the 1966 publication in English of *Phylogenetic Systematics.* One key point made by Hennig is that mere listing of similarities and differences among taxa is insufficient for making phylogenetic inferences. A phylogenetic relationship can only be identified if the member taxa share with one another the presence of some derived (apomorphous) character states not to be found among other related species. This does not

HOWARD DENE • Department of Anatomy, Wayne State University, School of Medicine, Detroit, Michigan 48201. MORRIS GOODMAN • Department of Anatomy, Wayne State University, School of Medicine, Detroit, Michigan 48201. WILLIAM PRYCHODKO • Department of Biology, Wayne State University, Detroit, Michigan 48202. GENJI MATSUDA • Department of Biochemistry, Nagasaki University, School of Medicine, Nagasaki, Japan.

rule out the possibility of the newly identified phylogenetic group sharing many similarities with less closely related taxa. These latter characters are those which are unchanged from an earlier ancestral (plesiomorphous) condition or which have become similar as a result of convergent evolution.

Although the terminology is his own invention, the recognition of a need to distinguish between apomorphous and plesiomorphous characters is not original with Hennig. The same idea may at least be inferred from the writings of other authors more familiar to English speaking systematists (i.e., Simpson, 1961; Mayr, 1963). What is unique to the Hennigian approach to taxonomy is the rigorous application of this concept to the construction of taxonomic groups. While the use of this approach has helped resolve certain taxonomic questions, inherent problems in identification of the apomorphous condition for morphological traits have in turn led to new controversies. In such situations, phylogenetic inferences drawn from the results of additional types of studies might help resolve some of these new controversies.

Students of ecology, behavior, embryology, physiology, and cytology have all contributed greatly to our understanding of systematic biology. In these fields, as well as in morphology and paleontology, Hennig's insights have provided a stimulus for renewed efforts toward phylogenetic studies (e.g. various chapters in Luckett and Szalay, 1975). Comparative studies of molecular variations among organisms have also contributed to our understanding of phylogeny and taxonomy. Although applications of molecular data to taxonomic problems were made early in this century (Nuttall, 1904), a concerted effort to use these data in such a way was not made until after the elucidation of the relationship between DNA and protein structure (Helinsky and Yanofsky, 1966). Since that time, various types of molecular data have been applied to systematic studies. Most extensive have been studies using various immunological approaches to detect differences between proteins of different species.

Works by Goodman based on the use of an immunological approach to the study of the systematics of Primates began appearing in the early 1960's (e.g. Goodman, 1962, 1963, 1965). Somewhat later, microcomplement fixation studies on this same group of organisms were published by Sarich and Wilson (1966, 1967, 1968). Both laboratories are continuing their studies and are expanding into other mammalian groups (i.e. Sarich, 1969; Hight et al., 1974; Shoshani et al., 1977).

Comparisons of amino acid sequences of evolutionarily related proteins from different species permit a more rigorous examination of molecular variability. As these data appear in the literature, they are collected and subjected to computer analysis, although such data cannot be accumulated as rapidly as can immunological data.

Data on the morphology of protein molecules, as is the case with all morphological data, must be analysed with the goal of distinguishing between apomorphous, plesiomorphous, and convergent similarities in homologous proteins of different species. Greatest progress in making such analyses has been with amino acid sequence data in which search procedures can systematically alter dendrograms expressing phenetic similarities in order to find the one most likely to be close to the true phylogenetic cladogram (based on maximum parsimony assumptions). From this tree and a knowledge of the genetic code, reconstructions of the ancestral condition can be attempted and identification of ancestral and derived similarities, based on these reconstructions, thus becomes possible.

2. *The Nature of the Molecular Evidence*

A. Immunological Data

Introduction of foreign material, such as injection of proteins of a different species, into the body of most vertebrates activates a series of mechanisms functionally designed to eliminate infections. Among these mechanisms is the production of a class of molecules called antibodies. Antibodies are produced in the host animal in response to the presence of those surface configurations on the injected foreign protein which differ from those present on proteins produced by the host's own body. The injected protein is called an antigen and those of its surface configurations which differ from those on the host's own proteins called antigenic sites.

Thus, to recapitulate, by injecting a protein preparation from one species (the donor) into a host of a different species, the host is stimulated to produce antibodies specific for antigenic sites which are on the injected proteins of the donor species but are not present on proteins of the host species. Serum obtained from the host species after immunization is called an antiserum because it contains the antibodies produced against the injected antigen. This antiserum can then be used to compare a range of different (heterologous) species to the original donor (homologous) species. If a heterologous species possesses on its proteins antigenic sites related to those of the homologous species, the antiserum will react with these antigenic sites. The extent of this reaction can serve as a measure of the degree of structural similarity between the proteins of the heterologous and homologous species. This, in turn, is a measure of the degree of genetic similarity between the two species (Goodman and Moore, 1971). Several ways of determining the degree to which antigens of one species will cross-react with an antiserum made to antigens of another species have been reported.

In the earliest published studies of this kind (Nuttall, 1904), antigen and antiserum were mixed and the antigen-antibody precipitates which formed in twenty-four hours were measured volumetrically after transfer to capillary tubes of standard inner diameter. The volume of material resulting when an heterologous species was used would then be compared to the volume resulting when the homologous species was used. With some considerable refinements, this method is still being employed (Fraguedakis-Tsolis, 1977).

The precipitin reaction is also the basis for the immunodiffusion procedure used by Goodman and co-workers. Here, two species are compared with a single antiserum in modified Ouchterlony plates. The plate is arranged in such a way that three wells are produced and separated from one another by an agar field. The two antigen preparations and the antiserum then diffuse through the agar field toward one another. A precipitin band forms when molecules of an antigen meet antibody molecules specific for antigenic sites on the antigen molecules. If the two antigenic preparations used are identical with respect to those antigenic sites which can be detected by the antiserum used, the bands formed by the two antigen-antibody reactions will fuse at their point of intersection and appear as an inverted "V".

If, on the other hand, one of the antigenic preparations possesses antigenic sites not found in the other preparation but, nonetheless detectable by the antiserum used, "spurs" will form in the plate. This means that one of the reaction bands will extend past its point of intersection with the corresponding band formed in the reaction of antiserum with the other antigenic preparation. The length of these spurs is dependent on the extent to which the two antigens differ with respect to detectable antigenic sites, the greater the difference, the longer the spur. As a result, spur sizes can be taken as an indirect measure of the extent of genetic divergence between the two forms being compared.

However, it is certainly possible that a given antiserum may be unable to detect apomorphous states of protein characters or that all character state changes occurred after the last shared ancestor of the two forms. Both of these possibilities tend to decrease the accuracy of any immunological measure as a detector for genetic changes. Nonetheless, the fact remains that the most reasonable assumption to make is that protein changes are of frequent enough occurrence that related forms will possess at least some apomorphous character states and that any given antiserum will be able to detect at least some of these states.

Computer analysis of immunodiffusion data was introduced in 1968 by Moore and Goodman. The computer algorithm (IMDFN) sums all spurs in both directions to determine a net spur value. Using these values for many comparisons, an immunological distance table is constructed containing the distances of each heterologous species compared from the homologous species for that particular antiserum.

For a variety of reasons previously discussed (Dene et al., 1976a), additional treatment of IMDFN results is felt to be desirable prior to using them for construction of dendrograms. A second computer algorithm, AJUST, is based on the assumption that the true evolutionary distance values between two species should be the same no matter which species is used to make the antisera for comparisons. AJUST uses a least squares procedure to produce a numerical factor for each antiserum such that, when all antisera results have been multiplied by their appropriate factors, values for all reciprocal comparisons made are as close as possible to their counterparts. This in no way affects the relative ordering of species on the various immunodiffusion tables as determined by IMDFN. These adjusted results are then used to produce an evolutionary tree (Baba et al., 1975; Dene et al., 1976a).

Assuming the existence of uniform rates of molecular evolution, many algorithms have been proposed which can produce dendrograms that correspond well to the true cladogram. Of these algorithms, the unweighted pair-group method of Sokal and Michener (1958) can best tolerate deviations from the uniform rate hypothesis and still produce a dendrogram approximating the true cladogram (Moore, 1971). As a result, this is the algorithm chosen for further treatment of the adjusted immunodiffusion results.

Dendrograms produced from immunodiffusion data are reflections of phenetic similarity. In order to make a distinction between apomorphous and plesiomorphous similarity, one must consider the original comparisons. When a heterologous species is compared directly with the homologous species, those detectable antigenic sites shared by the two species (symplesiomorphous sites) tie up antibody molecules directed against them and thus these antibody molecules can never contribute to formation of spurs, nor can they contribute to our calculated distance values. The spurs result from antibodies directed against sites found on molecules of the homologous species but not on those of the heterologous species. These sites are made up of apomorphous sites of the homologous species as well as plesiomorphous sites which have been lost by the heterologous species. These lost sites have been replaced by apomorphous character states in the heterologous species and thus, components of spurs resulting from these plesiomorphous sites of the homologous species are legitimate data in that they provide indirect information on undetectable apomorphous characters in the other species. These characters may be detected when antisera to the heterologous species are used for comparisons.

Spurs formed in comparisons between two heterologous species are made up of the same two components; one apomorphous and one plesiomorphous. Antibodies to symplesiomorphic sites shared by both heterologous species with the homologous one are once again prevented from contributing to spur formation. Because the phylogenetically closer heterologous species shared a longer period of common ancestry with the homologous species than did the more distant species, the closer species has had less time to lose plesiomorphic sites still found in the homologous species. The retained symplesiomorphic sites in this species which are lacking in the other heterologous species will contribute to the spurs formed in these comparisons, as will the synapomorphous sites shared with the homologous species but lacking in the more distant species. They thus contribute to the effectiveness of immunological techniques in detecting this period of common ancestry, even though not acceptable for such purposes according to Hennig. One could, of course, argue that highly irregular evolutionary rates on the line leading to the closer heterologous species might have been such that no apomorphous characters arose during its common ancestry with the homologous species, and increased rates after the split resulted in loss of plesiomorphous characters retained by the homologous species. Thus, when compared to the more distant heterologous species, the spur developed would be symplesiomorphous in nature and tend to lead to a fallacious grouping. This type of situation is certainly not unreasonable and could have occurred. However, it is unreasonable to believe that it is common.

Since antisera to as many species as is reasonably possible are used in immunological studies of systematic relationships, much of the plesiomorphic component of spurs using one antiserum will be complimented by a corresponding apomorphous component to spurs formed in the appropriate comparisons using other antisera. The net result is that most of the antigenic similarity indicated by dendrograms based on immunodiffusion data is the result of synapomorphy. A symplesiomorphic component is undeniably present in the relationships being depicted. However, this only means that caution is required in interpreting these dendrograms, not that they lack value in depicting cladistic relationships.

Immunological studies of taxonomic problems are also being done by V.M. Sarich and various collaborators. The immunological technique used by these workers, however, differs from the one used in Goodman's laboratory. Their technique, rather than depending upon production of a visible precipitin band, measures the degree of depletion of a complex of proteins called complement from their reaction mixtures as a result of an antigen-antibody reaction. This microcomplement fixation method is described by Sarich (1968).

Immunological distance, as measured by microcomplement fixation, represents the factor by which an antiserum concentration must be raised in order for its reaction with the protein of a particular heterologous species to give the degree of complement fixation equivalent to that given by the homologous protein. It should be pointed out that microcomplement fixation studies require that antisera be made to purified proteins. Thus far, extensive studies have been conducted using antisera to both purified albumin and purified transferrin.

Construction of phylogenetic trees by Sarich and coworkers follows an additive approach. Resulting trees apportion the total immunological distance between any two species along the two branches leading from their common ancestor in such a way as to make these values consistent with distances determined between these two species and more distant species. Their procedure has been described in detail previously (Sarich, 1969).

Both immunological methods are essentially measuring the same phenomenon. The difference in data lies solely in the method of measurement and the fact that microcomplement fixation data are restricted to two blood proteins at present while whole plasma (or serum) containing a variety of proteins has been used to collect immunodiffusion data.

B. Amino Acid Sequence Data

Immunological differences, whatever the method of detection, are rooted in amino acid differences of the protein antigens. While immunodiffusion and microcomplement fixation enjoy the obvious advantage of enabling the investigator to survey a wide range of organisms relatively rapidly, it is equally obvious that a more rigorous approach to molecular evolution as a whole would involve studies of the amino acid sequences of proteins from various organisms. Protein sequencing techniques are less readily adaptable to broad taxonomic surveys. It is only as a result of many years of work in many laboratories throughout the world that bodies of data for certain proteins have been built up which can offer useful information for taxonomic studies. The only protein sequences to date which relate directly to tree shrews are myoglobin (Romero-Herrera et al., 1976), α and β chains of hemoglobin (Maita et al., 1977), and lens α-crystalline protein (De Jong et al., 1977).

Data for these proteins have been analysed by a series of computer algorithms based on the maximum parsimony hypothesis. Initially, an unweighted pair-group method tree is analysed to determine its mutational or nucleotide replacement length. Then a branch-swapping algorithm is employed in which all nearest neighbor one step alterations in the topology of the tree for each of its adjacent pairs of nodes are tested for length. Of those topologies whose lengths prove to be less than that of the original tree, the one of lowest length is used for the next cycle of branch-swapping. This is continued until no further lowering of tree length is achieved by the algorithm. In order to test a wide range of phylogenetic possibilities, starting tree topologies showing greater alterations than those produced from the initial unweighted pair-group tree by nearest neighbor single step changes are tested by the branch-swapping algorithm. The search is terminated when it no longer appears to be possible to find a tree of lower length. By minimizing the tree's mutational length, the parsimony reconstruction maximizes the number of genetic identities within the set of protein sequences which can be attributed to common ancestry (Moore et al., 1973; Moore, 1976; Goodman et al., 1979).

The maximum parsimony method is entirely capable of drawing a distinction between the convergent evolution of similar character states and the inheritance of identical states (genetic homologies) from common ancestors. The time required for two already diverging proteins to develop, independent of one another, the same amino acid at a given position in a protein is of such an order of magnitude that additional divergent changes in amino acids at other positions would become fixed. To mistakenly group two species on the basis of some convergent state at one amino acid position would be expected to increase the overall length of the tree. As long as a majority of identities at sequence sites are indeed due to common ancestry rather than convergence, as predicted by evolutionary theory, the maximum parsimony reconstruction may be expected to be a close approximation to the actual genealogy of the sequences.

3. Immunological Results

A. Rabbit Antisera

The immunodiffusion procedures discussed earlier were used to make over 788 comparisons using antisera made in rabbits to the blood serum or plasma of the various species listed in Table 1. Results of these comparisons were analysed using the computer algorithms IMDFN, AJUST, and UWPGM. In addition, antisera to albumin and transferrin from a species of *Tupaia* unidentified in the literature have been used in microcomplement fixation studies (Sarich and Cronin, 1976).

Based on these data, all tree shrews are closer to one another than to any

Table 1. Homologous species used to make antisera

Tupaia chinensis	serum
Tupaia glis	plasma
Tupaia longipes	plasma
Tupaia montana	plasma
Tupaia minor	plasma
Urogale everetti	serum

Table 2. Immunological distance table for rabbit antisera to *Tupaia chinensis* serum

OTU	Distance value	OTU	Distance value
T. belangeri	0.000	*Nycticebus*	5.915
T. longipes	0.324	*Ursus*	5.921
T. glis	0.364	*Hemiechinus*	6.010
T. minor	0.692	*Loris*	6.051
T. montana	0.747	*Potos*	6.080
T. tana	0.766	*Manis*	6.213
T. palawanensis	*1.182*	*Canis*	6.241
Urogale	1.391	*Dasypus*	6.364
Tarsius	5.262	*Eumetopias*	6.417
Cynocephalus	5.398	*Petrodromus*	6.502
Macaca	5.465	*Rattus*	6.621
Aotus	5.534	*Atelerix*	6.679
Galago	5.544	*Tenrec*	6.688
Homo	5.549	*Cavia*	6.689
Ateles	5.579	*Rhynchocyon*	6.824
Lemur	5.592	*Didelphis*	6.863
Presbytis	5.600	*Tachyglossus*	6.863
Bradypus	5.639	*Myrmecophaga*	6.892
Cebus	5.669	*Suncus*	7.048
Loxodonta	5.701	*Nasilio*	7.096
Colobus	5.736	*Scapanus*	7.164
Perodicticus	5.740	*Eptesicus*	7.232
Lagothrix	5.760	*Sorex*	7.504

non-tupaiid species (Tables 2–7). Using antisera to *Tupaia chinensis* (Table 2) and to *T. montana* (Table 5), *Tarsius* is the closest primate species tested. However, with a third set of antisera to *Urogale everetti* (Table 7) several primates reacted better than *Tarsius*.

Examination of results with antisera to members of the genus *Tupaia* indicates the presence of two monophyletic groups within this genus. The first group consists of *Tupaia chinensis, T. belangeri, T. glis,* and *T. longipes* (Tables 2–4). In fact, using antisera to *T. chinensis* (Table 2), *T. belangeri* cannot be distinguished from the homologous species. The second group is composed of *T. montana, T. minor, T. tana,* and possibly *T. palawanensis* (Tables 5 and 6). *Urogale* is more divergent than any member of *Tupaia* except with anti-*T. glis* (Table 3) which indicates that *T. palawanensis* is more divergent than *Urogale*.

Members of the genus *Tupaia* give comparatively strong reactions using antisera to *Urogale everetti* (Table 7). With these antisera, as with all tree shrew antisera, primates and other non-primates give distinctly poorer cross-reactions than do tree shrews. Microcomplement fixation studies (Sarich and

Table 3. Immunological distance table for rabbit antisera to *Tupaia glis* plasma

OTU	Distance value	OTU	Distance value
T. belangeri	0.322	*Bos*	3.502
T. chinensis	0.534	*Presbytis*	3.529
T. longipes	0.626	*Cynomys*	3.551
T. tana	0.900	*Homo*	3.560
T. montana	0.912	*Ateles*	3.560
T. minor	0.997	*Colobus*	3.563
Urogale	1.046	*Saimiri*	3.604
T. palawanensis	1.119	*Aotus*	3.604
Lemur	3.182	*Potos*	3.623
Nycticebus	3.327	*Erinaceus*	3.628
Macaca	3.425	*Elephantulus*	3.896
Cebus	3.499	*Suncus*	3.956

Table 4. Immunological distance table for rabbit antisera to *Tupaia longipes* plasma

OTU	Distance value	OTU	Distance value
T. belangeri	0.609	*Presbytis*	4.167
T. chinensis	0.702	*Nycticebus*	4.230
T. glis	0.792	*Lemur*	4.231
T. tana	0.994	*Homo*	4.321
T. montana	1.085	*Saimiri*	4.354
T. minor	1.087	*Cynomys*	4.401
T. palawanensis	1.141	*Cebus*	4.537
Urogale	1.568	*Colobus*	4.603
Bos	4.101	*Potos*	4.723
Macaca	4.116	*Erinaceus*	4.762
Ateles	4.161	*Suncus*	5.292
Aotus	4.165	*Elephantulus*	5.397

Cronin, 1976) are in agreement with the above results regarding placement of *Urogale* with *Tupaia*. These studies also involve *Ptilocercus* which has not been included in immunodiffusion studies. *Ptilocercus* appears to be about one and one half times as divergent from *Tupaia* as is *Urogale*.

In almost all cases, primates tend to react better than other non-tupaiid groups. However, with anti-*Tupaia longipes* (Table 4), one ungulate (*Bos*) reacted slightly better than did primates. The dermopteran, *Cynocephalus*, reacted slightly better than did primates with anti-*T. chinensis* (Table 2) and at least as well as most other non-primates with anti-*T. montana* (Table 5) and anti-*Urogale* (Table 7). Only with anti-*Urogale* did an elephant shrew (family Macroscelididae) react as well as did most primates.

Table 5. Immunological distance table for rabbit antisera to *Tupaia montana* plasma

OTU	Distance value	OTU	Distance value
T. tana	0.431	*Cynomys*	4.025
T. minor	0.687	*Nycticebus*	4.042
T. palawanensis	0.829	*Perodicticus*	4.064
T. glis	0.893	*Cebus*	4.192
T. belangeri	0.945	*Cynocephalus*	4.196
T. chinensis	0.974	*Potos*	4.214
T. longipes	0.974	*Loris*	4.249
Urogale	1.225	*Canis*	4.262
Tarsius	3.732	*Galago*	4.270
Presbytis	3.748	*Dasypus*	4.289
Colobus	3.868	*Hemiechinus*	4.298
Aotus	3.873	*Erinaceus*	4.301
Homo	3.930	*Sus*	4.373
Lemur	3.943	*Suncus*	4.538
Ateles	3.962	*Elephantulus*	4.602
Lagothrix	3.977	*Echinops*	4.658
Macaca	3.988	*Didelphis*	4.778
Saimiri	3.993		

Table 6. Immunological distance table for rabbit antisera to *Tupaia minor* plasma

OTU	Distance value	OTU	Distance value
T. montana	0.562	*Cebus*	4.487
T. tana	0.592	*Lemur*	4.502
T. palawanensis	1.162	*Presbytis*	4.515
T. longipes	1.199	*Potos*	4.562
T. glis	1.391	*Cynomys*	4.574
T. chinensis	1.517	*Aotus*	4.587
Urogale	1.747	*Colobus*	4.607
Nycticebus	4.275	*Perodicticus*	4.674
Loris	4.373	*Erinaceus*	4.844
Galago	4.405	*Bos*	4.849
Ateles	4.436	*Elephantulus*	4.971
Homo	4.463	*Suncus*	5.351
Saimiri	4.487		

Table 7. Immunological distance table for rabbit antisera to *Urogale everetti* serum

OTU	Distance value	OTU	Distance value
Tupaia	1.423	*Potos*	4.135
Lemur	3.308	*Ateles*	4.144
Galago	3.589	*Colobus*	4.166
Macaca	3.652	*Cynocephalus*	4.184
Loris	3.793	*Perodicticus*	4.204
Tarsius	3.846	*Tenrec*	4.228
Petrodromus	3.860	*Cebus*	4.330
Saimiri	3.860	*Canis*	4.349
Homo	3.901	*Cavia*	4.365
Nycticebus	4.023	*Atelerix*	4.450
Dasypus	4.043	*Scapanus*	4.523
Presbytis	4.087	*Elephantulus*	4.677
Aotus	4.114	*Eptesicus*	4.851
Lagothrix	4.129		

Table 8. Immunological distance table for chicken antisera to an albumin-rich fraction of *Tupaia* plasma

OTU	Distance value	OTU	Distance value
Homo	4.7	*Lemur*	5.1
Echinosorex	4.8	*Nasilio*	5.3
Galago	4.9		

B. Chicken Antisera

Only one chicken antiserum to the partially purified proteins of an unidentified species of *Tupaia* has been analysed. Due to limited quantities of this antiserum and its relatively poor quality, only very limited comparisons were carried out. Results are shown in Table 8. Best reactions were with man and an insectivore (*Echinosorex*); poorest reactions were with *Lemur* and an elephant shrew (*Nasilio*). Chicken antisera to human and other primate albumins, however, demonstrate primate-like cross-reactions on the part of tree shrew albumin.

4. Amino Acid Sequence Data

A. Hemoglobin

In the combined α, β sequence tree (Fig. 1), the earliest eutherian branching separates ungulates from all other eutherians. Then, in the descent of the nonungulate eutherian stem, after the separation of Rodentia, a branch to Lagomorpha and Tupaioidea separates from one leading to Carnivora and Primates. The topology of the tree for α-globin sequences alone is almost

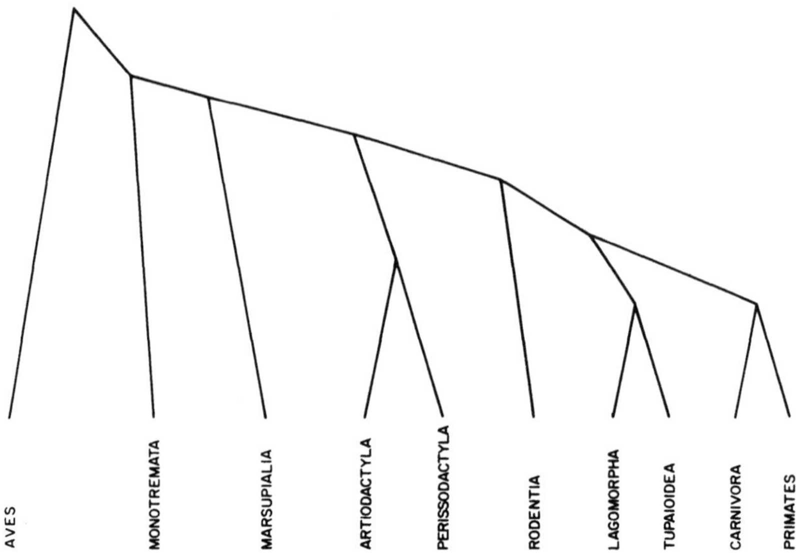

Fig. 1. Combined tree for α- and β-globin amino acid sequences.

identical to the combined tree. Differences are that the rodent branch is closer to the primate-carnivore branch than are either tree shrews or rabbits which separate from the trunk of the tree independently of one another, with the tree shrew branch closest to the carnivore-primate branch.

The β-globin tree has Proboscidea (elephant) separate from all other eutherians first. The remainder of the eutherian portion of the tree is somewhat similar to that shown in Figure 1. However, a tree shrew-rodent branch joins a rabbit-primate branch and then Carnivora either joins the enlarged primate-rabbit-tree shrew-rodent assemblage or the more distant ungulate group.

In an equally parsimonious β-globin genealogy, Carnivora separates first from all other eutherians. Next, the rodent-tree shrew branch separates from the remaining eutherians. Finally, an elephant-lagomorph branch separates from a branch which ultimately divides into primates and ungulates.

B. Myoglobin

The most parsimonious tree for myoglobin sequences is shown in Figure 2. Tree shrew, hedgehogs, and bats (*Rousettus aegyptiacus*) are seen to form a distinct branch on this tree while Primates is the most anciently separated order in Eutheria. However, a variety of alternative topologies should be considered, all of which are two mutations longer than this most parsimonious tree. Trees of the same length are found if the primate branch is placed between the carnivore and lagomorph branch, or between this last branch and

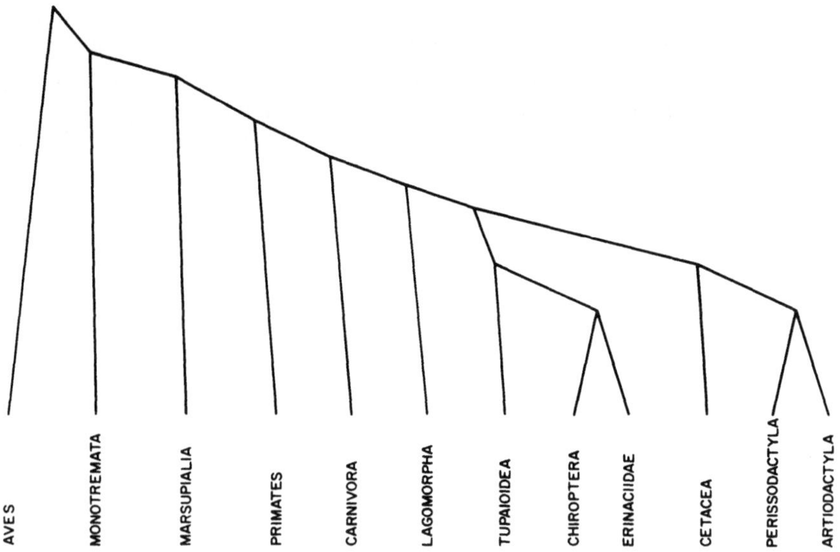

Fig. 2. Tree derived from myoglobin amino acid sequences.

the ungulate-cetacean branch. This slightly greater length is also obtained if primates are joined to the tree shrew-hedgehog-chiropteran branch.

C. Lens Protein

The most parsimonious tree for lens α-crystalline sequences (de Jong et al., 1977) has the first division of Eutheria separate subungulates (elephant and hyrax) from the other eutherians. These other eutherians are then divided into two major branches; one is composed of Carnivora and a branch made up of Cetacea and ungulates. The other branch has hedgehog, rat, and tree shrew diverge at the same point from a branch to rabbit and primates.

5. Systematic Implications of Molecular Evidence on Tupaiidae

A. Relationships within Tupaiidae

Of the lines of evidence presented, only immunological results using rabbit antisera are able to provide, at the present time, any insight into relationships within Tupaiidae. Extensive immunodiffusion results have been accumulated on seven species of *Tupaia* and for *Urogale*. Microcomplement fixation results (Sarich and Cronin, 1976) add *Ptilocercus* to the forms which may be discussed. These species are divided along traditional generic lines (Fig. 3).

These data support a division of the genus *Tupaia* into two subgenera. The subgenus *Tupaia* includes the forms *T. glis, T. chinensis, T. belangeri*, and *T. longipes*, all considered subspecies of a single species by Napier and Napier (1967). Of these forms, *T. chinensis* and *T. belangeri* are indistinguishable using antisera to *T. chinensis* (Table 2) and may , in fact, be subspecies of a single species. A similar conclusion was arrived at by Lyon (1913) based on morphological considerations. With this single exception, levels of immunological divergence in this group are comparable with those seen for separate species in other groups.

The second subgenus, *Lyonogale*, includes *T. montana, T. minor, T. tana*, and *T. palawanensis*. Only *T. tana* of these forms is included in this subgenus by Napier and Napier. As can be seen in Figure 3, however, *T. tana* and *T. montana*

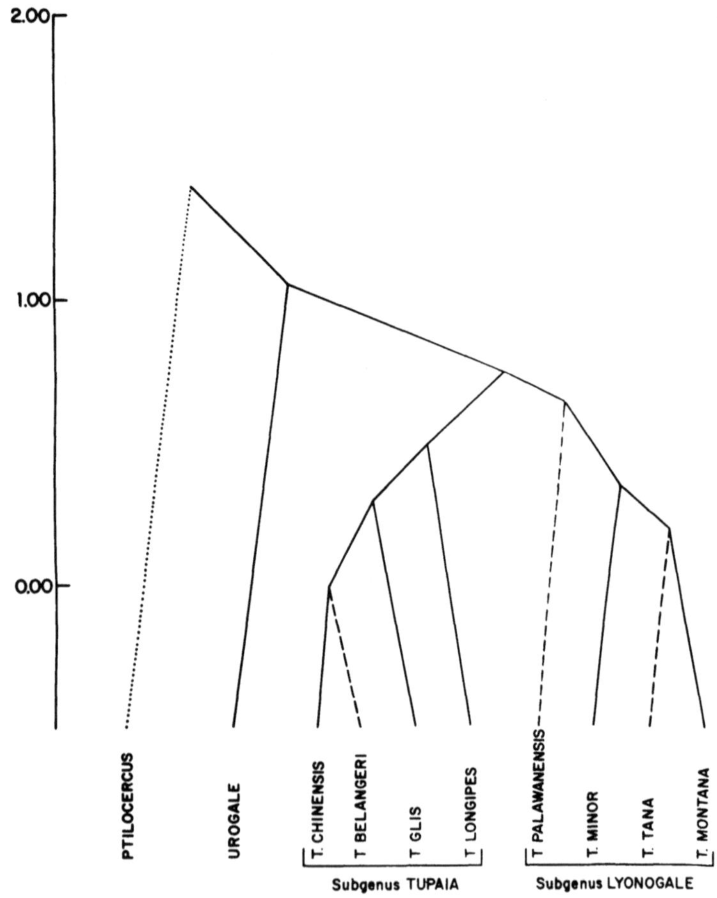

Fig. 3. Immunological divergence tree. Solid lines descend to forms used as homologous species to manufacture antisera used in immunodiffusion comparisons. Broken lines descend to forms used only as heterologous species in immunodiffusion comparisons. The dotted line descends to *Ptilocercus*, whose placement is based solely on information provided in Sarich and Cronin (1976).

are closely related. Placement of *T. palawanensis* in *Lyonogale* is less certain. Figure 3 is based on computer analysis of immunodiffusion results using antisera to tupaiid species only. If a wider range of antisera results are used, the other members of *Lyonogale* are closer to the subgenus *Tupaia* than to *T. palawanensis* which then becomes the most divergent member of the genus.

Examination of immunological distance tables shows that, using antisera to *T. montana* and *T. minor* (Tables 5 and 6), *T. palawanensis* reacts better with the homologous species than does any member of the subgenus *Tupaia*. In consideration of this fact, the grouping of *T. palawanensis* as presented here is favored.

Microcomplement fixation data of Sarich and Cronin (1976) indicate that *Urogale* is closer to *Tupaia* than is *Ptilocercus*. Placement of *Ptilocercus* in Figure 3 is based on this data. Immunological distance values from *Tupaia* assigned by Sarich and Cronin are 90 for *Urogale* and 131 for *Ptilocercus*. This is consistent with inclusion of *Urogale* in the same subfamily as *Tupaia* (Tupaiinae) while placing *Ptilocercus* in its own subfamily, Ptilocercinae (Lyon, 1913).

B. Menotyphlan Affinities of Tupaiidae

One recurring view in discussions of tree shrew taxonomy is that Tupaiidae possesses some special relationship to Macroscelididae (elephant shrews). This is expressed by their placement together in the order (or insectivore suborder) Menotyphla (Findley, 1967).

Unfortunately, no amino acid sequence data bearing on this point are yet available. Some immunological information has been accumulated, however (Dene et al., 1976b). These immunological data, which include results using antisera to elephant shrew species, give no indication of any particularly close relationship of Macroscelididae to Tupaiidae or to any other mammalian group. No support has thus far been found for the validity of Menotyphla as a natural taxon. Lack of any indication of a relationship of elephant shrews to various lipotyphlan insectivores suggests that the macroscelidids forms ought to be recognized as a distinct order, Macroscelidea (Goodman, 1975). A similar conclusion had been reached earlier based on morphological considerations (Patterson, 1965).

C. Affinities of Tupaiidae with Erinaceidae

Fossil forms morphologically related to modern Erinaceidae have, from time to time, been suggested as being ancestral to the order Primates. Results of immunodiffusion comparisons using antisera made in chickens to albumin-rich blood fractions from tree shrews have indicated a possible relationship between tree shrews and hedgehogs while rabbit antisera have not (Goodman, 1966).

For both myoglobins and for lens protein there are hedgehog and tree shrew amino acid sequences. In both cases, tree shrews and hedgehogs seem related. On the myoglobin sequence tree, tree shrews join the hedgehog-bat branch prior to joining the total tree (Fig. 2). On the lens protein sequence tree, tree shrews, hedgehog, and rodent branches come off the main tree at the same point.

Thus, where sequence evidence is available, it tends to support a tree shrew-hedgehog relationship although not necessarily a close one. In the case of myoglobin, tree shrews seem closer to hedgehogs than to primates, while for lens protein, the three groups are equally distant from one another. On the other hand, when immunological data using rabbit antisera are considered, no particularly close relationship between Tupaiidae and Erinaceidae is indicated.

D. Affinities of Tupaiidae with Rodentia

Both lens α-crystalline protein and β-globin indicate a possible relationship between tree shrews and Rodentia. This similarity is particularly intriguing in light of morphological suggestions of both tree shrew-primate affinities (Le Gros Clark, 1934; McKenna, 1975) and rodent-primate affinities (Wood, 1962). With rabbit antisera to both tree shrews and primates, rodents react more strongly, on average, than other heterologous orders except Dermoptera which also shows an affinity to Primates. However, such distant relationships are poorly depicted immunologically. Thus, the relationship between rodents and tree shrews implied by the two parsimony solutions mentioned above is an interesting possibility although poorly supported at present.

E. Primate Affinities of Tupaiidae

The one non-"insectivore" group most persistently associated with Tupaiidae is Primates. As a result of his very extensive morphological studies, Le Gros Clark (1934) not only felt that Tupaiidae ought to be included in Primates, he believed they showed a specific relationship to lemuriform primates. This viewpoint was modified somewhat in later publications in that, while primate affinities of tree shrews were still supported, tree shrews were considered a very ancient branch of this order (Le Gros Clark, 1971). Le Gros Clark's earlier point of view was expressed by Simpson in his 1945 classification of Mammalia. However, more recent considerations of tree shrew anatomy, reviewed by Campbell (1974), indicate that many similarities observed by Le Gros Clark were either retained primitive (plesiomorphous) features or convergent similarities in portions of the brain (also see Luckett, this volume).

None of the molecular data reviewed here can be taken to support placement of Tupaiidae with Primates. Rabbit antisera results suggest that tree shrews represent the closest non-primate order to Primates. However, all maximum parsimony trees of amino acid sequence data suggest a more distant relationship (Fig. 4).

Several of the amino acid sequences suggest a closer relationship of rabbit to either Primates or Tupaiidae than is commonly suggested by morphologists (i.e., McKenna, 1975). Should the rabbit be closer to one or the other of these groups as suggested by various of the maximum parsimony solutions (for

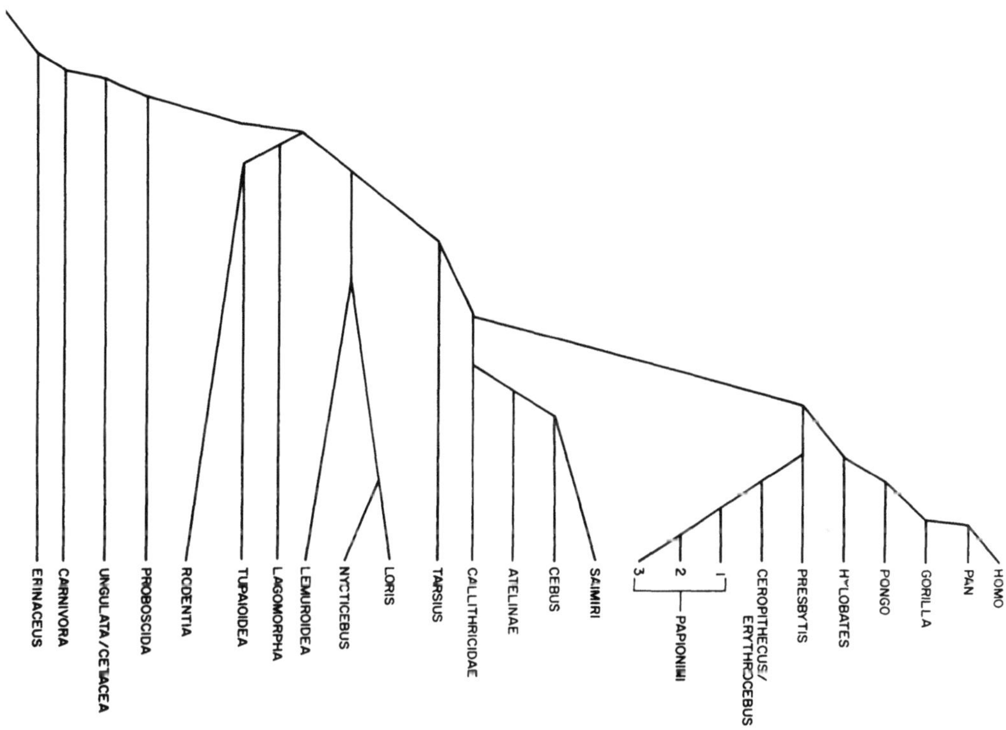

Fig. 4. Maximum parsimony analysis from combined amino acid sequence data for α and β hemoglobin, myoglobin, α lens crystallin, fibrinopeptides A & B, and cytochrome *c*. Not all taxa could be represented by all proteins: *Pan* and Atelinae are not represented by α lens crystallin sequences; *Nycticebus* is not represented by cytochrome *c:* Rodentia is not represented by myoglobin; Proboscidea is not represented by α hemoglobin and cytochrome *c; Gorilla, Pongo, Cercopithecus/Erythrocebus,* and one of the Papionini are not represented by cytochrome *c* and α lens crystallin; Lemuroidea, Tupaioidea, and *Erinaceus* are not represented by the fibrinopeptides and cytochrome *c; Hylobates* is represented by β hemoglobin, myoglobin, and the fibrinopeptides; *Cebus* is represented by the two hemoglobin chains and by fibrinopeptides; Callithricidae is represented by the two hemoglobin chains and myoglobin; *Saimiri* is represented by β hemoglobin and myoglobin; *Presbytis,* one of the Papionini, *Tarsius,* and *Loris* are represented by the two hemoglobin chains. Comparable trees have recently been constructed based on additional sequences. Although these trees differ in some details, they retain a lagomorph, rodent, tree shrew branch closer to Primates than to any other mammalian taxon.

combined α- and β-globins and for lens protein) discussed here, our use of rabbits as hosts for antibody production is undoubtedly introducing a bias into immunodiffusion results with respect to the possibility of a tree shrew-primate relationship.

One effect of this bias would be that, aside from convergent similarities, the only detectable similarities between tree shrews and primates would be those that were present in the common rabbit-tree shrew-primate ancestor and subsequently lost only in the rabbit lineage. The ability of a rabbit host to detect apomorphous characters between tree shrews or primates and any other mammalian order is thus open to question. If, on the other hand, the morphological assessment of rabbit affinities is correct (it is based on a wider sample of the total genome than is possible with amino acid sequence data at present), the rabbit data can reasonably be considered in determining tree shrew affinities with at least some other eutherian orders.

F. The Validity of the Superorder Archonta

Earlier discussions of immunological results with respect to both tree shrews and primates (Goodman, 1975; Dene et al., 1976a; Sarich and Cronin, 1976), have tended to stress the similarity between these two orders together with Dermoptera (flying lemur). As a result, a superorder Archonta has been supported to include the orders Primates, Scandentia (tree shrews), and Dermoptera. This superorder has been endorsed by morphologists periodically in the past. Recent advocates include Butler (1956) who would also include Macroscelidea and McKenna (1975) whose grandorder of the same name includes Chiroptera in addition to the three orders we have suggested.

Amino acid sequence data thus far collected are not in agreement with these morphological and immunological conclusions. No sequence data are yet available on Dermoptera or Macroscelidea and only for myoglobin are the three remaining potential members of Archonta (Primates, Scandentia, Chiroptera) represented by amino acid sequences. The most parsimonious myoglobin sequence tree (Fig. 2) places Scandentia and Chiroptera on the same branch but does not place Primates with them. It also places Erinaceidae in the same group, joining Chiroptera. However, it must be pointed out here that the bat myoglobin sequence was inferred from the amino acid composition of peptide fragments by homology with the hedgehog sequence (Castillo and Lehman, 1977). This would introduce a strong bias toward grouping bat and hedgehog myoglobins.

If primates are attached to the tree shrew-bat-hedgehog branch, the resulting tree in lengthened by two mutations. This group is, less Erinaceidae, almost identical to Archonta as recently proposed by McKenna (1975). Thus, although the most parsimonious tree for myoglobin data does not support the validity of Archonta, inclusion of an Archonta-like branching arrangement results in only a modest increase in the mutational length of the tree.

G. A Reply to Cronin and Sarich

Certain criticisms of our work have been expressed in another paper of this volume and require comment. Cronin and Sarich imply that for phylogenetic studies the microcomplement fixation technique (MCF) utilized by them is superior to other immunological methods. We question this claim. In MCF it is imperative that the antiserum be monospecific for a single type of antigen or, conversely, that the test antigens from all species compared be highly purified molecular species that are homologues of one another. When this is not the case the original MCF raw data may be seriously in error. Immunodiffusion comparisons, while admittedly requiring much higher concentrations of antibodies and antigen than MCF, are more robust in permitting use of antisera to mixtures of antigens. With care it is possible to visualize in immunodiffusion comparisons the separate precipitation reaction lines that form in the agar due to the different individual antigen-antibody systems. Thus the direction of the spurs that form in a network of comparisons of the homologous species to heterologous and the heterologous species to one another can readily be recognized. These spurs thereby produce an ordering of species progressing from those most closely related antigenically to the homologous species to those most distantly related. Our method of scoring spurs is defended in Dene et al. (1978).

In asserting that the original method of Nuttall is superior to the immunodiffusion method, Cronin and Sarich ignore some basic immunological principles. The problem with Nuttall's technique is the same problem that would be encountered in MCF if the antiserum was produced against a mixture of proteins, such as whole serum, and the test antigens employed were such mixtures, e.g. whole sera. Under these circumstances if one heterologous species appeared to cross react more strongly than another, it could be due to spurious quantitative differences in amounts of the antigens (cross reacting proteins) rather than due to true (qualitatively apparent) differences in antigenic specificities of these cross reacting proteins. Only the latter differences (and these are what spurs reveal in immunodiffusion comparisons) provide a measure of antigenic divergence or immunological distance.

With monospecific antiserum, of course, quantitative immunological techniques such as MCF do measure for the cross reacting antigen in a heterologous species its immunological distance from the homologous antigen. Nor do we deny that with such data calculations based on some additive model of evolution can be carried out to estimate for the protein under study amounts of antigenic change along the branches of a dendrogram. However we feel compelled to point out that while the "raw data" or the matrices involved in dendrogram construction have been published in immunodiffusion papers (e.g. Goodman, 1963a; Dene et al., 1976, this paper), Cronin and Sarich seem to publish only highly selected portions of their data and in the derivative form of I.D. values. Thus, in effect, the reader is being asked to accept on faith their dendrograms (e.g. Fig. 3 in Sarich and Cronin, 1976; Fig. 3, this volume).

The hypothesis of divergent evolution is, of course, not suited for calculating rates along different descending lineages but it does provide a simple and direct way to construct a dendrogram from immunodiffusion data. Both the tacit application of this hypothesis (e.g. Goodman, 1963a) and its explicit application through the IMMDF and UPGMA algorithms (e.g. the present study) to the spurs observed in immunodiffusion comparisons developed by chicken and rabbit antisera to a range of serum proteins have permitted us to draw phylogenetic conclusions that have not been possible for Cronin and Sarich, perhaps because their raw data with rabbit antisera have been focused on only one or two proteins (albumin and transferrin). A case in point is the phylogenetic position of the tree shrew (the very subject of this volume).

It was recognized early in immunodiffusion investigations involving tree shrew samples (Goodman 1963a,b) that while albumin placed tree shrews as close to higher primates as lorisoids were (and even slightly closer than Lemuroids), other proteins such as ceruloplasmin, gamma globulin, alpha$_2$ macroglobulin, and lens crystallins (Maisel, 1965) placed tree shrews further away than the two strepsirhines. That this greater antigenic distance of higher primates from tree shrews than from strepsirhines reflects genealogical distances was confirmed, decisively, by our analysis of the amino acid sequence data on myoglobin, alpha and beta hemoglobin chains, and alpha A lens crystallin (Fig. 4), as well as by DNA hybridization focused on the unique or so-called single copy genonomic nucleotide sequences (data gathered by Drs. Tom I. Bonner, Robert Heinemann, and George J. Todaro of the National Cancer Institute (T.I.B. and G.J.T.) and George Washington University (R.H.)). In contrast, depending on which outside reference species are used to construct the primate tree from the albumin and transferrin MCF data, the tree shrews can either emerge from the same node as strepsirhines and anthropoids (Fig. 3 of Sarich and Cronin, 1976) or from a tupaiid-dermopteran node just preceding the ancestral strepsirhine-anthropoid node (see Cronin and Sarich, this volume).

It also needs to be emphasized that Cronin and Sarich could be drawing unreliable conclusions on interordinal relationships among placental mammals from their rabbit antisera results. At issue is the evidence from amino acid sequence data that rabbits like tree shrews may be somewhat closer to Primates than to members of many other placental orders, and further than rabbits and tree shrews may even be slightly closer to each other than either is to Primates (e.g. our Fig. 4). Cronin and Sarich argue that this indicated relationship could not exist because if it did it "would make *Tupaia* proteins immunologically less, not more, similar to those of non-lagomorphs when using antisera made in rabbits." While that would be true for those particular protein surface configurations shared by rabbit and tree shrew, it is certainly not true for the many configurations that diverged between the two lineages since they separated from a Paleocene or Cretaceous ancestor. Certain of such configurations could still be shared by Primates and tree shrews and thus account for the relatively large tree shrew cross reaction compared to that of most other non-primate placentals. With this huge uncertainty concerning what particular portions of the albumin and transferrin amino acid sequences of rabbit are like those found in other placental mammals, we also have to question if the branch lengths calculated for Primates and tree shrews from MCF data by Cronin and Sarich using a carnivore or some other placental as the outside reference species have any meaning.

6. *Conclusions*

Extensive results of immunodiffusion comparisons using rabbit antisera indicate no relationship between tree shrews and lipotyphlan insectivores. In addition, these data fail to give an indication of any relationship between Tupaiidae and Macroscelididae. Thus, Menotyphla, as a natural taxon, is not supported by molecular data presently available.

On the other hand, sequence data do associate tree shrews with a lipotyphlan insectivore group, hedgehogs, using both proteins for which the two groups are represented, and a similar conclusion was reached on the basis of results using chicken antisera to tree shrew serum proteins. Even such evidence as can be found in currently available amino acid sequences (myoglobin) to support Archonta, also supports inclusion of hedgehogs in that superorder.

Tree shrew-primate relationships are supported by rabbit antisera results but not by those using chicken antisera, except when such antisera are directed against primate albumins. Similarly, amino acid sequence data fail to indicate a unique relationship between tree shrews and primates. On the other hand, lens α-crystallin protein and β-globin sequences suggest affinities between Rodentia and tree shrews. Thus, all of the available molecular data support the idea that tree shrews are a distinct order, Scandentia, as proposed by Wagner in 1855 (McKenna, 1975).

Archonta finds little support from currently available amino acid sequence data, although it is strongly supported by immunodiffusion results with rabbit antisera. Lack of sequence data for Dermoptera (flying lemur) is unfortunate. Data for this possible archontan form are desirable and could go far toward helping resolve the question of the validity of Archonta.

ACKNOWLEDGMENTS

We would like to thank Walter Farris, Jr. and Elaine Krobock for their help during the preparation of this paper. This work was supported by NSF grant GB36157.

7. References

Baba, M L., Goodman, M., Dene, H., and Moore, G.W. 1975. Origins of the Ceboidea viewed from an immunological perspective. *J. Human Evol.* 4: 89–102.

Butler, P.M. 1956. The skull of *Ictops* and the classification of the Insectivora. *Proc. Zool. Soc. Lond.* 126: 453–481.

Campbell, C.B.G. 1974. On the phyletic relationships of the tree shrews. *Mammal Rev.* 4: 125–143.

Castillo, O., and Lehman, H. 1977. The myoglobin of the fruit bat *(Rousettus aegyptiacus). Biochim. Biophys. Acta* 492: 232–236.

De Jong, W.W , Gleaves, J. T., and Boulter, D. 1977. Evolutionary changes of α -crystallin and the phylogeny of mammalian orders. *J. Molecular Evol.* 10: 123–135.

Dene, H.T., Goodman, M., Prychodko. W., and Moore, G.W. 1976a. Immunodiffusion systematics of the Primates.III. The Strepsirhini. *Folia primatol.* 25: 35–61.

Dene, H.T., Goodman, M., and Prychodko, W. 1976b. Immunodiffusion evidence on the phylogeny of the Primates, pp. 171–195. *In* M. Goodman and R. Tashian (eds.), *Molecular Anthropology.* Plenum Press, New York.

Dene, H.T., Goodman, M., and Prychodko, W. 1978. An immunological examination of the systematics of Tupaioidea. *J. Mammal.* 9:697-706.

Findley, J.S. 1967. Insectivores and dermopterans, pp. 87–108. *In* S. Anderson and J. Knox Jones (eds.). *Recent Mammals of the World: A Synopsis of Families.* Ronald Press, New York.

Fraguedakis-Tsolis, S.E. 1977. An immunochemical study of three populations of ground squirrel, *Citellus citellus*, in Greece. *Mammalia* 41: 61–66.

Goodman, M. 1962. Immunochemistry of the Primates and primate evolution. *Ann. N.Y. Acad. Sci.* 102: 219–234.

Goodman, M. 1963a. Serological analysis of the systematics of recent hominoids. *Human Biol.* 35: 377–436.

Goodman, M. 1963b. Man's place in the phylogeny of the Primates as reflected in serum proteins, pp. 204–234. *In* S. L. Washburn (ed.). *Classification and Human Evolution.* Aldine Press, Chicago.

Goodman, M. 1965. The specificity of proteins and the process of primate evolution, pp. 70–86. *In* H. Peeters (ed.). *Protides of the Biological Fluids.* Elsevier, Amsterdam.

Goodman, M. 1966. Phyletic position of tree shrews. *Science* 153: 1550.

Goodman, M. 1975. Protein sequence and immunological specificity: Their role in phylogenetic studies of Primates, pp. 219–248. *In* W.P. Luckett and F.S. Szalay (eds.). *Phylogeny of the Primates.* Plenum Press, New York.

Goodman, M., Czelusniak, J., Moore, G. W., Romero-Herrera, A. E., and Matsuda, G. 1979. Fitting the gene lineage into its species lineage, a parsimony strategy illustrated by cladograms constructed from globin sequences. *Syst. Zool,* 28:132-163.

Goodman, M., and Moore, G.W. 1971. Immunodiffusion systematics of the Primates. I. The Catarrhini. *Syst. Zool.* 20: 19–62.

Helinsky, D.R., and Yanofsky, C. 1966 Genetic control of protein structure, pp. 1–93. *In* H. Neurath (ed.). *The Proteins: Composition, Structure, and Function.* Vol. IV. Academic Press, New York.

Hennig, W. 1966. *Phylogenetic Systematics.* U. of Illinois Press, Urbana.

Hight, M.E., Goodman, M., and Prychodko, W. 1974. Immunological studies of the Sciuridae. *Syst. Zool.* 23: 12–25.

Le Gros Clark, W. E. 1934. *Early Forerunners of Man: A Morphological Study of the Evolutionary Origin of the Primates.* William Wood and Co., London.

Le Gros Clark, W. E. 1971. *The Antecedents of Man.* 3rd ed. Quadrangle Books, Chicago.

Luckett, W.P., and Szalay, F.S. (eds.) 1975. *Phylogeny of the Primates.* Plenum Press, New York.

Lyon, M. W., Jr. 1913. Tree shrews: An account of the mammalian family Tupaiidae. *Proc. U.S. Natl. Mus.* 45: 1–188.

Maisel, H. 1965. Phylogenetic properties of primate lens antigens, pp. 146–148. *In* H. Peeters (ed.). *Protides of the Biological Fluids*—12th Colloquium, 1964. Elsevier, Amsterdam.

Maita, T., Tanaka, E., Goodman, M., and Matsuda, G. 1977. Amino acid sequences of the α and β chains of adult hemoglobins of the tupaia, *Tupaia glis*. *Japanese J. Biochem.* 82: 603–605.

Mayr, E. 1963. *Animal Species and Evolution.* Harvard University Press, Cambridge.

McKenna, M.C. 1975. Toward a phylogenetic classification of the Mammalia, pp. 21–46. *In* W.P. Luckett and F.S. Szalay (eds.). *Phylogeny of the Primates.* Plenum Press, New York.

Moore, G.W. 1971. A mathematical model for the construction of cladograms. Institute of Statistics. Mineograph Series No. 731, Raleigh, N.C.

Moore, G.W. 1976. Proof for the maximum parsimony ("Red King") algorithm, pp. 117–137. *In* M. Goodman and R. Tashian (eds.). *Molecular Anthropology.* Plenum Press, New York.

Moore, G. W., Barnabas, J., and Goodman, M. 1973. A method for constructing maximum parsimony ancestral amino acid sequences on a given network. *J. Theor. Biol.* 38:459-485.

Moore, G.W., and Goodman, M. 1968. A set theoretical approach to immunotaxonomy: Analysis of species comparisons in modified Ouchterlony plates. *Bull. Math Biophys.* 30: 279–289.

Napier, J.R., and Napier, P.H. 1967. *A Handbook of Living Primates.* Academic Press, London.

Nuttall, G.H.F. 1904. *Blood Immunity and Blood Relationship.* Cambridge Univ. Press, Cambridge.

Patterson, B. 1965. The fossil elephant shrews (family Macroscelididae). *Bull. Mus. Comp. Zool. Harvard Univ.* 133: 295–335.

Romero-Herrera, A.E., Lehmann, H., Joysey, K.A , and Friday, A.E. 1976. Evolution of myoglobin amino acid sequences in Primates and other vertebrates, pp. 289–300. *In* M. Goodman and R. Tashian (eds.) *Molecular Anthropology.* Plenum Press, New York.

Sarich, V.M. 1968. Human origins: An immunological view, pp. 94–121. *In* S.L. Washburn and P.C. Jay (eds.). *Perspectives on Human Evolution.* Holt, Rinehart and Winston, New York.

Sarich, V.M. 1969. Pinniped origins and the rate of evolution of carnivore albumins. *Syst. Zool.* 18: 286–295.

Sarich, V.M., and Cronin, J.E. 1976. Molecular systematics of the Primates, pp. 141–170. *In* M. Goodman and R. Tashian (eds.). *Molecular Anthropology.* Plenum Press, New York.

Sarich, V.M., and Wilson, A.C. 1966. Quantitative immunochemistry and the evolution of primate albumins. *Science* 154: 1563–1566.

Sarich, V.M., and Wilson, A.C. 1967. Rates of albumin evolution in Primates. *Proc. Nat. Acad. Sci.* 58: 142–148.

Sarich, V.M., and Wilson, A.C. 1968. Immunological time scale for hominid evolution. *Science* 158: 1200

Shoshani, J., Goodman, M., Prychodko, W. , and Morrison, K. 1977. Relationships among the subungulates as demonstrated by immunodiffusion analysis of serum proteins. American Society of Mammalogists 57th Annual Meeting. Technical paper.

Simpson, G.G. 1945. The principles of classification and a classification of mammals. *Bull. Amer. Mus. Nat. Hist.* 85: 1–350.

Simpson, G.G. 1961. *Principles of Animal Taxonomy.* Columbia Univ. Press, New York.

Sokal, R.R., and Michener, C.D. 1958. A statistical method for evaluating systematic relationships. *Kansas Univ. Sci. Bull.* 38: .409–1438.

Wood, A.E. 1962. The early tertiary rodents of the family Paramyidae. *Trans. Amer. Phil. Soc.* 52: 1–261.

Tupaiid and Archonta Phylogeny: The Macromolecular Evidence

10

J. E. CRONIN and V. M. SARICH

1. Introduction

A. Tupaiid Affinities

The evolutionary affinities of the tree shrews, family Tupaiidae, have long been debated by mammalian systematists. Various hypotheses as to their evolutionary origin have been proposed since the initial description of the tree shrew in the 18th century (see Luckett, Chapter 1, this volume). In general, discussion of the affinities of this complex of "primitive" scansorial mammals has centered around postulated associations with (1) the Insectivora, specifically Macroscelididae, creating a group Menotyphla (see Luckett, Chapter 1, this volume); (2) Leptictidae (Van Valen, 1965; McKenna, 1966); (3) Mixodectidae (Szalay, 1977); and (4) Primates (Simpson, 1945; Le Gros Clark, 1971). Recently, McKenna (1975) has grouped the tree shrews with Primates, Dermoptera, and Chiroptera into the grandorder Archonta, with the tupaiids given an ordinal ranking as Scandentia (Butler, 1972). For primatologists, tree shrews take on special interest as they have been seen as a model for the basal primate morphotype (Le Gros Clark, 1971), although this model has been challenged on numerous grounds (Cartmill, 1972; Luckett, Chapter 1, this

J. E. CRONIN • Department of Anthropology, Peabody Museum, Harvard University, Cambridge, Massachusetts 02138. V. M. SARICH • Departments of Anthropology and Biochemistry, University of California, Berkeley, California 94720.

volume). Clearly, knowledge of the phylogenetic affinities of the tree shrews is crucial in fully testing this model and in dealing with the alternative proposals.

B. Molecular Systematics

Traditionally, the basic approaches to developing phylogenetic hypotheses have made use of dental, osteological, or soft anatomical traits. Since about 1960, a complementary approach which involves comparisons of proteins and nucleic acids among living species has become increasingly available. These include direct amino acid sequencing, immunological or electrophoretic comparisons of proteins, and annealing or hybridization (recently, also sequencing) of nucleic acids. The body of available molecular data is impressive and rapidly expanding.

Macromolecular comparisons of different species have been critical to the resolution of a number of longstanding controversies as to phyletic affinities over a wide taxonomic scale. For example, within Primates, *Theropithecus* (gelada baboon) is seen to be clearly more closely related to the savannah baboons of the genus *Papio* than are the forest baboons *(Mandrillus)* (Sarich, 1970; Cronin and Meikle, 1979). Similarly, the giant panda's closest affinities are with the ursids and not with the procyonids (Sarich, 1973). Other examples are the demonstration at the molecular level of the monophyly of anthropoid primates (Goodman, 1962, 1975; Wooding and Doolittle, 1972; Cronin and Sarich, 1975), and the association of all ratites into a monophyletic clade relative to other birds (Prager et al., 1976). Alternatively, previously unsuspected relationships have been discovered. These would include the polyphyletic origin of the mangabeys (Cronin and Sarich, 1976; Hewett-Emmett and Cook, 1978) and the surprising result that *Pan* and *Gorilla* appear to be no closer to each other than either is to *Homo* (Goodman, 1962; Sarich and Cronin, 1976; Sarich and Cronin, 1977; Zihlman et al., 1978).

Although there is, ultimately, one "true" phylogeny, and different molecules give highly congruent results (Prager and Wilson, 1976; Sarich and Cronin, 1976), there are cases of incongruence among different molecular trees. This can be caused both by discordance in the actual data and by differing interpretations of the same data. The phyletic position of *Tarsius* is one notable example (Dene et al., 1976; Cronin and Sarich; 1978; Sarich and Cronin, in press; Baba et al., in press; Goodman et al., 1979).

In this paper we make use of published protein sequence data and quantitative immunological information gathered in our laboratory to construct the most parsimonious tree positioning tree shrews among the mammals. Micro-complement fixation and precipitin analysis of albumin and transferrin proteins were the techniques we used to gather data, and in order to construct the phylogeny, we assume that the edentates are the sister group to all non-ungulate lineages. Given this assumption, we first ascertain basic intramammalian affinities. With these, we then attempt to position tupaiids. We conclude that tupaiids clearly link molecularly with "true" primates (*Tarsius*, anthropoids, and strepsirhines) and the flying lemur *(Cynocephalus)* into a monophyletic clade, Archonta (see also Goodman, 1975). The closest sister groups to this clade would appear to be bats and the lipotyphlan insectivores (shrews, moles, and hedgehogs). As yet, we cannot clearly differentiate molecularly any sequence of bifurcations among these lineages. Slightly more removed are macroscelidids and carnivores, and then we see (although on the basis of as yet very incomplete data) a basic placental mammal dichotomy into ungulates and nonungulates. Our results tend to support a modified grandorder Archonta, which includes the orders Primates, Dermoptera, and Scandentia but excludes Chiroptera (*contra* McKenna, 1975).

The positioning of both rodents and rabbits is problematical. The position of the latter, however, is crucial to evaluating results of immunological comparisons of mammalian proteins, as the rabbit is often the main source of antibody production. The possibility exists that the rabbit interjects a bias in the development of the "immunological picture."

2. *Nature of the Molecular Evidence*

A. Comparison of Macromolecular Techniques

The molecular data, particularly DNA and amino acid sequences, are especially suitable for a Hennigian approach to systematics (Hennig, 1966). Given a knowledge of the genetic code, probable ancestral sequences may be reconstructed for any particular node. Once this is accomplished, primitive sequence retentions (plesiomorphies), shared derived codons (apomorphies), and uniquely derived codons (autapomorphies) can be ascertained. The more sequences available in a taxon or clade from the contained species, the easier it is to define these character states. Tree construction can proceed without a specific knowledge of ancestral sequence (Fitch and Margoliash, 1967). However, most methodologies do take cognizance of such ancestral node sequence constructions and attempt to search for and derive the maximally parsimonious tree (see Dene et al., 1976, this volume; Goodman et al., 1979). An alternative method of phylogenetic reconstruction is the Wagner tree (Farris, 1970).

Morphological character state interpretations may be affected by parallel or convergent evolution in different lineages (Luckett, Chapter 1, this volume), and this is true at the molecular level as well. Globin gene evolution exemplifies this problem (parallel, back, and convergent mutations at a single codon along a lineage); the effect is a loss of optimal resolution of cladistic events. While such events are relatively infrequent, the problem grows larger as the taxonomic or temporal distances and the rate of evolution of the molecules involved increase (Holmquist, 1976).

Another inherent problem, again exemplified in globin evolution, is that of gene duplication and suppression. There is always the possibility that a particular protein in different lineages may be not homologous but paralogous; that is, the proteins may result from a duplicated locus and, although ultimately, are not strictly identical by descent. The two loci may follow different evolutionary pathways, with either being susceptible to suppression (becoming "turned off"). If such an event occurs in a particular clade, the resultant ancestral node sequence construction, and tree topology, become suspect. Such duplications can result only in exaggerated genetic distances as the two loci "diverge" in their sequence, while remaining part of a single lineage. Subsequent cladogenic events will add true phyletic substitutions to the total genetic distance. A relevant example is provided by Old World monkey hemoglobin evolution as interpreted by Hewett-Emmett et al. (1976); in this case there have been at least three duplications (2α, 1β) within the 42 chromosome clade alone.

Given these problems, we approach ancestral sequence reconstruction and cladistics, especially those derived from α and β hemoglobin and myoglobin, with some reservations. We should fully expect that as additional sequences become available, the topology of the tree may be substantially affected. Lineages with few substitutions will be particularly susceptible to such effects. Reassigning just a few mutations can alter significantly the sequence of bifurcations seen in the basal mammalian or primate radiations. Additionally, alternative methodologies of tree construction can yield different cladistics. For example, compare the alternative topologies of mammalian hemoglobin sequences in this paper and that by Dene et al. (this volume). To state it simply, the principle of parsimony does not yield results that should be taken as dogma, although we must note that while nature may not be parsimonious, science has to be.

One should approach other types of molecular data with all of the reservations listed above. Immunological and nucleic acid hybridization differences in macromolecules between organisms are reflections of underlying nucleotide base changes. Micro-complement fixation derived immunological distances are highly correlated with the sequence differences between the proteins compared (Prager and Wilson, 1971; Wilson et al., 1977). Differences in thermal stability in nucleic acid hybridization are highly correlated with per cent nucleotide base pair differences between the native strands (Benveniste and Todaro, 1976; Kohne, 1975). It is assumed that thermal stability differences between heterologous comparisons are also highly correlated with underlying nucleotide differences. However, as one cannot detect *specifically* what the changes are, one cannot strictly define apomorphic or plesiomorphic conditions. Yet, the detection of overall similarities or differences is a result of such underlying nucleotide substitutions. The more shared derived codon changes for a particular molecule, along a particular clade, the more similar the taxa within the clade will appear when measured by micro-complement fixation or by nucleic acid hybridization, when compared to clades which do not share these changes. As measured by these techniques, the more the synapomorphic base changes the greater the similarity between taxa and conversely, the more the apomorphic base changes the more the taxa will differ.

Thus while these methodologies may not yield results that are as "phyletic" as direct sequencing data, they are not purely phenetic. The process of the antigen-antibody interaction and, to a lesser extent, the formation of hydrogen bonds between bases, acts as a "phyletic" test, or screening device, of true synapomorphies at the molecular level.

One other molecular methodology deserves comment. This is the immunodiffusion approach, especially as it has been developed and used by Goodman and his colleagues. In this method, results are generated by a comparison of antigen-antibody precipitin lines in an agar matrix. Usually, antisera made against whole serum containing many components from a particular species, x, are reacted against the homologous species, x, and a heterologous species, y. The reactions are then graded on a scale of one to five, based upon subjective interpretations of the density of the precipitin lines or the length of the "spur" observed (Goodman and Moore, 1971). The molecular distance between two taxa is essentially a continuum, yet each antigen-antibody reaction is placed on an integral scale; the data are presented as though they are accurate to three decimal places, although the original readings may not be interpretable even at the integer value. In addition, the data as derived have never been analyzed in a cladistic manner. That is, since differential rates of protein change along a lineage can only be assayed by using an outside reference point (Sarich and Cronin, 1976), it is dendrograms constructed from immunodiffusion data which may lead to the *assumption* that there is a constant rate of change in the molecules—an *assumption* made by no other students of evolution (Sarich and Cronin, 1976). Also, it is not possible to evaluate a "goodness of fit" as one cannot go from the data presented to the tree, and vice versa, in any of the publications utilizing this methodology. In some ways, the technique lacks even the precision of the quantitative precipitin work performed by Nuttall in 1904. In fact, of all the molecular methodologies, immunodiffusion comparisons yield the *most phenetic* results; it can be described as nothing more than molecular numerical taxonomy but without the advantage of knowing what the input character states are that one has at the morphological level.

A final molecular technique is available which will revolutionize the field of molecular systematics and phylogenetics. In the last few years it has become possible to sequence DNA (Maxam and Gilbert, 1977). Specific genes have been isolated and the sequence of codons determined. One surprising result is that the region of the gene that codes for a particular protein is not in a continuous sequence. The gene is composed of segments that are transcribed (exons) and intervening sequences (introns) that, while translated, are not transcribed (Crick, 1979). These intervening sequences can be quite large, longer than the total length of the exons. Furthermore, the intervening sequences may have a faster rate of evolution than the exons. Thus it will be possible to gain substantial amounts of evolutionary information from such sequencing (Crick, 1979). This gain in information may make it possible to solve some of these long-standing phyletic questions, such as the positioning of *Tarsius* or *Tupaia*. Unfortunately, there are at present no relevant data from this approach.

B. The Utility of Molecular Data in Phylogeny Reconstruction

Finally, we discuss the utility of the macromolecular data in constructing phylogenies. A number of authors have commented that molecular data are phenetic in nature, less reflective of the effects of natural selection, and that each molecule is a single character equal in some sense to a tooth cusp or type of carotid circulation (see, for example, Schwartz et al., 1978). Thus they feel able to ignore, or give very low weight to, the molecular evidence. Clearly, we have a different perception of the situation and feel that most of these critics do not appreciate the nature of proteins and nucleic acids and their mode of evolution. The particular advantages of macromolecular data are actually readily enumerated. First is the fact that the unit of change at the molecular level (a single amino acid or nucleotide substitution) is known. Thus, differences between any pair of species can be measured in precisely the same terms. Second, the same molecules are available for comparison over very wide taxonomic ranges. It is, for example, quite possible to compare *Neurospora* and *Homo* at the molecular level; a comparison essentially meaningless at the morphological level.

Each nucleotide replacement in a codon is a unique character state and each molecule may have hundreds of such constituent character states. Furthermore, it has been suggested that proteins are not subject to the same rigors of natural selection as phenotypic morphological characters. It is obvious to geneticists that natural selection is working at the level of the structure of proteins. Note for example the differential fitness of individuals carrying the sickle cell allele (only one nucleotide different from the typical allele) in a malarial environment. Even the most ardent "neutralist" believes that a neutral allele has passed through a selective filter. This is readily seen by the fact that a particular amino acid is most likely replaced in evolution by one similar in size, configuration, and/or charge. The outside surface of a molecule seems to undergo more change than the inside and the active site less than other areas. Different substitutions obviously have differing effects if they alter the structure or function of the molecule. All of these observations are evidence that selection is working at the level of the gene. A protein does have a tertiary structure, that is, it is something more than just a linear string of amino acids, and thus has in some sense a morphology. But what molecular systematists attempt to ascertain is more than just overall phenetics or similarity; at each nucleotide position, the sequence of ancestral-descendent derived changes is determined.

As stated, the molecular data do have a distinct advantage in that each protein or gene has numerous character states. One can study many genes, or one gene at a time. Molecular character states have both a known homology and are not end products of complex polygenic developmental pathways. There is a distinct advantage in knowing that each of our character states has little environmental input into its final makeup. Even the simplest morphological trait may have a substantial environmental component influencing the final phenotype. Very few morphological characters have a heritability of one, that is, where environmental effects on the phenotype are nil.

Proteins and nucleic acids are finite structures and thus some parallelisms and convergences must occur either by chance or by natural selection. But with our knowledge of the genetic code, phylogenies can be constructed from molecular data and used as working hypotheses. We must strongly disagree with such statements as "molecular—and I would include karyological—studies do not, at present, offer the kind of data which are necessary to falsify hypotheses concerning the phylogenetic relationships of taxa" (Schwartz, 1978, p. 200). This statement demonstrates a fundamental misconception of the power of genetic data. It would be more accurate to state that the molecular data do falsify specific phylogenetic relationships of taxa (including some proposed by the above author, i.e., the position of *Daubentonia* and cheirogaleines, Schwartz et al., 1978). Molecular data confirm others (Luckett, 1975; Szalay and Drawhorn, this volume), are equivocal in some cases (i.e., *Tarsius*, Baba et al., in press; Sarich and Cronin, 1976, in press), and inconclusive in yet others (the position of lagomorphs, as discussed earlier and in Dene *et al.*, this volume). We strongly urge that all sources of information be used in assessing phylogenetic relationships, and that all data whether neontological, paleontological, or molecular—are suitable for corroboration or rejection of hypotheses. No one source of data should have unique or unequivocal preference over others, particularly since each line of evidence has been at some time either unavailable, misinterpreted, or uninformative. The point is that a synthetic approach is the most scientific, valid, and conscientious.

3. Molecular Data and Results

A. Introduction

Any independent positioning of a given lineage using comparative protein or nucleic acid data must involve an allocation of the measured differences among the species involved along a derived phylogeny that does not assume constancy of rates of change. In other words, two species may show protein or nucleic acid similarities either because of actual phylogenetic affinity or because of a relative lack of change along one or both of the lineages involved—and we have to be able to objectively choose the correct explanation. Ultimately the ability to make such a choice rests on the availability of "reference species" which can be assumed to be outside the clades of interest and which still show sufficient molecular similarities to the various species of interest within those clades. When we come to deal with questions of relationship above the ordinal level among mammals, this becomes a most restrictive requirement severely limiting the body of potential relevant data. Indeed, if one is to make any use of immunological data, one is, for all practical purposes, limited to albumin comparisons. Because of the relative paucity of other potentially useful information, in particular, amino acid sequence data for the majority of mammalian higher taxa, the albumin data do however have a unique current utility and are unlikely to become obsolescent until DNA sequence data become routinely available. In this contribution, then, we will emphasize the albumin immunological data insofar as they provide internally consistent placements of the various non-ungulate placental lineages of particular interest: tupaiids, dermopterans, bats, carnivores, and insectivores. We also assume that the South American edentates (sloths, anteaters, armadillos) bear no special relationship to any of the above (see McKenna, 1975) and thus can serve as our required "reference species" for beginning the assessment of amounts of albumin change along the various lineages of interest. One might wonder as to why we do not make use of various ungulate reference species in view of the fact that ungulate and non-ungulate albumins do cross-react. The answer is that antisera against ungulate albumins made in rabbits (for example, against those of *Elephas, Bos,* or *Equus*) do not discriminate among non-ungulate albumins known to be more or less changed. We assume this non-specificity stems from some immunological asymmetry of rabbit albumin relative to those of ungulates and non-ungulates and are currently actively investigating the matter (Chiment and Sarich, in preparation).

B. Tupaiid Molecular Systematics

The albumin and transferrin data provide the greatest resolving power here and so we begin with them. The albumin of *Tupaia glis* is an average of 104 immunological distance units distant from those of accepted primates (*Tarsius,* Anthropoidea, Lemuriformes, Lorisiformes) as shown in Table 1 (taken from Cronin and Sarich, 1975). This is quite similar to the 110 or so units observed when albumins from each of the four major primate lineages are compared to one another, and much less than the distance to *any* other albumin except that of the flying lemur, *Cynocephalus.* Thus, the question to be answered is whether this immunological similarity of *Tupaia, Cynocephalus,* and "true" primate albumins is due to markedly less change along the tree shrew and flying lemur albumin lineages or to the fact that the latter are sister lineages to the true primates? This question is dealt with in Figure 1, and the analysis there indicates that, relative to those of bats and edentates, the albumin of *Tupaia* is less changed, but not sufficiently so as to dissociate it from the

	P	T	C	B	E
Primates	0	106	105	154	168
Tupaia		0	103	139	148
Cynocephalus			0	155	158
Bats				0	172
Edentates					0

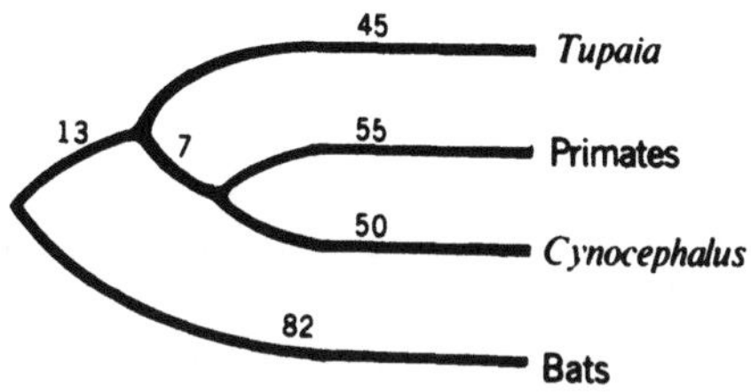

Fig. 1. The given albumin immunological distances were obtained using antisera to *Aotus, Lemur, Lepilemur, Nycticebus, Galago, Tupaia, Cynocephalus, Syconycteris, Tadarida, Bradypus, Tamandua,* and *Cabassous* albumins. It was assumed that bats and primates are at least as closely related to one another as either is to edentates. The immunological distances were then apportioned so as to minimize the disagreement between the input and output data. Here, it is a very low 1.8%.

primate clade. There is an indication, though, that other members of that clade, including the flying lemur, do share a period of common ancestry subsequent to the divergence of the tupaiids.

It should be noted here that we have in the past (for example, Sarich and Cronin, 1976, Figure 3) argued for an even closer affinity of the tree shrew and primate clades. These arguments were based on the data obtained using

Table 1. Immunological distances among primate and mammalian albumins and transferrins

	Anthropoidea	*Tarsius*	Tupaiidae	Dermoptera	Strepsirhini	Nonprimates[a]	
Anthropoidea	0	120	96	110	107	+19	A
Tarsius	164	0	93	109	109	+ 7	L
Tupaiidae	180	159	0	103	112	− 3	B
Dermoptera	169	161	183	0	111	+ 7	U
Strepsirhini	181	167	180	174	0	0	M
							I
	TRANSFERRIN						N

[a]Data gathered using antisera to the albumins of *Ursus, Genetta, Hyaena, Felis, Hipposideros, Pteropus, Tadarida, Antrozous, Bradypus, Cabassous, Tamandua, Scapanus,* and *Solenodon.* The numbers represent the amount the albumins are more or less distant from the outside reference points relative to the Strepsirhini mean.

a series of bat and carnivore reference points (Sarich and Wilson, 1973) which indicated only a moderate degree of conservatism in the *Tupaia* albumin lineage. However, this result was the average of the bat data, which showed less change along the *Tupaia* line, and the carnivore data, which showed significantly more. The subsequent addition of insectivore and edentate reference points gave results more consistent with the bat placement. Thus the weight of the evidence currently emphasizes the relatively conservative nature of *Tupaia* albumin evolution and, therefore, an earlier divergence for the tree shrew line. None of the albumin data, however, can be seen as consistent with the exclusion of the tree shrews from a close relationship with the primate clade.

This placement of the tree shrew and flying lemur close to primates is strongly supported by the transferrin evidence. Here we find appreciably stronger immunological cross-reactions between the proteins of the three groups (using antisera to their transferrins) than between their transferrins and those of any other mammal tested. In fact, in the course of our extensive investigations of transferrin evolution in mammals using antisera to those of numerous bat, carnivore, rodent, and artiodactyl species, any interordinal cross-reaction has been an extremely rare event—indeed, intraordinal cross-reactions often do not occur. We believe that the transferrin data would in themselves constitute extremely strong presumptive evidence for tree shrew-primate-flying lemur affinities; and together with the albumin data, they make for a most convincing association. We also reject the idea that these affinities could be an immunological artefact stemming from a rabbit—*Tupaia* cladistic association, as surely any such would make *Tupaia* proteins immunologically less, not more, similar to those of non-lagomorphs when using antisera made in rabbits. Nor can one appeal, in the case of the transferrins, to conservatism in the evolution of either the tree shrew or other primate molecules, as the result would then be immunological cross-reactivity with the transferrins of species in other orders—and this has not been observed, as noted above.

It is debatable whether the available amino acid sequence data make very much of a positive contribution in this area. The problems, as we see them, are two-fold. First, the majority of the relevant data come from proteins entirely unsuitable to the questions involved, as they evolve too slowly to provide adequate resolving power. For example, the myoglobin rate of substitution is about one per lineage per 5 MY. It is therefore to be expected that myoglobin lineages could exist for substantial periods of time—5, 10, 15 MY—without substitutions occurring along them, thus remaining invisible to us. Even more serious are the unanswered questions pertaining to the statistical analyses of the sequence data. Parsimony is certainly a necessary criterion, but the question of how many total substitution differences between two solutions are necessary before they can be considered statistically different remains open. It is obvious, though, that the amount of demonstrated homoplasy in the globin data should give one pause before attempting to choose among

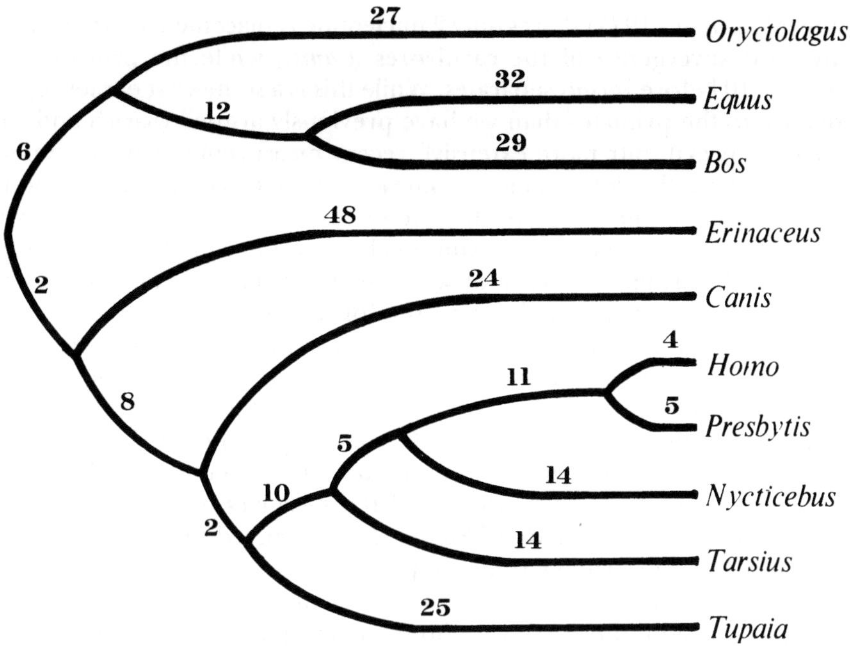

Fig. 2. This hemoglobin tree was developed by ancestral node reconstruction at the codon level using the parsimony criterion. The goal was to see if a cladistic arrangement which lacked certain of the discordant features of the Beard and Goodman (1976) trees, while remaining equally parsimonious, was possible. Discordance here is measured relative to our immunological data concerning interordinal relationships among the mammals. The analysis is based upon the following published sequences: *Homo, Gorilla, Macaca mulatta, M. fuscata, Cercopithecus, Presbytis, Cebus, Ateles, Lemur, Nycticebus, Tarsius, Tupaia, Erinaceus, Oryctolagus, Bos, Canis,* and *Equus*. The numbers represent the number of nucleotide replacements determined for each lineage. The specific nucleotide replacements are detailed in Sarich and Cronin (in press).

solutions that may differ from one another by only a few substitutions among hundreds.

In Figure 2 we present the most parsimonious solution that we have been able to achieve of the available hemoglobin sequence data—with the constraint that the solution also be congruent with other cladistic considerations. We obviously cannot claim that this tree reflects past reality; it is simply that one which best fits our immunological data, while remaining reasonably true to the hemoglobin sequence data by the parsimony criterion. Thus, we can recognize a basic dichotomy between ungulates and non-ungulates, with the rabbit falling on the ungulate (horse, cow) line, and the dog and tree shrew falling with the primates. The addition of the hedgehog (Maita *et al.*, 1979) near the basal node for placentals is again consistent with the general feeling that the basal placental stock was "insectivore" in character. Once the rabbit and dog are placed, the positions of *Tupaia* and *Tarsius* become much clearer. The *Tupaia* line (Maita *et al.*, 1977) shares only 2 nucleotide replacements with the primates after the divergence of the carnivores (*Canis*), while the primates share a further 10 before *Tarsius* separates. While this is a somewhat earlier separation relative to the primates than we have previously argued (Sarich and Cronin, 1976, in press), our more extensive recent measurements of the amount of change along the *Tupaia* albumin lineage (see Fig. 1) have also indicated an appreciable common ancestry for the primates subsequent to the separation of the tree shrew line. For maximum clarity we have deleted from Figure 2 many of the primate lineages for which sequence data are available. However, all sequences were included during the formulation of this tree.

Finally we note that the various lineages in Figure 2 have accumulated quite similar numbers of nucleotide replacements since the origin of the eutherians. The range is from lows of 33 and 34 for the rabbit and dog to 50 for horse and hedgehog. The mean is 40, and the primates are on this mean. No particular slowdowns and speedups are detectable here, and a figure of 2-2.5 MY per NR averaged across the α and β genes gives us a very reasonable 80-100 MYA for the basal node of Figure 2, as discussed previously (Cronin and Sarich, 1978; Sarich and Cronin, in press).

Although it might be thought that sequence data should produce an unequivocal solution (tree), very little handling of actual data is necessary to disabuse one of that notion. There are simply too many unavoidable homoplasies and no available criterion for choosing among several more or less equally good (parsimonious) solutions. By this we mean that a tree with 200 NR is surely better than one with 300, but is one with, say, 245, actually better, i.e., more representative of reality, than one with 248? This question has not been addressed within a proper statistical framework, and until it is, any choices made by strict deterministic applications of the parsimony criterion remain at best dubious. Our solution is reasonably parsimonious when compared to other published interpretations of much of the present data set (e.g., Beard and Goodman, 1976), and has the much more important further advantage of being more concordant with other data sets, both molecular and morphological. We have not included the rodents (*Rattus* and *Mus*) in this analysis as recent work (Zimmer, personal communication) suggests that the β globin gene of *Mus* may be paralogous to those of other placentals. Including *Mus* gives a divergence very near the basal node.

While our work within Tupaiidae has been limited, we have compared the albumins and transferrins of the three tupaiid genera available to us: *Tupaia, Urogale,* and *Ptilocercus.* The last is especially important as it was the form on which Le Gros Clark initially based his association of tupaiids and primates. The summed albumin and transferrin distances using antisera to the *Tupaia glis* proteins are: *Urogale,* 90; *Ptilocercus,* 131. Thus the living tree shrews share something close to 50 MY of common evolutionary development subsequent to their separation from any line leading to another living mammal and prior to the separation of the *Ptilocercus* line. In the absence of a relevant fossil record, this makes any reconstructive effort of tupaiid origins and evolution a most chancy proposition.

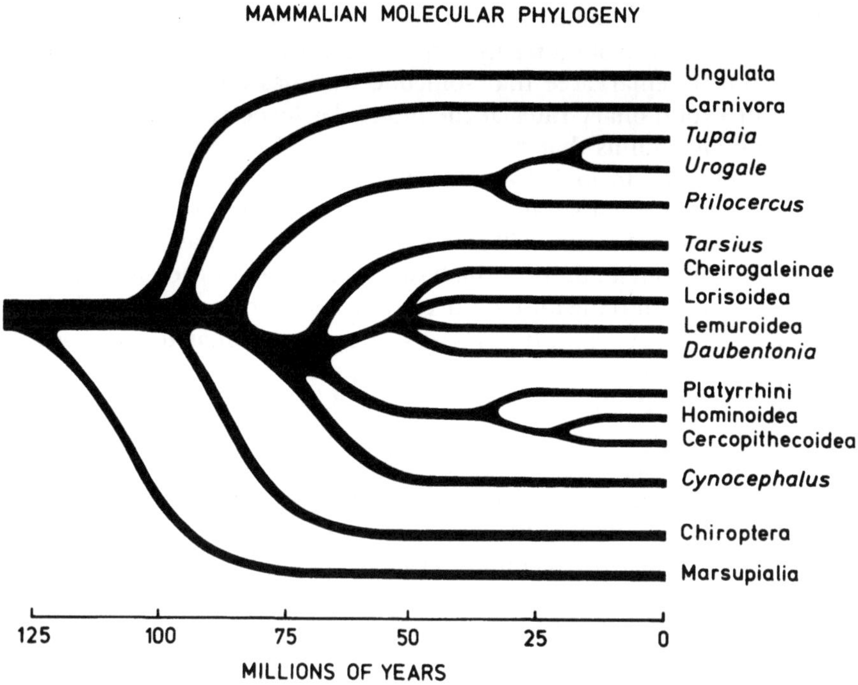

Fig. 3. This phylogeny represents an attempt to combine the available immunological and globin sequence placements of the various lineages involved so as to do minimal violence to the requirements of the various data sets involved. Obviously, the emphasis is on the primates and related lineages; Figure 5 attempts a more detailed cladistic look at a few other lines. Ungulata here includes three major clades: artiodactyls plus whales, perissodactyls, and elephants plus sirenians and hyraxes. The edentate and lipotyphlan insectivore lineages must have diverged very near to where we have placed the bat and carnivore nodes.

We include Figure 3 to emphasize the great antiquity of many divergence events in placental mammal evolution—an antiquity required by cladistic plus logical considerations. Simply stated, there is overwhelming evidence that protein evolution proceeds in a generally clock-like fashion; and rate tests tell us when it did not. Thus the fact that both the immunological and sequence data clearly resolve several internodal segments linking ordinal separations indicates a substantial existence in real time for those segments. Thus it is difficult to have a series of independent divergence events along the non-ungulate clade; for example: "true" primates, flying lemur, tree shrews, bats, carnivores, some insectivores, edentates, and rodents, without rapidly getting well back into the Cretaceous. We believe that any meaningful efforts at reconciling the paleontological, molecular, and comparative anatomical evidence concerning higher level mammalian systematics will have to seriously consider this extended time scale.

Finally, it might be appropriate here to extend our comments on the immunological work of Goodman and his colleagues. We believe that their obvious preference for amino acid sequences has served to perpetuate their inability to develop evolutionarily realistic analyses of their immunological data. We find it remarkable that someone who has always emphasized the inequality of evolutionary rates at the molecular level would continue to use tree-drawing algorithms that do not assess rates of change. It is also wholly illogical to continue to look for adaptive correlates of molecular evolution while at the same time assuming that looking at enough serum components with antisera to whole serum will somehow average out rate of change differences. Finally, of course, there is the basic insensitivity of the entire immunodiffusion approach and the false stature given to it by "computerizing" what are very poor data to begin with. It has been our goal to emphasize the production of reliable raw data and to analyze them using simple, rate-controlled, additive algorithms. We believe that we have succeeded in these efforts and there has certainly been no effective challenge to them. Thus we do not think that Goodman and his colleagues are justified in drawing *any* conclusions concerning the cladistic associations of tupaiids using their immunological data until they provide us with some estimates as to the uncertainties implicit in their *raw* data, and develop analytic techniques capable of coping with the question of rates of change.

C. *Chiropteran Affinities and Monophyly*

Some formulations of the Archonta include bats, along with the tree shrews, flying lemur, and primates in such a clade (McKenna, 1975; Szalay, 1977). In addition, questions have arisen as to the monophyly of bats (Smith, 1976). Our analysis of the albumin data, again emphasizing the use of edentates as the reference group, does not support the reality of a clade Archonta which includes bats and excludes the lipotyphlan insectivores (shrews, moles, hedgehogs). That is, we see no evidence of a closer association between bats and primates than between either and erinaceotan insectivores (tenrecs and elephant shrews probably diverged even earlier). At this point, however, the data are hardly such as to make this a definitive rejection, and we would not be surprised to find further molecular work delineating a relatively brief period of existence for a bat–primate clade.

The monophyly question is much more readily settled (Fig. 4). We saw in Figure 1 that about 80 units of albumin change had taken place along the bat lineages used in those comparisons since the separation of the proto-bat and proto-primate lines. Since the beginning of the bat adaptive radiation, however, those bat lineages have shown only about 55 units of change (Fig. 4). Thus all living bats share a period of common ancestry along which about 25 units of albumin evolution occurred subsequent to divergences of the lines leading to primates and, probably, lipotyphlan insectivores. This, along with the intra-bat distances, would place the beginning of the adaptive radiation leading to all modern bats at about 60 MYA, which is very close in time to the occurrence of

	Syc	*Hip*	*Art*	*Ant*	CP
Syconycteris	0	117	157	131	157
Hipposideros		0	155	103	141
Artibeus			0	92	187
Antrozous				0	147

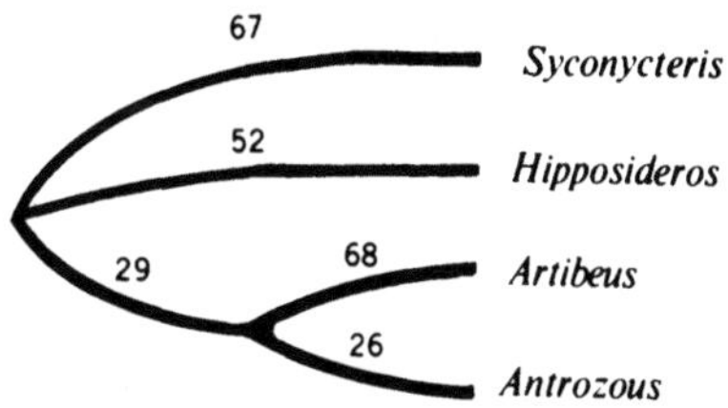

Fig. 4. Several bat taxa analyzed in a manner identical to that utilized in Figure 1. Here the difference between the input and output matrices is 2.6%, again a very low value. Extensive intra-bat work has shown that all pteropodids fall on the *Syconycteris* clade. all rhinolophoids on that of *Hipposideros,* and all other bats (emballonuroids, phyllostomatids, mormoopids, vespertilionoids) fall on the *Artibeus–Antrozous* clade.

the earliest, and clearly microchiropteran, bat in the Eocene of Wyoming (Jepsen, 1966). The recency of that adaptive radiation, and the lack of indicated monophyly of the Microchiroptera relative to the Megachiroptera (not to mention parsimony), then argue strongly for a single origin of bats. It should be noted that the markedly divergent nature of *Artibeus* albumin is characteristic of all phyllostomatids, mormoopids, and noctilionids. We would attribute this apparent acceleration to a single saltatory event (conformational change?) occurring immediately following the separation of the ancestral phyllostomatid-mormoopid stock from the lines leading to the vespertilionoids and emballonuroids.

D. *The Affinities of the Insectivores*

This data set is sketchy at best and most of our conclusions are drawn very tentatively. We simply have not been assiduous collectors of insectivore sera and have produced antisera only to the albumins of *Scapanus, Erinaceus,* and *Solenodon*. We are completely lacking representatives of some major groups such as chrysochlorids, and most others are represented by only one or two species. The anti-*Solenodon* sera react most strongly with the albumin of *Sorex* (90 units) and then with that of *Erinaceus* (140 units). *Scapanus, Tenrec,* and *Rhynchocyon* albumins are as distant as those of forms in other orders. The anti-*Scapanus* sera do not react well with anything besides the albumins of other moles. Thus we do not associate the moles with any other lineage. While the albumins of *Rhynchocyon* and *Tenrec* react no better with antisera to *Scapanus* and *Solenodon* albumins than do those of other non-ungulates, they do react more strongly with antisera to the albumins of the latter than do those of *Solenodon, Scapanus,* and *Erinaceus*. We take these results to imply relatively less change in the *Tenrec* and *Rhynchocyon* albumin lineages, and therefore conclude that they diverged at somewhat earlier times from the various erinaceotan insectivores than the latter did from one another or from other non-ungulates.

4. Summary and Conclusions

The albumin immunology (see above) and globin sequence data, when combined as discussed in this article, support a division of most placental mammals into two clades. The first of these contains those groups usually considered ungulates or ungulate-related: Artiodactyla plus Cetacea, Perissodactyla, and Proboscidea plus Sirenia. This latter unit most probably also contains Hyracoidea. The other clade (Fig. 5) is comprised of: Primates, Der-

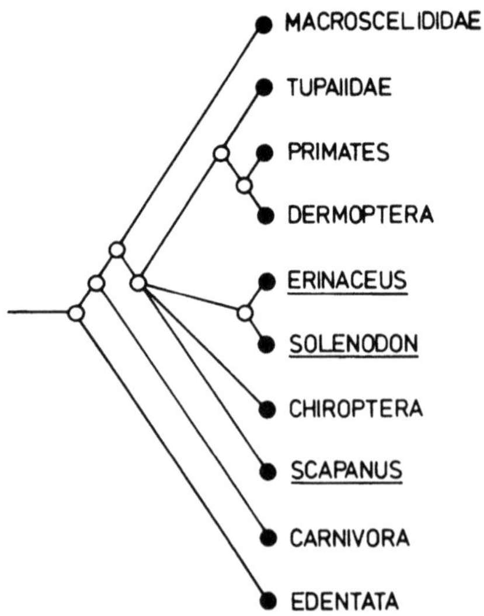

Fig. 5. Our current estimates as to the positioning of certain non-ungulate taxa not positioned in Figure 3. The conclusions are based on additional data obtained using antisera to the albumins of *Scapanus, Solenodon,* and *Erinaceus*. This is a cladogram and so the relative lineage lengths are not meant to be proportional to time. Again the assumption has been made that the edentates are a sister group to all the other taxa named. While this is morphologically and zoogeographically justified, it cannot be tested molecularly until more data become available. Immunological distances among the various non-edentates were apportioned using the edentates as the reference species.

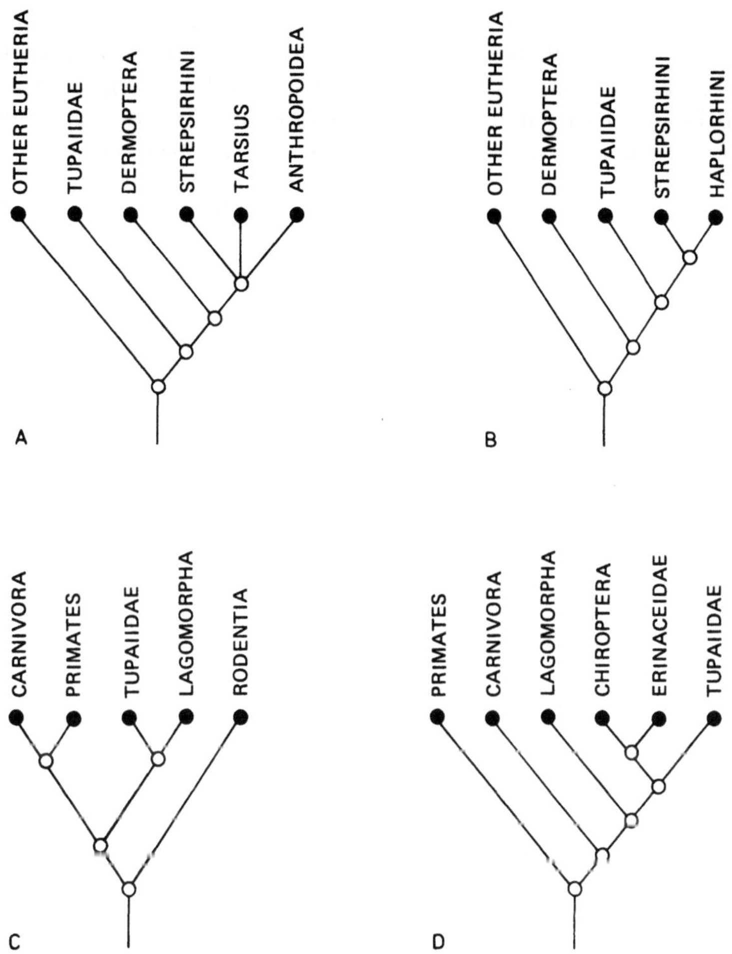

Fig. 6. A comparison of four different molecular cladograms. A. A cladogram derived through analyses of hemoglobin sequence differences and immunological distances (albumin and transferrin) among these taxa as discussed in this paper. B. An immunodiffusion cladogram as developed by Dene et al. (this volume). C. A hemoglobin cladogram as discussed by Dene et al. (this volume). D. A myoglobin cladogram as discussed by Dene et al. (this volume). We prefer cladogram A for reasons discussed in the text and suggest that this is the most parsimonious synthesis of the sequence and rate controlled immunological data.

moptera, Scandentia, Chiroptera, lipotyphlan Insectivora, Carnivora, Edentata, and, slightly less certainly, Rodentia. At this point we feel much less confident about lagomorphs, macroscelidids, tenrecids, pangolins, and the aardvark. The balance of a new, very limited, data set may even indicate an association of the first two on the ungulate clade (Chiment and Sarich, in prep.). We also refrain from any positive comment on the placement of the pangolins and aardvarks. However, they definitely are not associated with the South American edentates (Sarich and Cronin, in press; Sarich, in press).

The albumin and transferrin immunological data (Fig. 6) clearly associate the primates, flying lemur, and tree shrews into a clade for which the term Archonta might be retained (see also Dene et al., this volume). As discussed, we cannot see how this grouping could be the product of a possible rabbit antisera bias, and, clearly, if the tree shrews fall on the primate plus flying lemur clade, they cannot at the same time be lagomorph, macroscelidid, or erinaceid related. Each of these is definitely removed from any association with the primates by the immunological evidence. Finally, the association of the flying lemur and primates makes for a very strong case supporting the arboreal origin of the clade, while the tree shrews may represent a somewhat earlier and less specialized stage in the evolution of the group, much as Le Gros Clark (1971) discussed.

In conclusion, developing definitive answers to many of the questions with which this volume is concerned is beyond the scope of some of the various molecular approaches used to date. We have tried to indicate the limitations of the data whenever possible and suggest that a concerted effort be made to develop a program aimed at answering many of these questions with the appropriate methodology. This means, we believe, concentrating on and beginning a long-range program of sequencing appropriate genes (such as globins) from an adequate sample of mammal higher taxa. This does not mean that all other techniques will become obsolete. Immunological, protein sequence, DNA annealing, and electrophoretic comparisons will continue to have utility at appropriate taxonomic and temporal levels.

We now know a great deal concerning the applicability and reliability of various comparative macromolecular approaches to systematics, and we have a great deal of confidence in the results. We suggest that those who ignore such data do so at the risk of proposing easily falsifiable hypotheses.

ACKNOWLEDGMENTS

The authors wish to thank A. C. Wilson, of the University of California, Berkeley, for access to laboratory space. We would like to thank L. Brunker for help in many phases of the preparation and completion of the manuscript. C. Simmons and W. Powell provided the excellent illustrations. NSF provided grant support to one of us (V. S.). Lastly, we would like to thank W. P. Luckett for extremely helpful discussions and for his patience.

5. References

Baba, M. L., Darga, L. L., and Goodman, M. in press. Biochemical evidence on the phylogeny of anthropoids. *In* R. L. Ciochon and A. B. Chiarelli (eds.). *Evolutionary Biology of the New World Monkeys and Continental Drift.* Plenum Press, New York.

Beard, J. M., and Goodman, M. 1976. The hemoglobins of *Tarsius bancanus*, pp. 239–255. *In* M. Goodman and R. E. Tashian (eds.). *Molecular Anthropology.* Plenum Press, New York.

Benveniste, R. E., and Todaro, G. J. 1976. Evolution of type C viral genes: Evidence for an Asian origin of man. *Nature* 261:101–108.

Butler, P. M. 1972. The problem of insectivore classification, pp. 253–265. *In* K. A. Joysey and T. S. Kemp (eds.). *Studies in Vertebrate Evolution.* Oliver and Boyd, Edinburgh.

Cartmill, M. 1972. Arboreal adaptations and the origin of the order Primates, pp. 97–122. *In* R. Tuttle (ed.). *The Functional and Evolutionary Biology of Primates.* Aldine-Atherton, Chicago.

Crick, F. 1979. Split genes and RNA splicing. *Science* 204:264–271.

Cronin, J. E., and Meikle, W. E. 1979. The phyletic position of *Theropithecus:* Congruence among molecular, morphological and paleontological evidence. *Syst. Zool.* 28:259–269.

Cronin, J. E., and Sarich, V. M. 1975. Molecular systematics of the New World monkeys. *J. Hum. Evol.* 4:357–375.

Cronin, J. E., and Sarich, V. M. 1976. Dual origin of mangabeys among Old World monkeys. *Nature* 260:700–702.

Cronin, J. E., and Sarich, V. M. 1978. Primate higher taxa: The molecular view, pp. 287–289. *In* D. J. Chivers and K. A. Joysey (eds.). *Recent Advances in Primatology.* Volume 3. *Evolution.* Academic Press, New York.

Dene, H. T., Goodman, M., and Prychodko, W. 1976. Immunodiffusion evidence on the phylogeny of the primates, pp. 171–195. *In* M. Goodman and R. E. Tashian (eds.). *Molecular Anthropology.* Plenum Press, New York.

Farris, J. S. 1970. Methods for computing Wagner trees. *Syst. Zool.* 19:83–92.

Fitch, W. H., and Margoliash, E. 1967. Construction of phylogenetic trees. *Science.* 155:279–284.

Goodman, M. 1962. Evolution of the immunologic species specificity of human serum proteins. *Hum. Biol.* 34:104–150.

Goodman, M. 1975. Protein sequence and immunological specificity: Their role in phylogenetic studies of primates, pp. 219–248. In W. P. Luckett and F. S. Szalay (eds.). *Phylogeny of the Primates.* Plenum Press, New York.

Goodman, M., Czelusniak, J., Moore, G. W., Romero-Herrera, A. E., and Matsuda, G. 1979. Fitting the gene lineage into its species lineage, a parsimony strategy illustrated by cladograms constructed from globin sequences. *Syst. Zool.* 28: 132 163.

Goodman, M., and Moore, G. W. 1971. Immunodiffusion systematics of the primates. I. The Catarrhini. *Syst. Zool.* 20:19–62.

Hennig, W. 1966. *Phylogenetic Systematics.* Univ. of Illinois Press, Urbana.

Hewett-Emmett, D., Cook, C. N., and Barnicot, N. A. 1976. Old World monkey hemoglobins: Deciphering phylogeny from complex patterns of molecular evolution, pp. 257–276. *In* M. Goodman and R. E. Tashian (eds.). *Molecular Anthropology.* Plenum Press, New York.

Hewett-Emmett, D., and Cook, C. N. 1978. Atypical evolution of papionine α haemoglobins and indications that *Cercocebus* may not be a monophyletic genus, pp. 291–294. *In* D. J. Chivers and K. A. Joysey (eds.). *Recent Advances in Primatology.* Volume 3. *Evolution.* Academic Press, New York.

Holmquist, R. 1976. Random and nonrandom processes in the molecular evolution of higher organisms, pp. 89–116. *In* M. Goodman and R. E. Tashian (eds.). *Molecular Anthropology.* Plenum Press, New York.

Jepsen, G. L. 1966. Early Eocene bat from Wyoming. *Science* 154:1333–1339.

Kohne, D. 1975. DNA evolution data and its relevance to mammalian phylogeny, pp. 249–264. *In* W. P. Luckett and F. S. Szalay (eds.). *Phylogeny of the Primates.* Plenum Press, New York.

Le Gros Clark, W. E. 1971. *The Antecedents of Man.* 3rd Ed. Edinburgh University Press, Edinburgh.

Luckett, W. P. 1975. Ontogeny of the fetal membranes and placenta: Their bearing on primate phylogeny, pp. 157–182. *In* W. P. Luckett and F. S. Szalay (eds.). *Phylogeny of the Primates.* Plenum Press, New York.

Maita, T., Araya, A., Matsuda, G., and Goodman, M. 1979. Amino acid sequences of the α and β chains of adult hemoglobin of the European hedgehog, *Erinaceus europaeus. J. Biochem.* 85: 259–269.

Maita, T., Tanaka, E., Goodman, M., and Matsuda, G. 1977. Amino acid sequences of the α and β chains of adult hemoglobins of the tupaia, *Tupaia glis*. *J. Biochem.* 82: 603–605.

Maxam, A. M., and Gilbert, W. 1977. A new method of sequencing DNA. *Proc. Nat. Acad. Sci. U.S.A.* 74:560–564.

McKenna, M. C. 1966. Paleontology and the origin of Primates. *Folia Primat.* 4:1–25.

McKenna, M. C. 1975. Toward a phylogenetic classification of the Mammalia, pp. 21–46. *In* W. P. Luckett and F. S. Szalay (eds.). *Phylogeny of the Primates*. Plenum Press, New York.

Nuttall, G. H. F. 1904. *Blood Immunity and Blood Relationships*. Cambridge University Press, Cambridge.

Prager, E. M., and Wilson, A. C. 1971. The dependence of immunological cross reactivity upon sequence resemblance among lysozymes. *J. Biol. Chem.*246: 5978–5989.

Prager, E. M., and Wilson, A. C. 1976. Congruence of phylogenies derived from different proteins. *J. Mol. Evol.* 9:45–57.

Prager, E. M., Wilson, A. C., Osuga, D. T., and Feeney, R. E. 1976. Evolution of flightless land birds on southern continents: Transferrin comparison shows monophyletic origin of ratites. *J. Mol. Evol.* 8:283–294.

Sarich, V. M. 1970. Primate systematics with special reference to Old World monkeys: A protein perspective, pp. 175–266. *In* J. R. Napier and P. H. Napier (eds.). *Old World Monkeys*. Academic Press, New York.

Sarich, V. M. 1973. The giant panda is a bear. *Nature* 245:218–220.

Sarich, V. M. in press. Edentate systematics: The albumin immunological evidence. *In* G. Montgomery (ed.). Smithsonian Reports.

Sarich, V. M., and Cronin, J. E. 1976. Molecular systematics of the primates, pp. 141–170. *In* M. Goodman and R. E. Tashian (eds.). *Molecular Anthropology*. Plenum Press, New York.

Sarich, V. M., and Cronin, J. E. 1977. Generation length and rates of hominoid molecular evolution. *Nature* 269:354–355.

Sarich, V. M., and Cronin, J. E. in press. South American mammalian molecular systematics, evolutionary clocks and continental drift. *In* R. L. Ciochon and A. B. Chiarelli (eds.). *Evolutionary Biology of the New World Monkeys and Continental Drift*. Plenum Press, New York.

Sarich, V. M., and Wilson, A. C. 1973. Generation time and genomic evolution in primates. *Science* 179:1144–1147.

Schwartz, J. H. 1978. If *Tarsius* is not a prosimian, is it a haplorhine?, pp. 195–202. *In* D. J. Chivers and K. A. Joysey (eds.). *Recent Advances in Primatology*. Volume 3. *Evolution*. Academic Press, New York.

Schwartz, J. H., Tattersall, I., and Eldredge, N. 1978. Phylogeny and classification of the primates revisited. *Yearbook Phys. Anthrop.* 21:95–133.

Simpson, G. G. 1945. The principles of classification and a classification of mammals. *Bull. Amer. Mus. Nat. Hist.* 85:1–350.

Smith, J. D. 1976. Chiropteran evolution, pp. 49–69. *In* R. J. Baker, J. K. Jones, Jr., and D. C. Carter (eds.). *Biology of Bats of the New World Family Phyllostomatidae*, Part I. Spec. Publ. Mus., Texas Tech Univ., 10:1–218.

Szalay, F. S. 1977. Phylogenetic relationships and a classification of the eutherian Mammalia, pp. 315–374. *In* M. K. Hecht, P. C. Goody, and B. M. Hecht (eds.). *Major Patterns in Vertebrate Evolution*. Plenum Press, New York.

Van Valen, L. 1965. Treeshrews, primates, and fossils. *Evolution* 19:137–151.

Wilson, A. C., Carlson, S. C., and White, T. J. 1977. Biochemical evolution. *Ann. Rev. Biochem.* 46:573–639.

Wooding, G., and Doolittle, R. 1972. Primate fibrinopeptides: Evolutionary significance. *J. Hum. Evol.* 1:553–563.

Zihlman, A. L., Cronin, J. E., Cramer, D. L., and Sarich. V. M. 1978. Pygmy chimpanzee as a possible prototype for the common ancestor of humans, chimpanzees and gorillas. *Nature* 275:744–746.

Index

MIX
Papier aus verantwortungsvollen Quellen
Paper from responsible sources
FSC® C105338

If you have any concerns about our products,
you can contact us on
ProductSafety@springernature.com

In case Publisher is established outside the EU,
the EU authorized representative is:
Springer Nature Customer Service Center GmbH
Europaplatz 3, 69115 Heidelberg, Germany

Printed by Libri Plureos GmbH
in Hamburg, Germany